I0482756

NASA SP-4404

LIQUID HYDROGEN AS A PROPULSION FUEL, 1945-1959

JOHN L. SLOOP

The NASA History Series

Scientific and Technical Information Office   1978
NATIONAL AERONAUTICS AND SPACE ADMINISTRATION
Washington, D.C.

Contents

PART II: 1950–1957

Illustrations

Tables

Foreword

History is written and read for several reasons, and the NASA history program serves multiple purposes. John Sloop's history of liquid hydrogen as a fuel illustrates the most practical of those purposes: it is *useful* to current and future managers of high technology. Of course history does not repeat itself—there are too many variables. But similar situations often have similar results, and thoughtful study of the management of technology in the past can sometimes help us to recognize pitfalls in the present— pitfalls that managers can then act to avoid. We may also find ways to make desired outcomes more likely. In any event, study of history lets us see current problems more clearly.

For example, notice in this book how many times something had to be rediscovered. This has been a real problem, and a costly one, in the recent past; it is apt to get worse in the future. Are we in NASA doing all we reasonably can to manage this problem—not just making new technology available to industry, not just trying to stay current in our respective fields, but contributing something to the process by which the knowledge explosion can be made more tractable?

It is a truism that technology feeds on itself—that work in one area often is quickly applicable in an entirely different area. Perhaps the sharpest example in this book is the Air Force's building of plants for liquefaction of hydrogen and developing equipment and procedures for its handling. That program was cancelled short of completion, but the technology was on the shelf, already paid for, when NASA needed it for the Apollo program. Can we explain this process to Congress and to the taxpayers more effectively? The problem is similar to that of justifying basic scientific research. Can future NASA managers, in defending their programs, do so more effectively by elaborating that similarity?

A recurring theme in this book is the widespread fear of hydrogen, originating with the explosion of the *Hindenburg* and reinforced by the H-bomb. Proponents of hydrogen-fueled rockets had to overcome that prejudice. Are other technologies ignored today because of a bias against certain materials or processes? Engineers and scientists remain subject to the human condition; they, like the rest of us, need to be reminded from time to time to take a fresh look at old attitudes and familiar procedures.

The author illuminates the overlapping, often conflicting roles of the individual, who originates ideas, and of the group, which manages today's complex technology. Many worthwhile ideas have doubtless been lost, at least temporarily, because individuals were unable to convince committees. Hence how consensus is achieved within groups is worth studying. When agreement seems impossible, an individual is occasionally big enough, wise enough, to forego his preferred solution, so that a project may continue. In this regard, timing is critical. If the individual does not press his case hard enough, he is labeled irresolute; but if he says, in effect, "My way or none," he is obstinate. The story of the decision to use liquid hydrogen in the upper stages of the Saturn launch vehicles contains several accounts of individual-group interaction from which any manager can profit.

Finally, the book argues against the casual hindsight judgment of "the idea whose time had come." More than once participants were convinced—wrongly—that hydrogen's time had come. Its time came only after a number of disparate events gradually took on a pattern. If we are sometimes tempted to assume that a favorite project is inevitable, or that a solution to a sticky technical problem will inevitably be found, then we may be contributing to the failure of our own purposes.

This book is also a good story, with real drama, colorful men, and fascinating technology. If hydrogen comes to occupy an important place in the energy field, as some now predict, this book will take on an importance that cannot now be foreseen. But at a time when NASA is emphasizing the solution of workaday problems facing the nation and seeking early return on the taxpayers' investment, it seems appropriate to point out the book's practical significance.

JAMES C. FLETCHER
Administrator

April 1977

Preface

In 1957, when Russia launched the first satellite, the ability of the United States to respond depended on one small launch vehicle still under development, Vanguard, and modifications to ballistic missiles. The subsequent space race featured a rapid buildup of launch vehicle capability in this country during the 1960s, culminating with the giant Saturn V which launched the Apollo lunar expeditions beginning in 1968. A significant part of the increased launch capability resulted from technical decisions made in 1958 and 1959 to use liquid hydrogen in the upper stages of the Centaur and Saturn vehicles—and that story is not well known. The decision to use liquid hydrogen in developing the nation's largest launch vehicle was particularly bold, for many experienced engineers doubted the advisability of using a highly hazardous fuel associated with the *Hindenburg* disaster of 1937, a gas difficult to liquefy, a liquid so cold—close to absolute zero—that storage and handling are difficult, and so light—1/14 the density of water—that large tank volumes are required, with attendant problems of vehicle mass and drag. Hydrogen had been considered in astronautics and aeronautics several times before; but in each case, as the problems became better known, the attempt was abandoned. What was different in this case? Why was there so much confidence about hydrogen within the young space agency to warrant risking the success of the nation's manned spaceflight program? The decision, of course, turned out to be the right one. Subsequent advancements in the technologies of liquefying, storing, transporting, and using large quantities of liquid hydrogen made it just another flammable liquid that could be handled and used safely with reasonable caution.

The key role that liquid hydrogen played in the success of the Centaur and Saturn launch vehicles has long interested me. As a participant in research on hydrogen for rockets in the 1950s and a proponent for its use, I understood the potential as well as the risks and in recent years wanted to investigate more fully the circumstances leading to the 1958 and 1959 decisions.

In digging into the background for the decisions and the status of hydrogen technology that influenced those decisions, the question arose: how far back to investigate? The flammability of gaseous hydrogen has been known for centuries; its large heat content was measured in the 18th century; and it was liquefied by Dewar in 1898. Five years later, Tsiolkovskiy, the Russian rocket pioneer, proposed its use in a space rocket, as did Goddard in 1910. In the 1920s, Oberth correctly assessed the advantage of using hydrogen in the upper stages of space vehicles. None of these rocket pioneers experimented with hydrogen; other fuels appeared more attractive in the face of hydrogen's disadvantages, particularly its low density. One German experimenter,

Walter Theil, tried to use liquid hydrogen in a small rocket engine a few years before World War II, but numerous leaks and higher priority tasks ended the experiments. The first systematic investigations of liquid hydrogen to propel aircraft and rockets began in the United States in 1945 and although earlier developments undoubtedly had an influence, I have chosen to start this book at that point. A summary of the earlier story is in appendix A.

In describing the history of rocket technology, it is easy for an engineer-author to become immersed in the technical aspects that may be of little interest to some readers. I have tried to minimize mathematics, technical language, and other specialized details, but some are unavoidable if propulsion research is to be presented fairly and accurately. Adding to this problem has been the conversion of many familiar English units into the metric system. Those accustomed to thinking of rocket performance in terms of specific impulse will not find it here; instead, they will have to settle for its equivalent, exhaust velocity. Appendix B is provided as an aid in the technical aspects of propulsion, units, symbols, and abbreviations.

This work would not have been possible without the help of numerous participants in hydrogen and rocket research, who were generous with interviews and documents; the guidance of Monte Wright and Frank Anderson; the essential services of the NASA archivist, Lee Saegesser, and NASA librarians, particularly Mary Anderson and Grace Reeder, in obtaining many obscure documents; the aid of Col. John D. Seaberg (USAF, ret.) and Malcolm Wall, deputy command historian, AFSC, for securing invaluable Air Force documents; the enlightened attitude of Howard Maines, NASA security office; the encouragement of Gene Emme; and my wife, who served as editor as well as helpmate.

1
Introduction

In September 1944, a general and a professor met in an Air Force car parked at one end of a runway of New York's LaGuardia airport. General of the Army H. H. Arnold, chief of the Army Air Forces and on his way to a meeting in Quebec, had arranged the meeting with Professor Theodore von Kármán, famed aerodynamicist and jet propulsion pioneer at the California Institute of Technology. The two had first met in 1936; they had discussed auxiliary rocket thrust for bombers in 1938 and the design of a new research wind tunnel in 1939. Now Arnold wanted von Kármán to come to the Pentagon to draw up a plan for aeronautical research during the next twenty years.[1] Confident that the war was won, Arnold had turned to the future.

When the group of scientists von Kármán had organized for the task met in January 1945, Arnold stated his feelings bluntly: "I don't think we dare muddle through the next twenty years the way we have . . . the last twenty years . . . . I don't want ever again to have the United States caught the way we were this time."[2] Arnold was referring to technological superiority in the air.

When Arnold and the combined chiefs of staff met with Roosevelt and Churchill in Quebec, the tide of battle in Europe was decisively in favor of the Allies. Fleets of Allied aircraft were pounding Germany's industrial capacity into rubble. Eisenhower's armies were moving towards the Rhine and some units were on German soil near Aachen. In the Pacific, MacArthur was able to step up his plans for landing on Leyte by two months. U.S. production of aircraft and training of air personnel so far exceeded the demands of war that both were cut back in the fall of 1944 to save money.[3]

The air supremacy of the Allied European offensive in 1944 came not from technological superiority but sheer weight of numbers and better trained crews. Between May 1940 and September 1943, the United States alone produced 128 000 aircraft and 349 000 engines.[4] By 1944, however, there was ample evidence that piston-engine aircraft were rapidly becoming obsolete and that future military aircraft would be jet-propelled.

From the beginning, airplanes had been powered by the piston engine-propeller combination. Jet propulsion had been examined in the early 1920s but rejected as too inefficient at the prevailing aircraft speeds of 400 kilometers per hour. By the late 1930s, however, potential airplane speeds had doubled and this, along with other technical advances, made jet propulsion more attractive. Development began in Europe during the second half of the 1930s, but little work was done in the United States on gas turbine engines until 1939.* Even then, the U.S. military was lukewarm about the potential of jet aircraft.[5]

The two most serious disadvantages of early gas turbine engines for aircraft were their low thrust, which made long take-off rolls necessary, and high fuel consumption, which limited range. These disadvantages chilled Navy interest in gas turbines as primary propulsion systems until 1943. The Army showed greater interest in rocket propulsion for aircraft, rather than gas turbine engines. The Army became interested in rocket propulsion in 1938; in February 1941, when Arnold learned from intelligence reports that the Germans were using rocket propulsion, he asked the National Advisory Committee for Aeronautics (NACA) to study jet propulsion. The NACA, the government's aeronautical research organization, set up an advisory committee headed by 83-year-old Dr. William F. Durand, eminent aerodynamicist at Stanford University. Durand's interest in turbine machinery directed the NACA study almost entirely towards gas turbine engines. Representatives from three firms proficient in turbine machinery—Allis Chalmers, Westinghouse, and General Electric—served on the Durand committee, and their firms were given study contracts by the military services.[6]

Arnold visited Great Britain in the spring of 1941 and was impressed by the Whittle gas turbine engine. He arranged for General Electric to manufacture it in the United States. On 2 October 1942, the Bell P-59A, powered by a General Electric I-A gas turbine engine, became the first American jet-propelled aircraft to fly. The I-A produced so low a thrust, however, that performance was disappointing. Despite later installation of a more powerful engine, the I-16, the P-59A did not reach the production stage. The British developed the Meteor powered by a Rolls Royce W-2B gas turbine engine and used it in World War II, although its performance was little if any better than that of the P-59A. By 1944, General Electric had developed a much more powerful gas turbine engine, the I-40, which was used to power the Lockheed XP-80A fighter, developed by Clarence L. (Kelly) Johnson in just 143 days.[7] Production began before the war ended, but the P-80 did not reach tactical units until seven months after the war ended in Europe.

In mid-1944, the Allies confirmed that the Germans were using turbojet interceptors against Allied bombers. By January 1945, a special German squadron of sixteen ME-262 turbojet fighters, armed with twenty-four 55 mm rockets, operated against Allied bomber formations with high success. In early April 1945, a German pilot, tired of the war, landed an ME-262 at an Allied airfield. Arnold questioned the pilot about its capability and arranged for shipment of the aircraft to Wright Field for evaluation.[8]

Robert Schlaifer, who studied the development of aircraft engines through World War II, saw the lag of jet propulsion in the United States as a lesson in the importance of avoiding delays in adopting new technology:

---

*A gas turbine engine, the most common form of which is the turbojet, consists of an air inlet, a rotary fan or compressor, one or more combustion chambers, a turbine driven by hot, expanding combustion gases, and an exhaust nozzle. The turbine drives the compressor; the thrust comes from expanding and accelerating the hot air and combustion gases through the nozzle.

The most serious inferiority in American aeronautical development which appeared during the Second World War was in the field of jet propulsion. Had the Germans put their jet fighters in production a year sooner, as they were technically able to do, or had the Allied campaign in Europe come a year later, the use of jet fighters by the Germans might have had a most serious effect on the course of the war.[9]

What about the aeronautical research laboratories of the NACA and Air Force? Why had they not led in investigating advanced forms of propulsion? They had been slow in recognizing, during the second half of the 1930s, that the time of the gas turbine engine had come. A few investigators in NACA, particularly Eastman Jacobs and Benjamin Pinkel, began to realize this and were working on the problem by 1939, but progress was slow. The Durand committee provided new stimulus, but by that time war was close. The policy of mass production of piston engines led U.S. aeronautical laboratories to concentrate on solving urgent problems arising from their production and operation. Improvements were made in aviation fuels, in engine components such as the turbosupercharger, and in numerous operating problems. The laboratories of the NACA were at the disposal of the military services for this effort, giving first priority to war-related problems, leaving little time for long-range work on advanced propulsion systems.

In spite of concentration on piston engine problems, however, NACA continued some research on jet engines and rockets. In December 1943, both the Army and Navy asked the NACA to evaluate their jet engines developed under contracts originally recommended by the Durand committee. The first test was made in the unique altitude wind tunnel at NACA's engine laboratory in May 1944, and by fall the tunnel was used exclusively for jet engine research. The same year, NACA's director of research, George Lewis, authorized the engine laboratory in Cleveland to spend $43 000 for the construction of some simple rocket test stands; and about the same time, researchers at the NACA Langley laboratory began eyeing rockets as a means of propelling experimental models to transonic and supersonic speeds for aerodynamics and controls research.[10]

Late in 1944, the government aeronautical laboratories felt an easing of the pressure to concentrate on ad hoc problem solving, freeing men and funds for advanced concepts. The suppression of the long-felt desire by researchers to work on advanced propulsion was accentuated by reports of German accomplishments in jet propulsion and rockets, particularly the V-2. Teams of scientists and engineers were dispatched to obtain German technical data in the wake of advancing Allied armies and to interrogate German technical specialists. Plans were made to bring a group of German rocket experts to the United States. The mood in the government propulsion laboratories was the same as that expressed by Arnold to his advisory group—to catch up and not ever fall behind again in advanced propulsion.

Parallel to NACA research on aeronautics during the war was research and development in other fields of military importance by a large group of scientists and engineers, coordinated by the Office of Scientific Research and Development (OSRD). Among the many significant contributions OSRD made was rocketry. At the time of the Pearl Harbor attack, the U.S. military did not have a single rocket in service use:

but by the end of the war, $1.35 billion worth of solid-propellant rockets were being produced annually, mostly for the Navy. These were short-range, armament rockets. OSRD also sponsored work on liquid-propellant rockets for assisted take-off of aircraft. Information on the German V-2 was available to OSRD by mid-1943, but there were no plans for long-range rockets.[11] Like their fellow researchers in aeronautics, OSRD initially had their hands full with pressing war problems, with little time left for future systems. About 1944, however, an OSRD panel was formed on jet propulsion with Edwin R. Gilliland as its chief.*

Among the studies of the OSRD jet propulsion panel was a very significant one on fuels for jet propulsion reported by Alexis W. Lemmon, Jr., in May 1945.[12] The Lemmon report, or "blue book"—from the color of its cover—became a standard reference for researchers in jet propulsion and rocket fuels in the early postwar years. It marked the beginning of such research in the U.S.

For jet engines using the oxygen in air to burn the fuel, as in turbojet and ramjet engines, Lemmon considered eleven hydrocarbons and eleven high-energy fuels in the diborane and borohydride family.† High-energy fuels yield more heat in burning than conventional fuels, such as gasoline or kerosene, and therefore have the potential for greater performance. Lemmon concluded, however, that little change could be expected in fuels for jet engines using air and that "high density and high heat of combustion fuels will be used for minor applications but no major change from present fuel of gasoline or kerosene is probable."[13] In the years to come—extending into the second half of the 1950s—the government spent a quarter of a billion dollars investigating high-energy fuels containing boron and light metals for air breathing engines before abandoning them. Lemmon's early conclusion was right.

On rocket fuels, Lemmon presented the performance of 25 fuel-oxidizer combinations, 14 monopropellants, and 6 solid propellants.‡ Separate fuels and oxidizers, when mixed and burned, yield higher energy than either monopropellants or solid propellants. This advantage of higher energy is sometimes offset by the undesirable physical or chemical properties of fuel, oxidizer, or both. Of all the rocket fuel and oxidizer combinations that he considered, Lemmon found that the combination of liquid hydrogen–oxygen gave the highest performance, but he rejected it. "Although the liquid hydrogen–liquid oxygen system has by far the highest specific impulse performance of any system considered in this report, the low average density of the fuel components almost completely eliminates this system from all but very

---

*Other members: Neil P. Bailey, Howard E. Emmons, Ernst H. Krause, Alexis W. Lemmon, Jr., Lloyd W. Morris, John C. Quinn, Edward M. Redding, Theodore H. Troller, Merit P. White, Glenn C. Williams, and Harold A. Wilson.

†A ramjet engine uses atmospheric air but no mechanical compressor or turbine. Essentially an open duct, the ramjet depends upon high-speed flight and ram air for compression. Fuel is injected and burned and the hot gases expand through a nozzle to provide thrust.

Diborane (B_2H_6) and pentaborane (B_5H_9) were of great interest in the late 1940s and the 1950s. Lemmon listed as borohydrides compounds containing light metals such as sodium, beryllium, aluminum, and magnesium.

‡A fuel-oxidizer combination, also called bipropellant, is a fuel and an oxidizer which are injected and burned in the rocket combustion chamber; a monopropellant decomposes and gives off heat in the process; a solid propellant contains both fuel and oxidizer elements and burns to yield heat.

minor applications."[14] Low density meant that large tanks were required, which added mass and drag to the vehicle. Lemmon went on to point out that the development of equipment to produce liquid hydrogen would be difficult, the cost high, and handling hazardous.

On the practical application of liquid hydrogen to flight, Lemmon was proved wrong. In 1958 and 1959, decisions were made to use liquid hydrogen in the upper stage of the Centaur launch vehicle for unmanned space missions and the upper stages of the Saturn launch vehicle for manned voyages to the moon. Both decisions turned out to be sound; both vehicles were remarkably successful. Liquid hydrogen–oxygen emerged as the first high-energy rocket propellant combination to find practical application among many candidates investigated. To explain why and how this happened, and why it took so long, is the purpose of this book.

Part I

1945–1950

Part I
1945–1950

Up to 1945, gaseous hydrogen had been considered many times as a fuel for internal combustion engines, particularly for dirigible engines. It caused engine knock (detonation) and was limited to experimental investigations or as a component in a gaseous fuel mixture.

Three early rocket pioneers—Tsiolkovskiy, Goddard, and Oberth—all proposed to use liquid hydrogen with oxygen in a rocket engine for space travel, but none tried it experimentally. The Germans, who made the greatest advances in rocketry up to 1945, experimented with liquid hydrogen in a small rocket engine prior to World War II, but numerous leaks and other problems made the fuel appear impractical.*

The development of jet engines and rockets during World War II opened up a new line of propulsion systems and with them, new considerations of fuels. The Germans, following the lead of Goddard, showed that a cryogenic fluid—liquid oxygen—could be used in a practical propulsion system, the V-2. From this cryogenic fluid to another—liquid hydrogen—was a big step, but it was inevitable that propulsion engineers would take another look at hydrogen's possibilities. That step came a month after the close of World War II in Europe, when the Air Force contracted for a general investigation of liquid hydrogen as a fuel for aircraft and rockets. The Navy was not far behind in becoming interested in liquid hydrogen, but contracted for a specific application. These military contracts resulted in the first experiments in the United States with liquid hydrogen as a propulsion fuel and were responsible for advancing the technology of liquid hydrogen during the second half of the 1940s.

---

*For a summary of hydrogen properties and technology through World War II, see appendix A.

2
Air Force Research on Hydrogen

The origins of Air Force interest in liquid hydrogen as a fuel are obscure, but researchers were well aware of hydrogen from general studies and from occasional external suggestions. One of the latter came to Robert V. Kerley on a warm July day in 1942 and, at the time, made no sense to him. As chief of Wright Field's fuels and oil branch, Kerley was the Air Force's leading expert on aviation fuels and its representative on the fuels and lubricants subcommittee of the National Advisory Committee for Aeronautics (NACA). The subcommittee, under the chairmanship of Professor W.G. Whitman of the Massachusetts Institute of Technology, played a key role at the beginning of the war by coordinating aviation fuel needs and stressing the imperative of increased production.*

Kerley's branch at Wright Field had a long tradition of leadership in improving aviation fuels. Although fuels research can be traced to the establishment of the aeronautical engineering laboratory at McCook Field in 1917, the first systematic fuels research program dates from 1928 when studies were started to determine the relationship between fuel composition, engine performance, and knock. As engine designers sought increased power output per unit volume of engine piston displacement, fuels had to be improved to keep pace. During the 1930s, the fuels and oil branch at Wright Field was the recognized leader in promoting research and production of improved aviation fuels. As a result, the United States was the only one of the Allies at the beginning of the war having a significant capacity for producing high-performance aviation fuel.

Kerley was up to his ears in practical problems of increasing aviation fuel production and operating problems in July 1942 when he was requested to comment on a British suggestion forwarded by the NACA. It was a ten-page proposal by F. Simon to use—of all fuels—liquid hydrogen as a means for increasing aircraft range. Kerley knew that hydrogen produced knock; further, hydrogen liquefaction capacity in the United States was on the order of a few hundred liters per day, and those plants were in scientific laboratories. If the exasperated Kerley considered Simon a nut and his suggestion ridiculous, it would be understandable. Although the suggestion was

---

*In July 1940, when President Roosevelt announced a goal of 50000 airplanes, the subcommittee estimated that current production of 100-octane aviation fuel must be increased twelvefold but could not convince the military services, who agreed only to a fourfold increase. By war's end, Allied production of 100-octane aviation fuel was 40 times greater than in 1940. Sam D. Heron, "Development of Aviation Fuels," in *Development of Aircraft Engines and Fuels* (Elmsford, NY: Maxwell Reprint, 1970), pp. 631–34.

impractical at the time and indicated Simon's naiveté with respect to fuel production and aviation, he was anything but a nut. F. Simon was Franc Eugen Simon (1893–1956), a thermodynamicist and ingenious experimenter with liquid hydrogen at Oxford University. He had earned his doctorate under the famous Nernst and worked in Germany on low-temperature phenomena until 1933 when, disturbed by rising Nazi power, he accepted an invitation from F.A. Lindemann (Lord Cherwell) to come to Oxford. Simon managed to bring a hydrogen liquefier with him and was instrumental in building an outstanding low-temperature laboratory at Clarendon; in August 1940, he was placed in charge of isotope separation research in Britain's nuclear fission effort.[1]

Kerley immediately recognized the utter impracticability of Simon's suggestion to use liquid hydrogen, but was not so pressed that he could not respond with a bit of humor:

> Now F. Simon went a-hunting
>   With a lot of gaseous pride
> We say his purpose does appear
>   To take US for a long sleigh ride
>
> Hydrogen is a knocking fuel
>   And is plenty good for heating
> But what good is a B T U
>   When horsepower goes a-fleeting?

After several more verses, the doggerel ended:

> If morals one must always sing
>   To Ally Simon we'd sing thus
> 'Keep working on a simple thing
>   And shut off all this goddam fuss!'[2]

Simon, who had an impish sense of humor and laughed at jokes on himself, would have been delighted with the verses, if not the disposal of his suggestion.

Simon was not alone in considering hydrogen for aviation fuel. Much earlier, P. Meyer had written an article entitled "Is There Any Available Source of Heat Energy Lighter than Gasoline?" which the NACA translated as Technical Note 136 in the early 1920s. Meyer noted that hydrogen had a greater heat content than any other known fuel. Apparently considering it only in gaseous form under pressure, he also noted that the containers had to be strong and heavy, which counterbalanced the energy advantage.

Both Meyer and Simon, therefore, found that hydrogen in any form was an aviation fuel whose time had not come. Interest in hydrogen, however, was not lost entirely and surfaced when war pressures eased in late 1944 and 1945 and the men at Wright Field began to think again about future projects. Opie Chenoweth, chief civilian engineer of the power plant laboratory, suggested that research be sponsored on increasing the energy content of aviation fuels.[3] Hydrogen was not a good fuel for piston engines because of the tendency to knock, but what about using it in jet engines? Over at Ohio

State University in nearby Columbus was a professor who had built a cryogenics laboratory during the war and was one of a few experts in liquefying hydrogen and studying its properties. Why not have him study liquid hydrogen for aircraft and rockets? The professor's name was Herrick L. Johnston.

The Cryogenics Laboratory at Ohio State University

Soon after his arrival at Ohio State University in 1929 as an assistant professor of chemistry, Herrick L. Johnston (1898–1965) prepared plans for a cryogenics laboratory to match that of his preceptor, William F. Giauque of the University of California at Berkeley. This was ambitious planning, for Giauque's laboratory and one at the Bureau of Standards in Washington were among the very few in the country capable of research at the temperatures of liquid hydrogen. Giauque and Johnston had just published their revolutionary discovery that atmospheric oxygen contains atoms of mass 17 and 18, as well as 16, a discovery that set into motion a chain of experiments leading to the discovery of heavy hydrogen by Harold C. Urey in 1939 (appendix A-3).

Unfortunately for Johnston, his move to Ohio took him out of the mainstream of the swiftly moving research of low-temperature phenomena, and his dream of a cryogenics laboratory lay dormant a decade for lack of funding. In 1939, Johnston, a full professor and still pushing for his cryogenics laboratory, got a big break. The year before, William McPherson, a former head of the chemistry department, was called out of retirement to be acting president of the university. The first annual alumni development fund drive in 1939 included plans for a cryogenics laboratory and McPherson personally contributed the first $1000. This amount was augmented by $5000 from the university budget and Johnston was quick to start spending it. He ordered a hydrogen compressor and other equipment needed for a liquefier but soon encountered another obstacle—no space for the equipment. This problem was solved when federal funds—part of a plan to involve universities in war research—became available for a building. It was bluntly named the War Research Building. Johnston was initially allocated part of the first floor for a cryogenics laboratory, but later he took over the first two floors.[4]

Construction of the building began about mid-1942, but before the foundation and framework were completed, another crisis threatened to shatter Johnston's dreams for a laboratory. The government had decided to push forward the research necessary to build and test an atomic bomb. Part of the urgently needed research was for more information on hydrogen and deuterium as likely moderators. The university received word that its low-temperature equipment was needed elsewhere for war research, and Johnston was requested to set up and direct a cryogenics laboratory in the East. By some remarkably fast footwork and persuasion, university officials and Johnston managed to get the government to locate the cryogenics laboratory at Ohio State University. By mid-November, Johnston had a research contract.

Johnston worked best under pressure and short deadlines. He quickly recruited a staff including Gwynne A. Wright, an engineer who was to remain with him for 16 years, and Dr. Thor A. Rubin, a research chemist and another pupil of Giauque. Wright was placed in charge of installing the liquefier equipment. Typically, Johnston drove himself and his men hard. During December and most of January they worked in overcoats, for the building was still under construction and without heat. On 2 February 1943, Johnston and Wright produced their first batch of liquid hydrogen (fig. 1) and Rubin lost no time in making use of it in an experiment.[5]

Fig. 1. Gwynne A. Wright, left, operating Professor Herrick L. Johnston's first hydrogen liquefier at Ohio State University, as Johnston observes, ca. 1943. (Courtesy of G. A. Wright.)

By the end of his Manhattan Project research contract in 1946, Johnston had a fine cryogenics laboratory. Included were air, hydrogen, and helium liquefiers and other low-temperature equipment. He organized five sublaboratories—calorimetric, high pressure, spectroscopic, electrical and magnetic, and high temperature. The last, capable of reaching temperatures up to 2700 K, indicates that Johnston viewed a cryogenics laboratory in very broad terms.

The hydrogen system comprised five major components. Gaseous hydrogen was generated by electrolysis of water and the equipment was capable of producing 2 cubic meters per hour. The hydrogen was purified by a series of steps including heating to 570 K to remove oxygen, chilling to remove moisture, and use of a liquid-air trap to remove other condensable impurities. The third component was a three-stage compressor with an output of 1.7 cubic meters per minute at 300 atmospheres pressure. The hydrogen liquefier, a group of heat exchangers, was capable of 25 liters of hydrogen per hour. A large vacuum pump, capable of handling 5 cubic meters at a vacuum of 0.03 atmosphere, comprised the last component.

The hydrogen liquefier was modeled after the one developed by Giauque which was, in turn, a refinement of the basic process of regenerative cooling used by James Dewar in the first liquefaction of hydrogen in 1898. The process consisted of cooling high-pressure gaseous hydrogen as close as possible to the boiling point of liquid hydrogen (20.3 K) and then expanding the gas through a valve. Expansion provided the final cooling needed to liquefy part of the gas. Dewar used boiling liquid air for part of the hydrogen cooling and passed the cold, expanded hydrogen gas through a coil containing the incoming high-pressure gas on its way to the expansion valve. Giauque and Johnston did the same, although they used a total of eight heat exchangers to increase liquefaction efficiency. The liquefier (fig. 2) was diagramed and described by Johnston in 1946.

In steady-state operation, liquid hydrogen was in the left column (fig. 3). This column had four heat exchangers: the three upper ones, A', B, and F, used escaping cold gaseous hydrogen as a coolant; the bottom heat exchanger, G, was immersed in the liquid hydrogen, which served as a coolant. The right column also had four heat exchangers; the two upper ones, A and C, used escaping cold gaseous nitrogen and oxygen, from liquid air boiling under reduced pressure, as coolants. The two lower heat exchangers, E and D, were immersed in liquid air as coolant. The liquid air was in two containers connected by a float valve, to ensure that the escaping gases were nitrogen-rich. (If the gas were oxygen-rich, it would burn when in contact with the oil of the pump.)

Incoming hydrogen gas at room temperature and a pressure of about 125 atmospheres was split between the two columns and received its first cooling in heat exchangers A and A'. The two hydrogen streams then combined and passed, successively, through heat exchangers B, C, D, E, F, and G, getting progressively colder until (at G) the gas was near the boiling point of liquid hydrogen, 20.3 K. Finally, the high pressure, cold hydrogen gas expanded through valve H and about 20 percent of it liquefied. The rest passed up through the heat exchangers and cooled the incoming high pressure hydrogen as previously mentioned. The liquefier produced about 25 liters per hour.

Fig. 2. The several heat exchangers (shell removed) of H. L. Johnston's hydrogen liquefier, ca. 1946.
(Courtesy of W. V. Johnston.)

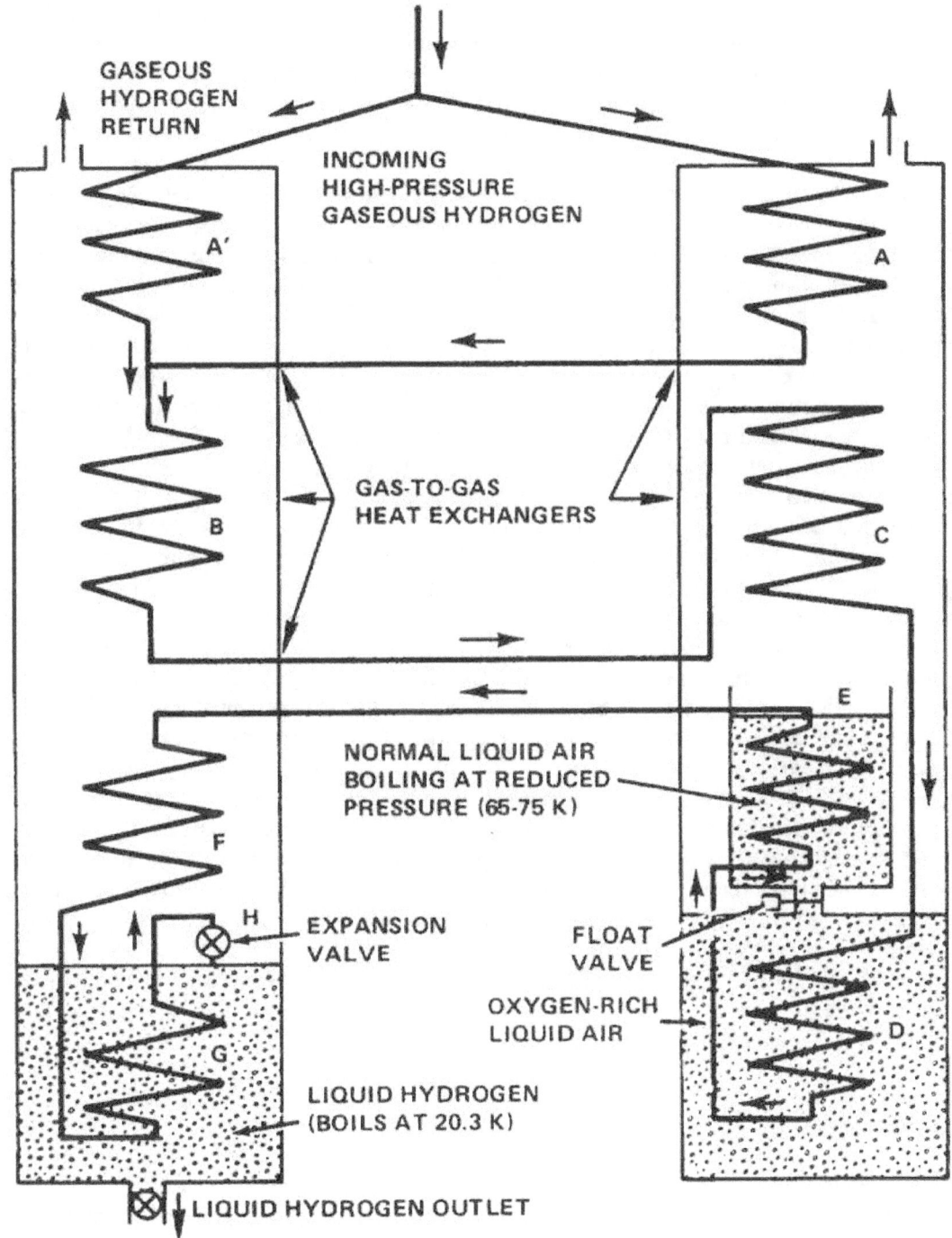

Fig. 3 Diagram of hydrogen liquefier, Cryogenic Laboratory, Ohio State University, 1946. Professor H. L. Johnston modeled this liquefier after one developed by Professor Giauque of the University of California.

Hydrogen for Aircraft and Rockets

By early 1945, the pace of war-needed military research had slackened. The government's laboratory at Los Alamos, New Mexico, was preparing for the first atomic bomb test in July. Johnston needed new support for his cryogenic laboratory and was receptive when the men in charge of fuels research at Wright Field approached him in the spring; agreement was soon reached on a contract, the first on hydrogen for aircraft and rockets in the United States.

Starting on 1 July 1945, the contract covered two major types of investigations. The first was hydrogen as a fuel for aircraft and rockets and was essentially engineering research. The second dealt with measurements of the physical, chemical, and thermochemical properties of hydrogen and the effect of very low temperatures on the properties of metals. This was science, the kind of work Johnston was most familiar with and which provided the research opportunities academicians seek for their graduate students. In 1948, both types of work were continued but under separate contracts. The fuel contract ended in December 1951, but the scientific properties contract continued. The contracts required bimonthly progress reports and annual summaries. In addition, special reports were written and the scientific work appeared in numerous doctoral theses and papers in scientific journals.[6]

The properties research contributed to the propulsion research by providing basic data needed for the theoretical aspects of propulsion research, such as thermochemical calculations of performance at various fuel/oxidant mixtures and combustion pressures, the composition of the exhaust gas and its properties for heat transfer calculations, and the properties of liquid hydrogen as a coolant.

Johnston devoted most of his time to his specialty, low-temperature equipment and properties research. The propulsion work was delegated largely to a group of engineers and technicians assisted by engineering students, all in the charge of a chief engineer. Three chief engineers served during the course of the propulsion work: Marvin L. Stary from early in the contract until 1949; Willard P. Berggren from 1949 to 1950; and William L. Doyle from 1950 to 1951. The rocket work involved, at one time or another, 18 research engineers, 21 students, 13 technicians, 7 administrative personnel, and 3 consultants. Figure 4 is a photograph of the rocket laboratory staff about 1950 and shows a typical mix of skills: 3 engineers, 3 engineering students, and 5 technicians.

Many aspects of the hydrogen work at Ohio State are beyond the scope of our subject, and only the work directly related to propulsion will be described. This is divided into five topics: hydrogen-air experiments, hydrogen-oxygen rocket performance, hydrogen-oxygen rocket cooling, pumping liquid hydrogen, and hydrogen-fluorine rocket performance.

Hydrogen-Air

In the initial days of the contract, the studies of hydrogen as a fuel related to its ignition and burning in air for possible application to jet engines. The work began with 67 tests of gaseous hydrogen injected, ignited, and burned in an air stream. No data were published, but presumably there were no problems. In the next series of experiments, liquid hydrogen was injected in open air ahead of a stream of air from a small pipe (about 7.5 cm in diameter), and later the liquid hydrogen was injected in the air flowing within the pipe. The liquid hydrogen flow was very small (about 2 grams per second) and when the liquid hydrogen was directed as a straight jet, an icicle—from moisture in the air—formed on the injector and impeded the flow. A splash plate was placed in front of the liquid hydrogen jet to spread it radially into the air stream and this gave better results. Combustion was maintained over a wide range of hydrogen-air mixture ratios.[8]

An unsuccessful attempt was made to use hydrogen in a pulsejet engine, the type of engine used in the German V-1. The inlet of a pulsejet consists of a number of flapper, or one-way, valves. When air enters through these, fuel is injected and the mixture ignited. The rise in pressure from combustion closes the flapper valves and the hot gases flow rearward through the nozzle, producing thrust. When operated on gasoline, the rapid series of explosive bursts was very noisy—as anyone can attest who experienced them popping along overhead in World War II. The Ohio State investigators obtained a tiny pulsejet engine marketed for model airplanes and substituted hydrogen for the fuel. It would not work because the very wide flammability limits of hydrogen resulted in continuous, rather than intermittent, burning. The investigators concluded that the narrow range of flammability of gasoline was responsible for establishing cyclic combustion by flaming out at lean and rich mixtures; it was these characteristics that made the pulsejet work.

Fig. 4. Staff of Ohio State University's rocket laboratory, ca. 1950. First row, L to R: Lester Cox, shop supervisor; Darwin Robinette, student; James Pierce, student; Arthur Brooke, test mechanic; William L. Doyle, chief engineer; Philip Petre, student. Standing: Lawrence Anthony, rocket shop supervisor; Ross Justus, machinist; Harold Smeck, engineer; James Sweet, test mechanic; William Strauss, engineer; unidentified. (Courtesy of Arthur Brooke.)

In the latter part of 1948, large-scale equipment was built to investigate hydrogen as a fuel for ramjets. A few tests were made, but were discontinued when the facilities were needed to test a liquid-hydrogen pump.

To sum up the hydrogen-air burning experiments, they were qualitative observations and verified only what was already well known—hydrogen burns in air over a wide range of conditions. The nature of the experiments and their cessation in favor of another project indicated a lack of interest in hydrogen as a fuel for air-breathing engines.

Other hydrogen-air experiments were made to assess the hazards of handling hydrogen. Tests of hydrogen-air explosions were made using a liter of liquid hydrogen in an open-mouth dewar. Ignition of the evaporating hydrogen resulted in a quiet flame, whereas hydrogen containing 10 percent solid air exploded with violence. Johnston was well aware of these characteristics as the following incident, part of the legend about him, illustrates.

Johnston supplied liquid hydrogen not only for his own experiments but also for the low-temperature experiments of other groups on the campus. One day a fire broke out at the top of a liquid hydrogen dewar of about 25 liters capacity being used for some materials testing. The fire department was called and the dewar was hurriedly rolled out into a parking lot. The firemen and a crowd were standing in a circle about the dewar, obviously puzzled about what to do next, when a passing car suddenly stopped in the middle of the street and a man got out. He pushed through the crowd, approached the dewar, pulled out his handkerchief and used it to snuff out the flame. He returned to his car and departed without having said a word. None in the crowd recognized Professor Johnston.[8]

Hydrogen-Oxygen Rocket

Experiments with hydrogen and oxygen in a rocket began at Ohio State University on 2 April 1947 and ended 29 May 1950. Similar tests were also underway at Aerojet General Corporation in California from 1945 to 1949 and at the Jet Propulsion Laboratory of the California Institute of Technology from 1948 to 1949, which will be described in the next chapter.

At Ohio State, the first twelve tests were made with liquid hydrogen and gaseous oxygen, because the installation of a liquid-oxygen tank at the test cell had been delayed. On 13 June 1947, Stary and his staff made the first rocket engine test in the United States using liquid hydrogen and liquid oxygen. The engine produced 471 newtons (106 lb of thrust) at a chamber pressure of 21.1 atmospheres with an oxygen-to-hydrogen mass ratio of 4.2. Exhaust velocity was 2405 meters per second, or 82 percent of the theoretical performance for that ratio (according to the theoretical performance given in the Lemmon report). Following this, an additional 118 runs were made with the same engine, and beginning in September 1948, 38 runs were made with an engine five times larger. One of the most significant accomplishments was a series of 37 runs at the smaller thrust using an engine regeneratively cooled with liquid hydrogen, starting of 26 August 1949. These will be discussed later.

An early problem for all rocket experimenters was satisfactory instrumentation to measure thrust, mass flow rates of fuel and oxidizer, and combustion pressure. From these the exhaust velocity at a given mixture ratio can be obtained and compared with theoretical calculations.* Of these measurements, the mass flow rate of liquid hydrogen was of most concern. It was determined by measuring the pressure differential across a sharp-edged orifice—a time-honored method of measuring flow rates. The accuracy depends upon the density of the fluid being measured; for liquid hydrogen, the large density changes with temperature are lessened at the high pressures used in rocket experiments. Measurements of hydrogen properties by Johnston, David White, and others were going on in parallel with the rocket work. Using Ohio State's temperature-density data, Johnston and Doyle reported that if the temperature of liquid hydrogen in a tank increased from its normal boiling point to the critical point at 25 atmospheres, the flow measurement would be approximately 10 percent too high. For this reason and because improvements in measurements were made as the tests progressed, strict comparisons of the various runs were not made, but qualitative comparisons were made to show trends.

The major design element affecting high performance is the propellant injector. Stary came to Ohio State from Aerojet General Corporation where impinging jet injectors had been successfully used with other propellants. In this type of injector, streams of propellant are directed so as to impinge on each other to break up the liquid stream into fine droplets and mix the fuel and oxidizer (fig. 5). This was the prevailing design philosophy of the period, and it is not surprising that major emphasis was placed on this type of injector for hydrogen-oxygen at Ohio State. In fact, 18 out of 20 injectors at Ohio State used impinging jets for at least one, and usually both, propellants. The two exceptions, a concentric tube and a "showerhead," were not tested. Ironically, these were later found by the Aerojet team and other investigators to be best for liquid hydrogen.

Rocket-Engine Cooling

In rocket experiments, the measurement of the heat flowing from the combustion gases to the engine walls and the use of this information to devise satisfactory cooling of the engine are second only to obtaining maximum performance. Without cooling, a flight-weight rocket engine would be heated to its melting point in a second or two. Major factors affecting this heat transfer are gas temperature, density, and velocity; all three of these are much higher in rocket engines than in other internal combustion engines. These factors, plus gas composition, are functions of the propellants, engine design, and operating conditions. The particular fuel and oxidizer, the proportions used, combustion pressure, and combustion efficiency determine gas composition, temperature, and density. Injector design, propellant proportions, mass flow, and combustion chamber design affect gas velocity. The rocket engineer seeks a design giving both high performance and a cooling method for steady-state operation. He is aided by combustion characteristics, for peak performance usually occurs at a fuel-rich mixture where the heat transfer is lower than at a leaner mixture.

---

*Rocket experiments used specific impulse (thrust divided by total propellant flow rate) for determining performance, which is equivalent to exhaust velocity used in this text (appendix B).

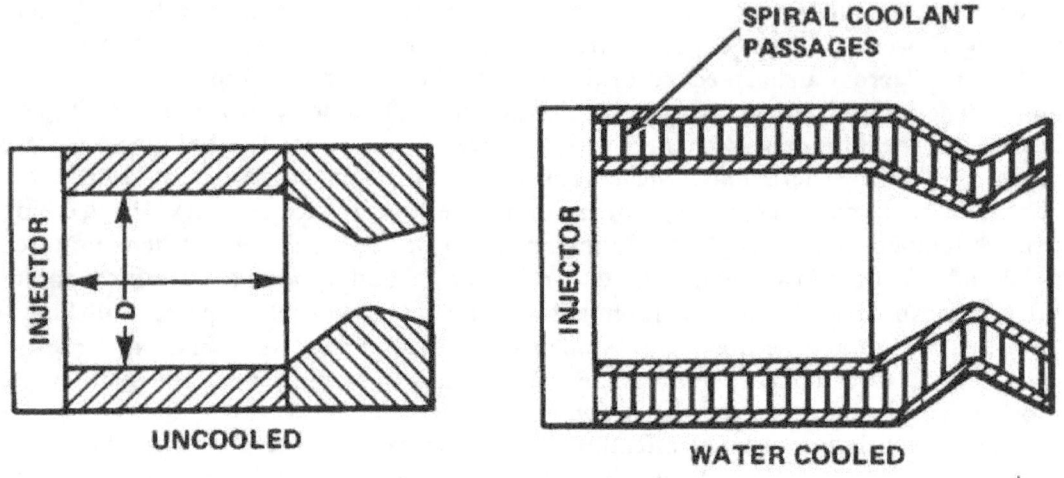

	445 N	2.2 kN
D, cm	4.4	8.7
L, cm	6.4	17 (UNCOOLED)
		10.3(COOLED)

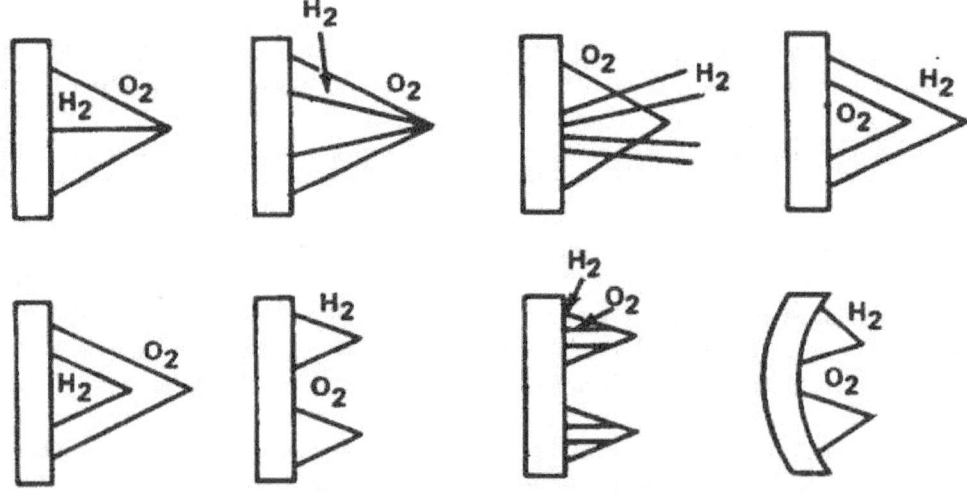

VARIOUS HYDROGEN-OXYGEN IMPINGING-JET INJECTORS

Fig. 5. Experimental rocket engines using hydrogen-oxygen, Ohio State University. 1947–1950.

Heat transfer measurements at Ohio State used two techniques common in rocket experiments. In the "heat-sink" method, the combustion chamber and nozzle are made from a high-conductivity material, usually copper, in which a thermocouple to measure temperature is buried in the thick, uncooled wall. During rocket operation, the high thermal conductivity of the copper keeps the inside wall from melting as the heat rapidly flows into the interior of the mass. This allows a rocket to operate for a few seconds, and sometimes as long as 30 seconds. After the run, the temperature of the copper mass comes to equilibrium and by measuring this temperature, the total amount of heat absorbed can be calculated from the known mass and specific heat of the copper. In the second method, a water jacket surrounds comparatively thin engine walls and a high-velocity water flow keeps the walls cool. The average heat transfer can be obtained by measuring the water flow and its temperature rise. Using these methods, Ohio State measured average heat transfer rates of about 1.6 joules per second per square meter (1 Btu/sec-sq in) for the combustion chamber and about twice that for the nozzle. These values were on the same order as found in high-performance rocket engines using other propellants, but are several times higher than heat transfer rates in other types of internal combustion engines and are, for example, from 20 to 200 times higher than in steam plants.

In mid-1948 a mechanical engineer from Aerojet, Irwin J. Weisenberg, joined the Ohio State rocket staff under Stary and specialized in heat transfer and cooling experiments. The first attempt to use hydrogen as a coolant was to employ a porous combustion chamber wall and force hydrogen through the wall into the combustion chamber.[10] This type of cooling, called transpiration or "sweat" cooling, was popular at the time and work with it was under way at several other rocket laboratories.

In the first part of 1949, another engineer at the Ohio State rocket laboratory, Clair M. Beighley, made a theoretical analysis in which a temperature ratio involving combustion gas temperature, wall temperature, and coolant temperature was related to dimensionless flow parameters. A porous combustion chamber was tested later and the experimental data agreed with the theoretical predictions. Porous wall chambers with uniform permeability were difficult to make, however, and the Ohio State rocket engineers turned to regenerative cooling when an analysis showed it to be feasible. In this method, hydrogen is circulated in coolant passages surrounding the engine prior to injection and burning.

In the midst of preparations to try it experimentally (in June 1949) Stary returned to Aerojet and still another Aerojet engineer, Dr. Willard P. Berggren, arrived at Ohio State as the new chief engineer for rocket experiments.[11]

The experimental thrust chamber for regenerative cooling was designed to produce 445 newtons at a chamber pressure of 20.4 atmospheres (fig. 6). Liquid hydrogen in the coolant jacket would be well above this value and hence far above its critical pressure of 12.8 atmospheres so that no boiling could occur in the coolant passages. The first successful regenerative cooling run was on 26 August 1949, when the thrust chamber operated for 60 seconds at an oxygen-to-hydrogen mass ratio of 4.1 and produced an exhaust velocity of 3190 meters per second—about 93 percent of theoretical performance.

In all, 33 successful runs were made, over half of which operated for 60 or more seconds; one operated for 159 seconds. The runs covered a range of mixture ratios and

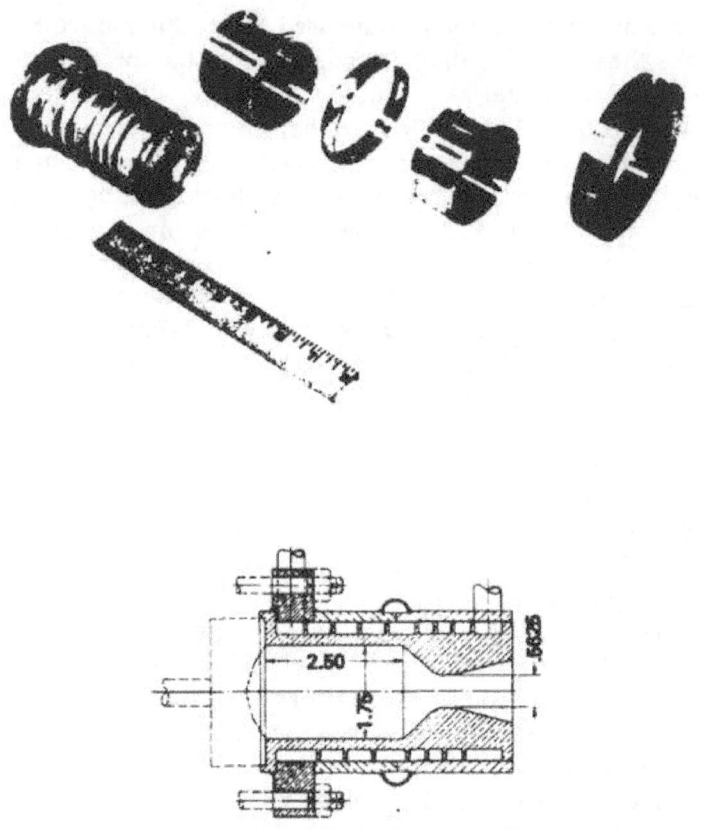

100 LB. THRUST REGENERATIVE MOTOR

Fig. 6. Rocket thrust chamber of 445 newtons designed to use liquid
hydrogen-oxygen and be regeneratively cooled by the liquid hydrogen,
Ohio State University, 1949. Scale and dimensions are inches.
(Courtesy of I. J. Weisenberg.)

the maximum exhaust velocity for the series was 3270 meters per second.* In general,
performance with the regeneratively-cooled engine was considerably higher than that
obtained with the water-cooled chambers. The experimenters attributed this not only
to the elimination of heat losses, but also to a lower-density hydrogen entering the
combustion chamber, which produced improved mixing and higher combustion
efficiency. Figure 7 shows the regeneratively-cooled rocket operating in December
1949 during the series of tests. The frost on the chamber indicates that it was well
cooled.[12]

---

*The highest performance run lasted 90 seconds at a fuel-rich mixture (O/F.4.7), 21 atm, and a relatively
low overall heat transfer rate of 2.1 $J/s \cdot m^2$. In contrast, the longest run (159 sec.) was at the stoichiometric
mixture (O/F,8), 19.6 atm, much lower exhaust velocity (2800 m/s), but almost triple the overall heat
transfer rate (5.2 $J/s \cdot m^2$). The comparison illustrates that peak performance does not come at the same
operating conditions as maximum heat transfer. It also shows that hydrogen cooling handled the higher heat
load.

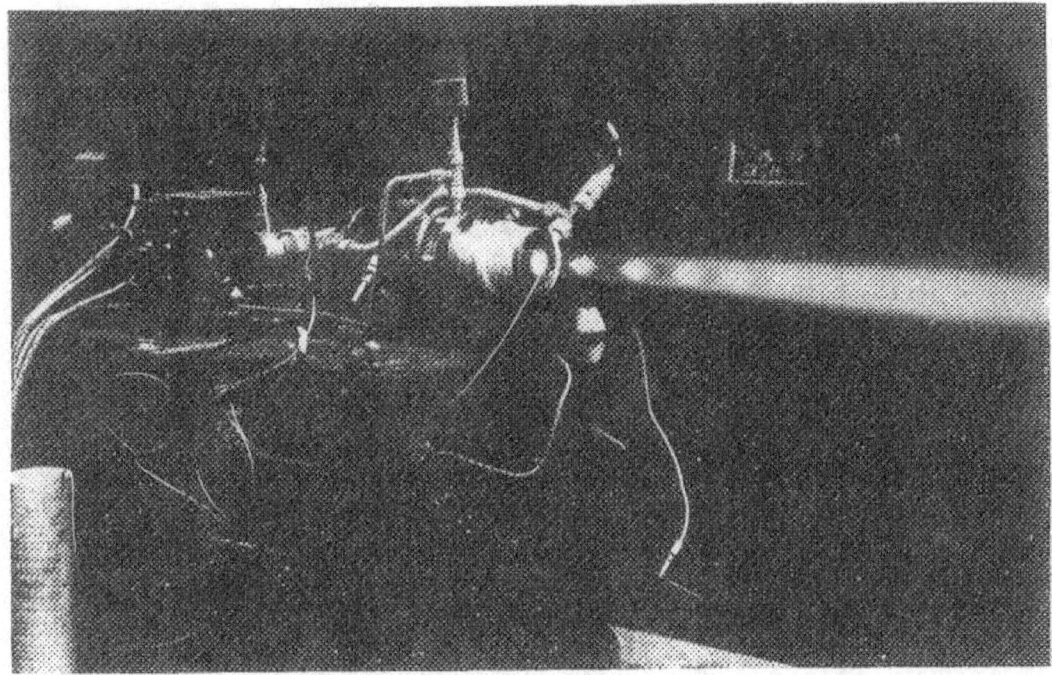

Fig. 7. Liquid hydrogen–oxygen rocket engine regeneratively cooled by the hydrogen, Ohio State University, December 1949. Note the frost on the outside of the rocket chamber and the shock diamonds in the exhaust. (Courtesy of I. J. Weisenberg.)

Pumping Liquid Hydrogen

By 1947, the scope of rocket research at Ohio State had broadened to study pumping of liquid hydrogen, which was carried out by Leroy F. Florant with the assistance of another engineer, Harold F. Snider. They were aware of the German development of a pump for liquid oxygen and of parallel work on liquid hydrogen at Aerojet General beginning in 1948, but their research was the most comprehensive analytical and experimental investigation of liquid hydrogen pumping of the period. They built two facilities—one for using fluids normally liquid at room temperature and the other for low-temperature fluids. They used water and isopentane in the first and liquid nitrogen and liquid hydrogen in the second. In the low-temperature facility the tanks, lines, and valves were vacuum jacketed for insulation. After initial troubleshooting, the system worked well, although there were high liquid hydrogen losses from the conversion of orthohydrogen to parahydrogen (pp. 266–67).

Florant and Snider designed, built, and tested two types of centrifugal liquid hydrogen pumps. They also investigated bearings and seals at speeds up to 10 000 RPM. They concluded that centrifugal pumps were a desirable and practical way of pumping liquid hydrogen for rocket engines. They found that water was a satisfactory test fluid for determining and verifying pump design parameters for liquid hydrogen pumps. This facilitated testing and greatly reduced its cost.[13]

One of the most significant results the two investigators obtained was that properly mounted, precision ball bearings would operate satisfactorily in liquid hydrogen at speeds up to 10000 RPM. The bearings were cooled by the liquid hydrogen and required no lubrication. This useful information was to be rediscovered by Richard Mulready of Pratt & Whitney in 1958 in developing the first flight-model liquid hydrogen–oxygen rocket engine.

Hydrogen and Fluorine

Early in 1949 William L. Doyle, a chemist engaged in rocket propellant research at North American Aviation, made a deal with Herrick Johnston. Doyle would come to work at the Ohio State University Rocket Laboratory if given a free hand to investigate the performance of liquid hydrogen with his favorite oxidizer, liquid fluorine.*

In February 1949, Doyle reported for duty at the Ohio State laboratory. He did not like the experimental equipment, the operations, or the procedures, so he began to make changes.

William Doyle was a dynamic young man who knew what he wanted and just how to do it. The antithesis of the desk-bound supervisor and paper shuffler, he liked to be part of the action. He found his right environment at Ohio State where a senior engineer was responsible for his entire project—from inception, through design, fabrication, installation, operation, data analysis, and writing up the results. Doyle found this situation ideal and he made the most of it.

Doyle's interest in the hydrogen-fluorine combination was natural. It represented the combination of the ultimate fuel and the ultimate oxidizer, with a higher theoretical performance than hydrogen and oxygen. In addition, the mixture of 6 percent hydrogen and 94 percent fluorine by weight not only resulted in near-maximum performance, but also meant higher average propellant density for the combination. Doyle visited the men in the fuels and oil branch at Wright Field and convinced them to modify the Ohio State contract to include the work he wanted to do.

One of Ohio State's rocket test facilities was rebuilt to handle liquid hydrogen and liquid fluorine. The hydrogen flow system was encased in a vacuum jacket for insulation. A series of problems with maintaining the vacuum were solved. The flow of liquid hydrogen was measured by a dual system: the conventional way of measuring the pressure differential across a sharp-edged orifice as well as continuous measurement of the hydrogen tank mass. Once the hydrogen system was functioning, it gave little more trouble, but many problems were encountered in the fluorine system. The fluorine gas, procured commercially as a compressed gas, was condensed in the

---

*Doyle had become interested in fluorine at North American Aviation when he was assigned to take over an experimental rocket project. Doyle redesigned the equipment and the test rig to his own liking and proceeded to investigate the burning of hydrazine and fluorine. On 8 Nov. 1947, Doyle successfully operated the first rocket to use fluorine, the most powerful of all stable oxidizers. He found that fluorine quickly decomposed Teflon, at the time the favorite "inert" material for gaskets and sealants in rocket experiments. Fluorine will also combine with moisture and impurities and once a reaction starts, the heat generated makes it quick to attack and burn metals. It is also highly toxic, and Doyle's parking-lot operation became a matter of some concern by the time he was ready to head for Ohio. Interview with William L. Doyle. Redondo Beach. CA. 26 Apr. 1974.

propellant tank by immersing it in a liquid air bath. Liquid fluorine flow was measured by the same methods used for hydrogen.

Doyle made his first liquid hydrogen–liquid fluorine run on 15 June 1950. He first operated the injector alone to see if the hydrogen and fluorine would ignite readily and spontaneously, which they did. He followed this experiment with rocket engine tests. By the first part of August, nine runs had been made and Doyle felt confident enough to invite his sponsors from Wright Field to witness a test. Judging from the mishaps reported for the first eight runs, Doyle was displaying a considerable amount of confidence. Don Kennedy arrived in response to the invitation and witnessed the tenth test on 11 August 1950. The run was perfect in Doyle's view, with a measured exhaust velocity of about 3600 meters per second at 20 atmospheres. Kennedy was greatly impressed and reported the results to his boss, Weldon Worth. Doyle continued the experiments and in mid-January 1951, Kennedy informed him that a group of high officials at Wright Field would visit Ohio State to witness a run with hydrogen-fluorine. Soon after the call, Doyle made a run at a high pressure (38 atmospheres) and measured an exhaust velocity of over 4300 meters per second. On 29 January, 14 people from Wright Field's Power Plant Laboratory arrived in terrible weather—a sheet of ice compounded by mist and drizzle. Icing difficulties delayed the run for an hour, but it was a success, lasting over a minute. Performance, however, was lower than obtained in earlier runs.[14]

One measurement necessary to determine performance—fluorine flow—had bothered Doyle from the start. Whereas the two flow measurements for liquid hydrogen checked with each other, the fluorine flow as measured by the orifice was lower than that measured by weighing the propellant tank. The difference was consistent—about 18 percent lower for the orifice. Five design changes were made to improve the orifice measurement, but the discrepancy remained.

Doyle was not the only experimenter having difficulty measuring the flow rate of liquid fluorine. Aerojet was having the same difficulty and investigators there began to suspect that the density of fluorine might somehow be wrong. This was heresy, for a number of eminent scientists had measured the density of fluorine and they all agreed. James Dewar and Henri Moissan had first measured it in 1897 and found it to be close to 1.14 grams per cubic centimeter at 83 K. The value in use in the 1950s was 1.13 grams per cubic centimeter at 77 K, determined by E. Kanda in 1937.

Near the end of April 1951, Kennedy telephoned Doyle that Aerojet, using a hydrometer, found that the density of liquid fluorine was 1.55 grams per cubic centimeter, considerably higher than the published value. Doyle used the Aerojet value with his orifice measurements and found that the 18 percent discrepancy with the weighing measurement disappeared! The greater density of liquid fluorine was an exciting discovery to rocket engineers, for it meant the oxidizer was even more attractive than first realized.*

---

*The specific gravity of liquid fluorine, 1.54 at 77 K, reported by Kilner, Randolph, and Gillispie (*J. Am. Chem. Soc.* 74:1086) in 1952 was verified by Elverum and Doescher the same year (*J. Chem. Physics* 20:1834). See also National Bureau of Standards Technical Note 392, rev., 1973. Some contributors to chemical handbooks were slow in noting the change.

Doyle made his 48th and last hydrogen-fluorine run in mid-April 1951 and turned his attention to the ammonia-fluorine combination. This ended the Ohio State rocket experiments with liquid hydrogen, although the properties work continued, as well as some small-scale combustion research of a fundamental nature.

Significance

The first experimental investigation of liquid hydrogen as a fuel for aircraft and rockets was started in 1945 by the research arm of the Air Force, as part of a long tradition of searching for new and improved fuels. Hydrogen, the ultimate fuel in energy content, needed to be investigated for its potential application in air-breathing and rocket engines. The availability of hydrogen liquefaction equipment and the experts at Ohio State University provided the catalyst for starting the experimental investigation.

The Ohio State research on hydrogen for air-breathing engines never progressed beyond a few exploratory experiments. These showed that hydrogen burned readily over a wide range of conditions—a result that could have been predicted from earlier work. That more was not done with hydrogen for air-breathing engines could have come from one or more of the following: (1) hydrogen's low density, long its outstanding disadvantage for aircraft applications, as pointed out by Tsiolkovskiy in 1930; (2) rising interest in boron compounds as high-energy fuels for ramjets, as sponsored by the Navy; (3) greater interest in rocket applications by Wright Field; and (4) lack of equipment needed for research on air-breathing engines.

Ohio State University investigators focused their engine research on rocket engines and made many contributions to liquid hydrogen technology. The high performance potential of liquid hydrogen–liquid oxygen was verified, and it was also found that liquid hydrogen was a satisfactory regenerative coolant. Research established that centrifugal pumps were capable of pumping liquid hydrogen to the high pressures needed for rocket engines. It was also found that ball bearings for pumps would operate satisfactorily when immersed in liquid hydrogen without the usual oil lubrication, showing that design of practical pumps was feasible. Pump tests with water produced data that were valid in predicting performance with liquid hydrogen— a decided convenience in determining several design parameters. Finally, it was shown that the performance of liquid hydrogen–liquid fluorine was higher than for liquid hydrogen–liquid oxygen, and density was higher also.

With such significant results with liquid hydrogen, then, why did Air Force interest in sponsoring further research begin to wane in the late 1940s? Several possible reasons come to mind, one being the shift in Air Force interest from rockets to air-breathing propulsion in the late 1940s. Another possibility is that the Air Force managers may have felt the exploratory research had fulfilled all of its objectives and without an application, there was no need for further work. There is, also, the ever-present possibility that the sum of all of hydrogen's disadvantages—formidable for military applications—may have overwhelmed Wright Field's attraction to the high energy of hydrogen in the same manner experienced earlier by both Tsiolkovskiy and Goddard.

The scientific and technological progress made at Ohio State with liquid hydrogen served as the foundation for contributions by other groups. Running parallel to Air

Force interest in hydrogen as a fuel was Navy interest, which also faded by the end of the 1940s. Unlike the Air Force, however, the Navy had a specific application in mind and its efforts to secure approval to develop a hydrogen-oxygen rocket will be discussed next.

3

Hydrogen-Oxygen for a Navy Satellite

The Navy's Bureau of Aeronautics became interested in liquid hydrogen as a rocket fuel in the second half of 1945 in connection with its early satellite proposals. Unlike the Wright Field contract with Ohio State University, which was research-oriented with no specific application in mind, the Navy interest was, from first to last, linked directly to its proposal to use a single-stage rocket to boost a satellite into orbit. For this reason, the effort is best viewed within the broader context of the Navy's early interest in missiles and satellites.

Origins of Navy Interest in Satellites and Hydrogen

Considering the Navy's involvement in solid rocket research and development during the war, the rising interest in jet propulsion as German developments became known, the Navy's sponsorship of OSRD's Jet Propelled Missiles Panel, and the Lemmon report on jet propulsion fuels (p. 4), the interest in hydrogen would appear to be an evolutionary step. In fact, these prior events had little influence. The proposal to use liquid hydrogen to place a satellite into orbit with a single-stage-to-orbit rocket came from Comdr. Harvey Hall, a Navy physicist who had educated himself quickly in jet propulsion, had not heard of the Wright Field contract with Ohio State University on liquid hydrogen (p. 18), and had not read the Lemmon report. Neither was he acquainted with the proposals of Tsiolkovskiy, Goddard, or Oberth to use hydrogen in rockets (appendix A-2); but like Tsiolkovskiy, he had gone to chemistry textbooks in search of the most energetic fuel. Not surprisingly, Hall found and selected liquid hydrogen, and in his quest for more information on its use in rockets, he met Robert Gordon of the Aerojet Engineering Corporation, who also had gone to his textbooks and was thinking about hydrogen at about the same time.[1]

The train of events that led to the Navy's interest in satellites and the use of liquid hydrogen as a fuel in the booster rocket was triggered by information brought to the Bureau of Aeronautics in July 1945 by a young Marine officer, Lt. Abraham Hyatt. The Bureau of Aeronautics was aware of German developments in jet propulsion and rockets from intelligence reports during the war. Hyatt had been part of a technical intelligence team in Europe following closely in the wake of the advancing armies early

31

in 1945 to interrogate German scientists and technicians and gather documents. Among the Germans interrogated in May 1945 were Wernher von Braun and his associates who had developed the V-2 at Peenemünde. Among the documents Hyatt brought to the Bureau of Aeronautics in July 1945 was a summary by von Braun of liquid propellant rocket developments in Germany and his view of future prospects. Von Braun listed five future possibilities: (1) rocket-propelled transports for intercontinental travel; (2) multi-stage, piloted rockets orbiting the earth; (3) a large space station orbiting the earth; (4) a large orbiting mirror to concentrate solar energy and beam it to the earth for various purposes, including weather control;* (5) travel to other planets but "first of all to the moon," possibly by harnessing atomic energy. Von Braun saw the rocket as having the same impact on future scientific and military activities as the airplane.[2]

Among those in the Bureau of Aeronautics who were most excited over the potential of satellites were Lt. Robert Haviland and Comdr. Harvey Hall. By the first part of August, Haviland had written an internal memorandum proposing that the Navy initiate a program leading to a manned space station. He developed the Tsiolkovskiy equation† relating vehicle velocity to exhaust gas velocity and mass ratio, but referred only to available fuels, with no mention of hydrogen. A British report of March 1945 on the mass of various components of the V-2 was used by Haviland to calculate the terminal velocity of a two-stage rocket based on these masses. The result was disappointing; the second stage velocity was too low to achieve orbit. To get out of this dilemma, Haviland drew on a 1934 publication of E. Sänger to assume that an exhaust gas velocity of 3500 meters per second was achievable with gasoline and oxygen.‡ This is highly optimistic even at altitude: the V-2 exhaust gas velocity, using alcohol-oxygen, was only about 2/3 that value. However, his conservative mass and optimistic rocket performance assumptions led him to the correct conclusion that a satellite could be launched with a two-stage rocket booster using gasoline-oxygen. He wisely included a recommendation that more research be undertaken to secure a high energy fuel. As a further assurance of success, he suggested that the launch be made from a mountain top, to gain altitude, and in the direction of the rotation of the earth, to gain rotational velocity.[3]

In spite of his excitement over satellites, Hall took a slower and more deliberate course than Haviland. For one thing, he was not well acquainted with jet propulsion, but having a doctorate in physics, he went to basic concepts to work out the flight and energy relationships for himself. In the process, he also obtained the Tsiolkovskiy

---

* The same month that Hyatt brought von Braun's predictions to the Bureau of Aeronautics, *Life* published an article on the German plans for a large orbiting mirror which was also a manned satellite. The article stated that the Germans had planned to use the mirror to focus the sun's rays into a beam to scorch the earth. *Life* 19 (23 July 1945): 78-80.

† The Tsiolkovskiy equation is $V = V_r \ln(M_o/M_e)$ where V is the maximum velocity of the rocket in gravity-free, drag-free flight; V_r is the rocket exhaust velocity; ln is the natural logarithm; M_o is the initial, full, or gross mass of the rocket; and M_e is the final or empty mass of the rocket. The two masses differ by the amount of propellant expended. More details are given in appendix A-2.

Haviland used Willy Ley, *Rockets, the Future of Travel beyond the Stratosphere,* 1944, for tabulated values in the Tsiolkovskiy equation.

‡The NACA translated and published the Sänger paper in 1942 as Technical Memorandum 1012.

equation. He then began to explore the extremes of its two variables—exhaust gas velocity, determined by the energy of the reactants and expansion through the nozzle; and mass ratio, determined by the structure. He could have obtained excellent data on exhaust gas velocities of various propellants from the Lemmon report which had been issued in May, but it had not come to his attention. Instead, he simply went to his chemistry textbooks in search of the most energetic fuel he could find to use as a yardstick in comparing the performance of various fuels. There he found the hydrogen-oxygen combination, whose heat of combustion had been measured numerous times since Lavoisier and Laplace first measured it in 1783. Hall was totally unaware that he was repeating the same steps Tsiolkovskiy had taken almost a half century earlier (appendix A-2).

In considering the ratio of initial to final mass, Hall thought of very light structures, somewhat analogous to Oberth's, and his structural design was as optimistic as Haviland's was conservative. Hall's calculations led him to believe that, using liquid hydrogen and oxygen and very light structures, he could put a payload in orbit with a single-stage vehicle, eliminating the complications of multiple staging.

Hall wanted to discuss his calculations with rocket experts, so he visited the Jet Propulsion Laboratory (JPL) of the California Institute of Technology where he met with Martin Summerfield, Frank Malina, and Homer Joe Stewart.* Encouraged by his visit, Hall went on to the Aerojet Engineering Corporation to talk about rocket propellant experiments.

Aerojet Propellant Research, 1944–1945

At the end of 1944, the Aerojet research group, headed by Fritz Zwicky, noted astrophysicist at the California Institute of Technology, had completed a Navy contract to investigate high-energy solid and liquid propellants. The results led the investigators to monopropellants; they were enthusiastic over the possibilities of using nitromethane, which has a theoretical exhaust velocity of 2200 meters per second. Zwicky was aware of other Navy-sponsored work on boron hydrides that had potential exhaust velocities of 2800 to 3100 meters per second—considerably higher than nitromethane but also much further from practical utilization.[4] At the time of Hall's visit, Aerojet was in the second phase of investigating nitromethane—determination of its experimental performance and handling characteristics. David A. Young and his new assistant, Robert Gordon, were in charge of the work, and Hall asked them about the combustion properties of hydrogen and oxygen.

Gordon had worked on aircraft engines at the power plant laboratory at Wright Field for several years and later was a navigator with the Eighth Air Force in Europe. There he had acquired a first-hand awareness of German competence in jet propulsion.

---

*Rocket research began at the Guggenheim Aeronautical Laboratory of the California Institute of Technology (GALCIT) in 1936 and was known as the GALCIT Rocket Project. GALCIT became the undisputed leader in rocket research during the 1940s. In 1944 the project was reorganized and named the Jet Propulsion Laboratory, GALCIT; it is now called the Jet Propulsion Laboratory of the California Institute of Technology or simply JPL. R. Cargill Hall, "GALCIT-JPL Developments, 1926–50, a Chronology," 8 Sept. 1967, NASA History Office.

He took part in bombing Peenemünde, observed the launching of a V-2, and was attacked by ME-163s—the first rocket-powered aircraft. Gordon joined Aerojet in July 1945 and during his orientation, Young introduced him to the fundamentals of rocket theory. Gordon then began calculating theoretical rocket performance of various propellant combinations using the heat of formation of exhaust products. This lead to the consideration of the simplest and most energetic reaction—hydrogen and oxygen—and he asked Young to let him try hydrogen-oxygen in a rocket experiment.[5]

Aerojet's First Series of Experiments, 1945–1946

Hall's visit to Aerojet was fortunate in its timing. He believed that he brought a new idea to Young and Gordon because none of Aerojet's previous work or proposals on propellants mentioned liquid hydrogen as a fuel. To Gordon, however, here was his boss's boss—the Navy—voicing ideas similar to his own and he was eager to get started. After Hall returned to Washington, Aerojet was authorized to experiment with hydrogen and oxygen as part of their nitromethane contract. In less than a month, Young and Gordon operated the first recorded run of a hydrogen-oxygen rocket in the U.S. on 15 October 1945.* The run ended after 15 seconds when the uncooled thrust chamber burned out, but not before a thrust of 200 newtons (45 lb) and a chamber pressure of 25.5 atmospheres were recorded. From these, the experimenters estimated the exhaust velocity to be 2600 meters per second.[6] They were undaunted by the burnout and began preparations to use a water-cooled thrust chamber. In the next test, they obtained a lower exhaust velocity, and in spite of water cooling, the chamber showed signs of overheating.†

From the first test in October until the end of the first phase of the work in June 1946, about 50 rocket runs were made at thrusts of 445 and 1780 newtons (100 and 400 lb) and chamber pressures of 20.4 and 34 atmospheres. The experimenters found it relatively easy to achieve high performance (3050 meters per second). Much of the work was concentrated on cooling and several methods were tried. One was a porous chamber through which water was forced as a form of transpiration or "sweat" cooling. Another was gaseous hydrogen flowing through the porous combustion chamber. The main method, however, remained water cooling.

As had other experimenters since the eighteenth century, the Aerojet research team found that hydrogen and oxygen ignite very readily and burn over a wide range of mixture ratios. Rapid burning meant that the combustion chamber could be small, and this led Young to his idea of the ultimate small thrust chamber—the "flared tube."

---

*Richard B. Canright operated a gaseous hydrogen-oxygen rocket at JPL about 1943, but no reports on this work have been found. The first JPL laboratory was referred to as the "Gashouse" and apparently Canright used gaseous hydrogen and oxygen, mainly for their convenience and availability. Howard S. Seifert, "Twenty-Five Years of Rocket Development," *Jet Propulsion* 25 (Nov. 1955):595; telephone interview with Howard Seifert, 22 Aug. 1973; Seifert to Sloop, 29 Nov. 1973; telephone interview with Richard B. Canright, 21 Aug. 1973; interview with Richard B. Canright, Camp Hill, PA, 7 Mar. 1974. The Germans operated a hydrogen-oxygen rocket during 1937–1940 (appendix A-3).

†Average heat transfer rate was 5.7 $J/s \cdot m^2$; this value and the relatively low exhaust velocity are approximately the same as Ohio State obtained later at the stoichiometric mixture (p. 24, n.).

Essentially it was a straight wall tube for the combustion chamber with a flare for the expansion portion of the nozzle, as shown at the top of figure 8. Young experimented to find the minimum size tube chamber and soon became confident that he could use from 1/10 to 1/20 the volume normally used for rocket thrust chambers. This was a great step forward, for a tiny combustion chamber meant less mass for the vehicle and less surface area to cool—both big advantages. He became a missionary for the idea and set forth to sell the Navy an expanded program.

The Hall Committee

Haviland's August memorandum proposing a manned space station (p. 32) was convincing to his supervisor, Comdr. J. A. Chambers, head of a special weapons section, who saw among its advantages the possibility of a worldwide navigation and communication system on high frequencies—free from horizon limitations and sky-wave errors. He endorsed it and passed it up the line. It also received support from Capt. Lloyd V. Berkner, head of the electronics materiel branch. During this time, Hall was arguing his case for the single-stage rocket, and he must have been persuasive because on 3 October 1945, Capt. R. S. Hatcher, deputy director of engineering in the Bureau of Aeronautics, established the Committee for Evaluating the Feasibility of Space Rocketry. Its purpose was "to investigate the presently available materials and techniques and to arrive at some estimate of the possibility of attaining a velocity of liberation from one stage of operation." Hall was made chairman and the first meeting was held five days later.* Both Haviland and Hall explained their ideas, and it was revealed that detailed calculations for an earth satellite were under way in another branch of the Bureau of Aeronautics.[7]

The second meeting took place on 15 October 1945, and the subject was experimental data on some fuels and theoretical estimates on others. Lt. Comdr. F. A. Parker presented experimental data on only two propellant combinations: mixed nitric and sulfuric acid with methyl-ethyl-aniline, and alcohol with liquid oxygen. He gave their exhaust velocities at sea level as 1950 and 2300 meters per second, respectively. He thought that any hydrocarbon-oxygen system would likely have an upper limit near that of the alcohol-oxygen value. Parker estimated that increasing the combustion pressure to practical limits would increase exhaust velocity about 15 percent. A greater increase would be possible by increasing the area ratio of the exhaust nozzle. The upper limit on this appeared to be an increase in velocity of about 40 percent over sea level values. The theoretical performance of hydrogen and oxygen was given as 3000 meters per second at sea level and 4300 at altitude. The performance of diborane and oxygen was unknown, but was estimated (optimistically) to be about the same as for hydrogen and oxygen.

The Hall Committee concluded that a single-stage rocket for boosting a satellite to orbit would need an exhaust velocity on the order of 4300 meters per second and recommended that the performance of both hydrogen and diborane be investigated.

---

*Other members: Comdr. C. D. Case, Lt. (jg) K. W. Max, Lt. R. P. Haviland, Lt. Comdr. F. A. Parker, Lt. L. A. Hansen, Comdr. O. E. Lancaster, and J. R. Moore.

theoretically and experimentally.* The same day, Aerojet made their first experimental rocket test with hydrogen and oxygen. The exhaust gas velocity during the run was estimated at 2600 meters per second, which meant that 3600 would be attainable at altitude with proper design. No one had tried diborane, but Hall was attracted to it as an alternate to hydrogen. At the next meeting, on 22 October 1945, he discussed diborane and estimated that it could produce an exhaust velocity of 5500 meters per second, a value far greater than that for hydrogen-oxygen.[9] Diborane therefore appeared to be the dream fuel, but Parker pointed out that boron oxides, formed during combustion of diborane and oxygen, might solidify when expanded to the lower temperatures in the nozzle, and this would lower performance.* L. A. Hansen raised the problem of dissociation, where energy is absorbed in breaking molecules apart, which would further reduce the exhaust velocity. In spite of these cautions, the committee accepted the 5500 meters per second theoretical value for diborane-oxygen and estimated that actual performance would probably be close to the desired 4300. Hall recommended that: (1) an experimental program be initiated leading to a satellite orbiting the earth at an altitude of 1850 kilometers; (2) engineering layouts be made on the basis of an exhaust velocity of 4300 meters per second and a mass ratio of 10, and an empty mass of at least 4500 kilograms; (3) the vacuum performance of the most promising fuels having estimated exhaust velocities of 4300 meters per second be tested; and (4) diborane and similar compounds be studied. With this proposal, Hall—the original proponent of hydrogen-oxygen—was now referring to that combination only indirectly in terms of performance and was urging the study of diborane as a fuel. The committee agreed with Hall's proposal for an engineering design layout with his guidelines, but made no reference to diborane.

By the fourth meeting, on 29 October, the committee amended the minutes of the previous meeting to agree with Hall's higher estimate for the performance of diborane. Both Lancaster and Haviland, however, had analyzed boosters, and they continued to prefer hydrogen and oxygen. The two analysts differed in their mass assumptions. Lancaster found an initial-to-final mass ratio of 10 impractical, but Haviland did not. The committee found both sufficiently close to the desired goal to be promising and recommended that a more detailed study be made.[10] This was carried out by Lt. Comdr. Otis E. Lancaster and J. R. Moore.

In November 1945, Lancaster and Moore reported their study: "Investigation on the Possibility of Establishing a Space Ship in Orbit above the Surface of the Earth." Using the basic energy relationships and a simplified formula for estimating structural weight, comparisons were made of the minimum mass ratio needed for rockets to orbit at various altitudes with the mass ratios attainable with several propellant combinations. Liquid hydrogen–oxygen was considered the best on the assumptions

*Parker was right. In 1948, at the NACA Conference on Fuels, Flight Propulsion Research Laboratory, Cleveland, the author, P. M. Ordin, and V. N. Huff reported results from rocket experiments in which boron oxides were deposited on the nozzle, verifying Parker's speculation. In the 1950s the Navy and Air Force mounted a major effort to use boron hydrides in turbojet engines and failed, largely because boron oxides clogged the turbine blades. *Hearings on Boron High Energy Fuels before the Committee on Science and Astronautics*, U.S. House of Representatives, 26–27 Aug. and 1 Sept. 1959.

of a jet velocity of 4300 meters per second in a vacuum, which was realistic. The structural formula, however, made mass ratio results very pessimistic. Lancaster and Moore concluded that an initial-to-final mass ratio of from 10.9 to 12.1 was needed to orbit at a high altitude. Since the structural mass formula indicated that for a ratio of 10, a very large rocket (one with a mass of some 2270 metric tons) would be necessary without considering payload, the authors concluded that a single stage to orbit was not feasible.* A multiple stage rocket using alcohol and oxygen, however, could orbit a satellite.[11]

The analysis was a blow to Hall's single-stage-to-orbit concept, and he proposed that JPL conduct an independent analysis.

JPL Study

The rocket experts of the Jet Propulsion Laboratory of the California Institute of Technology began their study of single-stage rockets for the Bureau of Aeronautics in December 1945 and completed it by July 1946† They wrote six reports, with the earliest appearing on 3 January 1946. The study was based on three assumptions: (1) the orbiting vehicle would be a single-stage liquid propellant rocket, (2) the propellants would be liquid hydrogen and liquid oxygen, and (3) the exhaust velocity of the rocket would be 3240 meters per second at sea level and 4320 at very high altitude. The rocket performance values were furnished by David Young of Aerojet. With these assumptions, the JPL men sought to determine the most suitable trajectory and designs for minimum initial-to-final mass ratio.

The final report, appearing in July, stated that if the single-stage rocket was launched from sea level, the initial-to-final mass ratio must be 8.70; if launched from a high mountain (4300 m), the mass ratio could be decreased slightly.[12] These results made clear to the Bureau of Aeronautics what the next steps should be: expand Aerojet's work on the experimental performance of hydrogen and oxygen and get improved weight estimates for rocket engines and vehicle structures. The latter called for the experience of an airframe manufacturer. The JPL-GALCIT study also pointed out that the mass ratio requirements for orbiting a satellite could be greatly reduced if multiple-stage rockets were used.

Apparently as a derivative of these classified military studies, Frank Malina and Martin Summerfield reported on the problem of escape from earth by rocket in August 1946, and Malina presented the results at the Sixth International Congress for Applied Mechanics at Paris in September. They made a strong case for using hydrogen-oxygen. A multistage rocket using nitric acid and aniline (a combination in use at that time) was considered too large to be practical even for a 5-kilogram payload. They concluded

---

*Lancaster and Moore doubted the accuracy of the structural masses they were using and recommended that a detailed structural design study be made. They also recommended intensifying the research program on rocket fuels and engines to find fuels with higher exhaust velocities and to develop larger engines.

†Participating in the study were W. Z. Chien, Lt. Comdr. E. C. Sledge, Lt. Comdr. G. G. Halverson, J. V. Charyk, and H. J. Stewart. Stewart wrote the final report.

that a multistage rocket of reasonable size using liquid hydrogen and liquid oxygen could carry a payload of 45 kilograms and was within engineering feasibility. They assumed an exhaust velocity of 3660 meters per second for hydrogen-oxygen, five stages, and a gross mass of 37600 kilograms. The authors also pointed out the advantages of using hydrogen as the working fluid with heat supplied by a nuclear reaction. Potential exhaust velocities were as high as 11400 meters per second—close to the vehicle velocity needed for escape from the earth's gravitational field.[13]

Attitudes towards Missiles and Satellites

While the advocates of satellites in the Bureau of Aeronautics were pursuing their technical studies, they were also attempting to obtain high-level support. They estimated that five to eight million dollars would be needed, but in the budget competition, they faced an uphill struggle. Ironically, their sister service, the Army Air Forces, had support at the top but little initiative at the working level. During September, the AAF's Scientific Advisory Committee, headed by Dr. Theodore von Kármán, issued the first volume of its series, *Towards New Horizons*—a bold assessment of future developments.[14]

On 12 November 1945, in his Third Report to the Secretary of War, Gen. H. H. Arnold predicted that strategic bombers would eventually be replaced by long-range ballistic missiles that would need to be launched "from true space stations, capable of operating outside the earth's atmosphere."[15]

If the Bureau of Aeronautics men were heartened by Arnold's statement, they must have been dismayed the next month at the lack of support from the top scientist in the government. In December, Vannevar Bush, Director of the Office of Scientific Research and Development, appeared before the Special Senate Committee on Atomic Energy and stated:

There has been a great deal said about a 3000-mile [5600 km] high-angle rocket. In my opinion such a thing is impossible and will be impossible for many years.[16]

Bush was not alone. The following April, the chairman of the National Advisory Committee for Aeronautics, Jerome C. Hunsaker, echoed the same view in an address before the National Academy of Sciences:

One is tempted to speculate about the possibilities of an improved rocket of this type [V-2]. An engineer cannot see much prospect for an improved propellant nor for much better materials of construction. It is unlikely that a ratio of starting weight to empty weight of much more than three can be obtained. It, therefore, appears that the range of 200 miles is near the maximum for the type.[17]

By the first part of 1946, the funding prospects for the satellite project were well below what its supporters in the Bureau of Aeronautics considered to be a minimum. They decided that drastic action was needed to save the project and contacted the AAF regarding a jointly supported satellite project. A meeting on satellites was held on 7 March 1946, with Hall speaking for the Bureau of Aeronautics on the proposed joint effort. The initial reaction was favorable and Hall was elated. However, his joy was shortlived; in less than a month he was called to the office of Lt. Gen. Curtis E. LeMay, the AAF deputy chief for research and development, and told that the AAF would not support the Navy proposal. LeMay did leave the door open for future discussions on earth satellites.[18] Almost coincident with the meeting on 22 March a JPL–Army Ordnance WAC rocket became the first American rocket to go beyond the earth's atmosphere. It reached an altitude of 93 kilometers.

The Air Force's Interest in Satellites

With Arnold an outspoken proponent for long-range missiles and satellites, the Air Force was not about to take a back seat to the Bureau of Aeronautics on the subject. An organization well staffed to study the potentialities of military satellites had just been formed—a "think tank" known as Project RAND.*

Soon after the meeting with Hall, LeMay instructed the Douglas Aircraft Company, RAND's parent organization, to give priority to a design study of a satellite vehicle. He wanted the basic study in three weeks "to meet a pressing requirement."[19] Douglas assigned the top manpower of its Santa Monica engineering department to this task and stopped all other RAND studies and several important Douglas design projects.

At the peak, over fifty of the best scientists and engineers of the firm were on the study—including Louis Ridenour and Francis H. Clauser, both of whom had been in the team that interrogated Wernher von Braun in 1945. The result of the study, "Preliminary Design of an Experimental World-Circling Space Ship," was hand-carried to Wright Field on 12 May 1946. Project RAND stopped further work while the Air Force evaluated the report and decided what further studies were wanted.

World Circling Spaceships

In their first quick look, the RAND group faced the same problems as the earlier investigators at the Bureau of Aeronautics and JPL. Simple physics gave the required orbital velocity and the Tsiolkovskiy equation gave the vehicle velocity without drag or pull of gravity. The major unknowns, other than those velocity losses, were the structural weights and the performance of propellant combinations. The velocity losses were not difficult to assess. The V-2 data furnished a guide for structural mass estimates as well as the actual performance of the alcohol-oxygen propellant combination. RAND considered 39 different fuel-oxidizer combinations and found that hydrogen-oxygen ranked highest (the same result as the Lemmon report, p. 4).

---

*Project RAND was the brainchild of Frank Collbohm, an engineer working for the Douglas Aircraft Company. In late 1945, he talked to government officials about forming a postwar scientific organization to work on problems of national security. He got plenty of expressions of interest but no action until he met General Arnold in October 1945. Arnold liked the idea and implementation began the same day. On 2 March 1946, the Douglas Aircraft Company was given a letter contract for $10 million to set up an autonomous group of engineers and scientists, Project RAND. On 12 May 1947, RAND became an independent corporation. William Leavitt, "RAND—The Air Force's Original Think Tank," *Air Force/Space Digest,* May 1967, p. 100.

Hydrogen's low density, low temperature, and wide explosive range would cause problems, but RAND decided to accept it for design studies anyway. A parallel design study used alcohol and oxygen. A satellite with a mass of 227 kilograms was selected as the payload to orbit at 556 kilometers.[20]

The RAND study gave the V-2 structural mass as 18 percent of its initial mass, estimated that 16 percent was as good as could be obtained, and used the latter for propellants not involving hydrogen. The larger tank needed for low-density hydrogen would probably increase the structural mass proportion to 25 percent.* This, of course, greatly offsets the advantages of the high exhaust velocity of the hydrogen-oxygen combination. Not surprisingly, RAND concluded that a vehicle using either hydrogen-oxygen or alcohol-oxygen could not reach orbital velocity with a single stage—a repudiation of the Navy proposal.

The RAND study found that with multistage rockets, however, orbital velocities could be reached with either hydrogen-oxygen or alcohol-oxygen, but the designs would differ considerably. The alcohol-oxygen vehicle required 4 stages with an initial mass of about 100 metric tons. A 2-stage vehicle using hydrogen-oxygen, but having a third more mass, could do the same job. A 3-stage hydrogen-oxygen rocket would reduce the initial mass to below that of the alcohol-oxygen vehicle. RAND concluded that hydrogen-oxygen should be given serious consideration in any future study. The cost of constructing and launching a satellite was estimated at $150 million over a 5-year development period.

The RAND study gave the AAF a strong position in discussing satellite proposals with the Bureau of Aeronautics. The War Department had a mechanism for coordinating similar programs between the air services—the Aeronautical Board, created during World War II. In June 1946, the board considered the satellite studies of the two services and took the neutral position that both should continue their studies independently.[21] Both the Bureau of Aeronautics and the Air Force moved to strengthen their positions.

The Air Force instructed RAND to start a 6-month study to provide a design sufficiently complete that development contracts could be negotiated for a vehicle capable of launching a satellite. For their part, the Bureau of Aeronautics contracted with North American Aviation for a 90-day study of the feasibility of their proposal, using the GALCIT structural limits as a guide. For a more detailed study of the structural aspects, the Navy also contracted with the Glenn L. Martin Company for a 12-month study, using the same guidelines as the North American contract. To supply data on rocket power plants, the Navy contracted with Aerojet for the detailed design study of a 1.33 meganewton (300000 lb thrust) engine suitable for a vehicle of 45400 kilograms initial mass. The Navy called its vehicle the High Altitude Test Vehicle, or HATV.

---

*Structural mass is generally assumed to be the final mass less payload and engine; the RAND structural figures are not convertible directly into initial-to-final mass ratios.

North American Aviation Study

On 26 September 1946, North American Aviation reported the results of its study. The Navy had specified an initial mass of 45 360 kilograms with a 454 kilogram payload. From the GALCIT report series, an initial-to-final mass ratio of 9.09 was assumed (which meant a propellant mass of 40 370 kg and an engine and structural mass of 4536 kg). Aerojet was asked for an estimate of the rocket engine mass and gave a range of 1361 to 2268 kg. North American used the higher number, leaving an equal mass for the structure—comprising the tanks, supporting structure, external vanes, controls, and skin. R. G. Wilson, the principal structural analyst, found that a structure with a mass of 2903 kg—635 over the limit—was the lightest that could be designed for the propellant mass specified. This increased the initial mass of the vehicle (to 45 995 kg), and Wilson concluded that the use of a single-stage rocket to achieve orbit was not possible with the specifications given. This not being what the Navy wanted, Wilson added that if the initial mass was increased to 59 000 kg and rocket burning time to 165 seconds, the vehicle could achieve orbit with a single stage.[22]

The North American Aviation study reloaded the Navy's guns. A 59 000 kg vehicle could place 454 kg in orbit with a single-stage vehicle, whereas the Air Force with the RAND study needed from 2 to 4 stages and initial masses 1½ to 2 times greater to place half as much payload in orbit. One reason for the light North American design was pressure-stabilized tanks with a common bulkhead separating the liquid hydrogen and oxygen. Pressure-stabilized tanks are thin-walled vessels without bracing which depend upon internal pressure for rigidity in the same manner as does a balloon. The technique had been proposed by Oberth in 1923 (p. 262) and was a controversial design in the 1940s and early 1950s.

Concurrent with the North American study, RAND was proceeding with its second phase of satellite studies scheduled for completion by the end of January 1947. The RAND engineers selected vehicle mass, volume, and complexity as criteria for evaluating a number of propellant combinations. Hydrogen-oxygen was still the best on the basis of initial mass, but considering all three criteria, RAND liked hydrazine-fluorine better. The study was far from complete when the North American Aviation report came out. In the interim, James Lipp of RAND wrote a special report on the advantages of satellites. Using an estimate of $50–150 million to orbit a satellite in the 1950s, Lipp urged a quick start so that the United States could maintain superiority over possible enemies. He recommended that the AAF be given priority for a research program leading to a satellite.[23] This recommendation was strengthened a week later when Army Ordnance launched a V-2 from White Sands. The missile reached an altitude of 120 kilometers and took motion pictures of 100 000 square kilometers of the earth's surface. Lipp's arguments, however, fell on barren ground, for the country was complacent in its atomic bomb superiority—a complacency that was to last until the Russians exploded their bomb in 1949.

Fading of Satellite Proposals

The competition between the Air Force and the Navy's Bureau of Aeronautics over satellites might have grown keener had it not been overshadowed by national and international events. In the fall of 1946, President Truman's administration faced formidable problems at home and abroad. The railroad strike in August had threatened to paralyze the nation's transportation system, and Truman had countered by threatening to take over the railroads and draft its workers into the military. On the international front, the hoped-for mutual understanding with the Russians became less likely as Stalin became increasingly more hostile. With the United States facing increasing obligations abroad, preparations for the next year's budget brought decisions to restrict long-range research and development programs in favor of expenditures promising more immediate benefits. By December, such strictures had ended the prospects for satellites as a military weapon. The Air Force ordered RAND to shift emphasis from satellites to airplanes and ramjet vehicles.

During the first quarter of 1947, Project RAND wound up its first satellite study and published a final report in April. The favored configuration was a 3-stage rocket using hydrazine and oxygen, with an initial mass of 38 600 kilograms and an orbit altitude of 648 kilometers. The cost for a satellite in orbit was estimated to be $82 million.[24] If there was no immediate result, RAND's dozen reports on satellites had an important side benefit. The RAND staff had become thoroughly versed in rocket vehicles and their potential. Although the new Air Force directive emphasized air-breathing engines, RAND continued to consider the possibility of long-range rockets. In effect, this marked the beginning of the RAND-Air Force work on intercontinental ballistic missiles—the great driving force for rocket developments in the 1950s.

The Navy, however, took a different tack. The previous August, the Naval Research Laboratory had been authorized to develop a high-altitude test vehicle for scientific research. The satellite supporters in the Bureau of Aeronautics saw science as the savior of their project and began emphasizing this aspect. In November, the Bureau of Aeronautics requested the Naval Research Laboratory to study the use of satellites for scientific research and allowed the Martin and Aerojet contracts to continue.

Aerojet and Martin Design Studies

Aerojet's contract that began July 1946 called for furnishing detailed design information to the North American Aviation and Glenn L. Martin design study groups on a hydrogen-oxygen rocket engine suitable for their vehicles. The thrust of the rocket engine was specified as 1.33 meganewtons (300 000 lb), the exhaust velocity 4165 meters per second, and the mass not more than 1814 kilograms. Aerojet chose a combustion pressure of 34 atmospheres and a hydrogen-to-oxygen molar mixture ratio of 3 to 1. The combustion chamber and nozzle were to be made of porous stainless steel for transpiration cooling. Young's flared tube design concept (fig. 8) was to be used. A greater unknown than the thrust chamber was the turbopump design, and Aerojet concentrated its initial effort there. By mid-October, pump characteristic curves had been determined and a pump speed of 10 000 revolutions per minute selected. Although larger than any previously designed for a rocket engine, the pump would be about the size of the turbines in turbojet engines of the period and not beyond current technology.

The Aerojet design study was completed and reported by the end of March 1947—in time for use in the Martin study but too late for the North American analysis. The

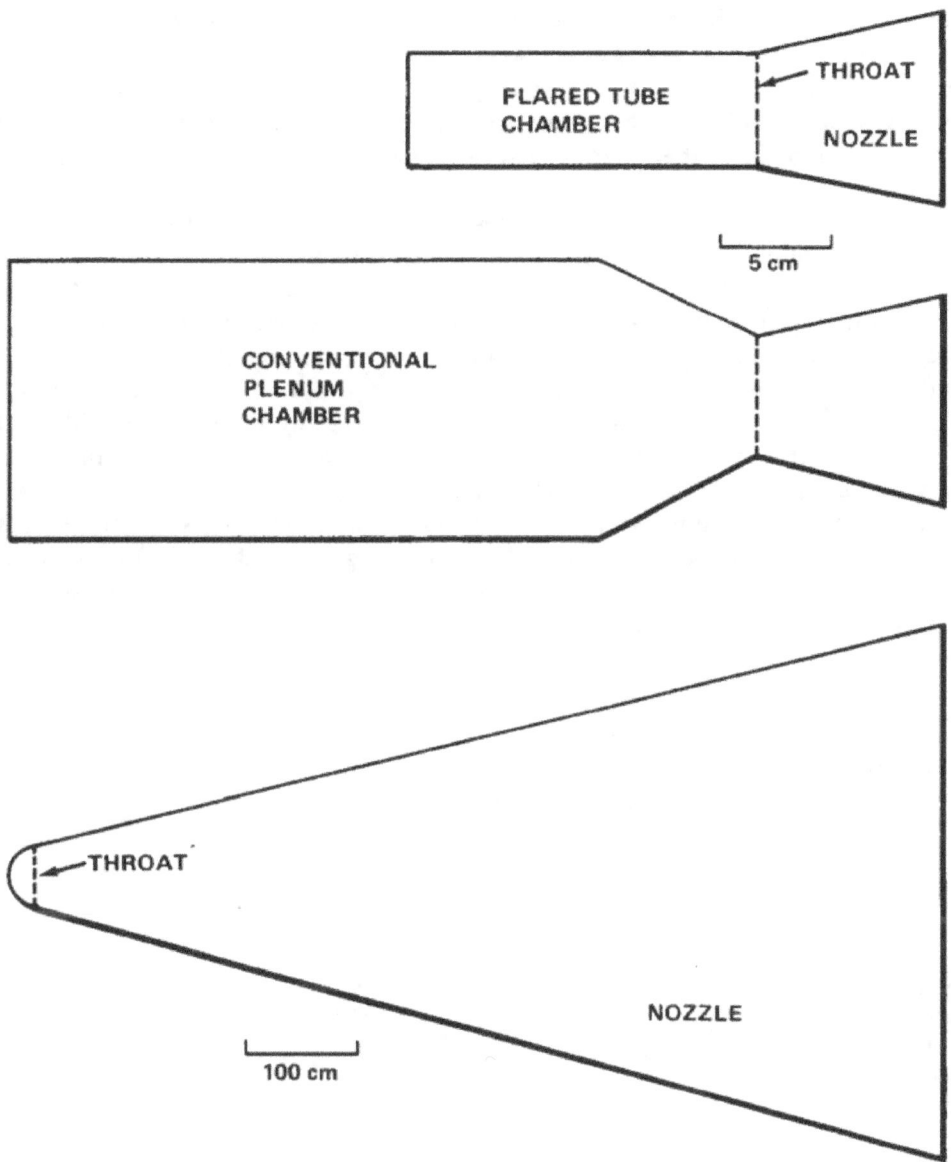

Fig. 8. Aerojet's experimental flared-tube engine (top) had less than a tenth the combustion volume of a conventional plenum chamber engine (middle) of the same size throat, nozzle, and thrust (4.5 kN or 3000 lb). Below: Aerojet's application of the flared tube concept to the design of a large engine (1.3 MN or 300 000 lb thrust) where the nozzle dwarfs the combustion chamber. Note difference in scales.

thrust chamber resembled a huge ice-cream cone some 7 meters long; the combustion chamber at the small end was dwarfed by the large conical nozzle (fig. 8, bottom). The inner wall, porous stainless steel, was cooled by hydrogen flowing through it into the combustion chamber. The mass of the chamber, turbopump, and assorted valves and lines added up to 1762 kilograms, comfortably within the specifications.[25]

The Glenn L. Martin Company had the same general guidelines as North American Aviation for designing a single-stage rocket to orbit a satellite, but they too found that it could not be done within these guidelines.* In striving to do so, Martin's structural designers developed a remarkably ingenious and lightweight structure using pressure-stabilized, thin-wall tanks. With initial vehicle mass only 5 percent greater than specified in the guidelines, they managed to increase the payload by 50 percent over that specified.[26]

A comparison of the North American and Martin designs is given by table 1. Martin increased the wall thickness of Aerojet's thrust chamber and used a heavier engine than Aerojet furnished. In addition to the thin-wall, pressure-stabilized tanks, the Martin design made the large thrust chamber an integral part of the aft liquid-hydrogen tank, and added four small auxiliary rockets around the nozzle exit for stability and control. The small rockets eliminated the need for external aerodynamic stabilizer fins and movable fins in the hot exhaust stream for thrust-vector control. The idea of surrounding the thrust chamber with the tankage was remarkably similar to Tsiolkovskiy's hydrogen-oxygen spaceship of 1903 (fig. 9).

Using the same basic design, Martin analyzed a family of vehicles with initial mass from 13 600 to 72 600 kilograms with payloads varying from 136 to 780 kilograms. With these the Bureau of Aeronautics had a range of vehicle sizes for possible development.

TABLE 1.   *Comparison of Single-Stage-to-Orbit Rocket Designs*

Item	North American	Martin
Guidelines	kg	kg
Initial mass (Navy)	45 360	45 360
Payload (Navy)	454	454
Engine (Aerojet)	2268	1762
Results		
Initial mass	59 000	47 468
Propellant	52 510	42 484
Final mass	6490	4984
Mass ratio (initial-to-final)	9.09	9.52
Payload	454	658
Engine	2268	2044
Structure	3768	1791
Instruments for control		491

Aerojet's Second Series of Experiments, 1946–1947

In addition to the rocket engine design study, Aerojet's contract that began in July 1946 called for experiments with a gaseous hydrogen–liquid oxygen thrust chamber. The thrust was 4.5 kilonewtons (1000 lb) and the minimum exhaust velocity was specified as 2940 meters per second. Moreover, the engine was to operate continuously

---

*Martin used the same JPL satellite study as North American but chose an initial-to-final mass ratio of 9.52, rather than the 9.09 used by North American.

for three minutes. The chief experimenters were Robert Gordon and Herman L. Coplen, reporting to David Young. By the end of the twelve month period they had met the specified performance.

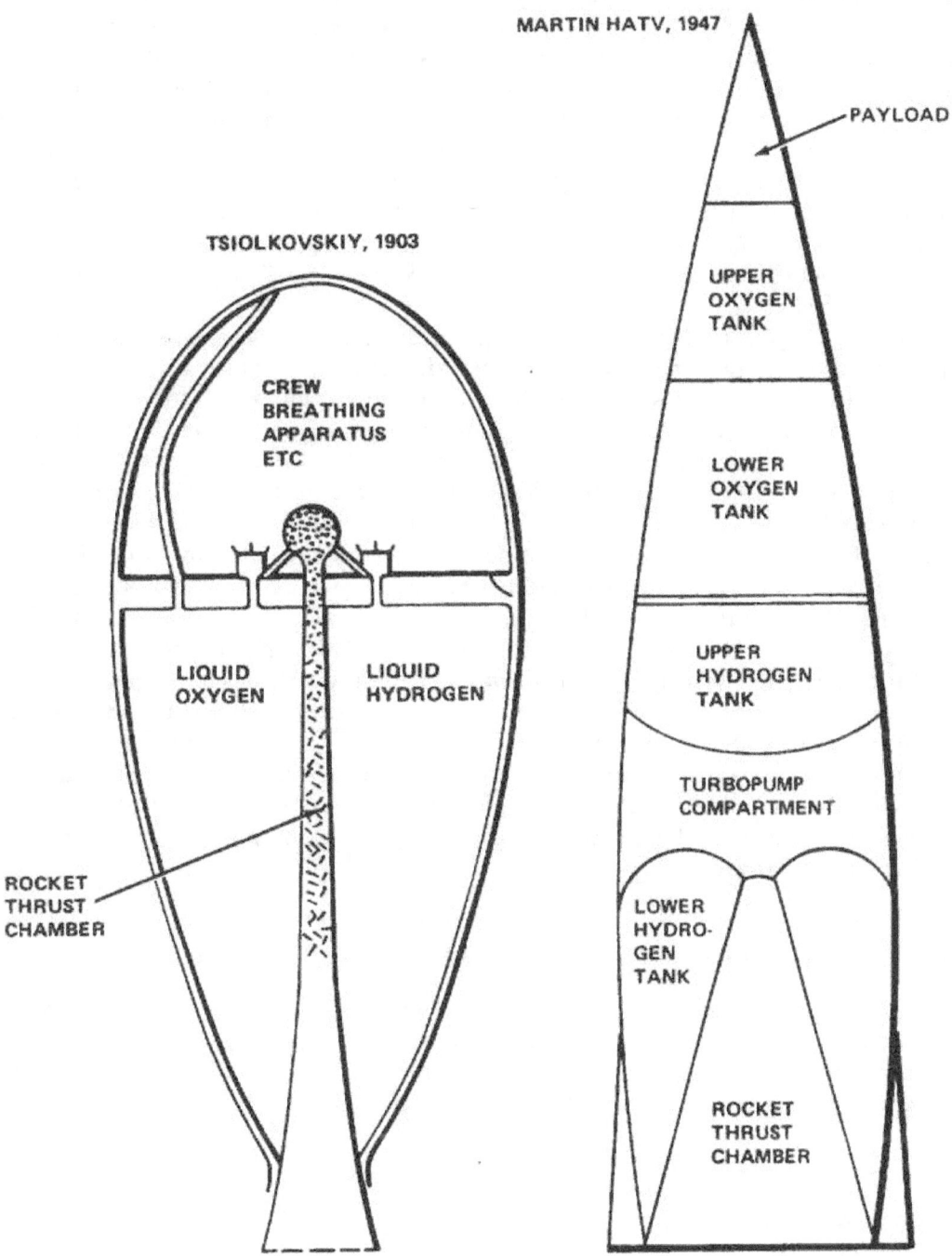

MARTIN HATV, 1947

PAYLOAD

TSIOLKOVSKIY, 1903

CREW BREATHING APPARATUS ETC

LIQUID OXYGEN

LIQUID HYDROGEN

ROCKET THRUST CHAMBER

UPPER OXYGEN TANK

LOWER OXYGEN TANK

UPPER HYDROGEN TANK

TURBOPUMP COMPARTMENT

LOWER HYDRO-GEN TANK

ROCKET THRUST CHAMBER

Fig. 9. Comparison of Tsiolkovskiy rocket concept (1903) and Martin HATV (1947). Note similarity of integral tanks and thrust chambers in the aft sections.

The thrust chamber had a water jacket and an inner liner of porous material through which the water seeped and evaporated on the inner surface for cooling. The shape was the flared tube design (fig. 8), having in this case a chamber diameter of 5 centimeters and overall length of 21. The gaseous hydrogen was injected through a series of holes to form a cone in the chamber, and the gaseous oxygen was injected radially inward to intercept the hydrogen cone. The combination of this injector and the flared tube design produced very high heat transfer rates—several times higher than normally experienced in rocket experiments. This led to a separate investigation of the characteristics of the flared tube by Gordon using a smaller engine independently cooled with water. Gordon found that high performance (95 percent of theoretical) could be obtained with the design, but the combustion pressure was not constant as in a conventional plenum chamber; it dropped rapidly throughout the length of the flared tube chamber.[27] The average heat transfer rates were much higher than those of a plenum chamber.*

Instead of reconsidering their basic engine design, the Aerojet men focused most of their attention on cooling. They tried a dozen different porous materials. Porous nickel made by the Amplex Division of the Chrysler Corporation proved to be the best. An attempt was made to match the water flow through the porous liner with the large variation of heat transfer rate along the combustion chamber and nozzle, but this was only partially successful. The best they could do was to use almost twice as much coolant as they had originally calculated to be necessary. This was a matter of some concern, as the water entering the combustion chamber diluted the propellant and lowered performance, for its mass had to be considered in determining thrust per unit mass flow or its equivalent, exhaust velocity. (One percent increase in water flow decreased the exhaust velocity by 0.75 percent.) To make up for the drop in performance, the combustion pressure was increased, which increased gas expansion and exhaust velocity. On 26 June 1947, four days after expiration of the contract, the performance objective was achieved on the 46th run, which lasted over three minutes.[28]

With these experiments, Young, Gordon, and Coplen were still confident that their 1.3 meganewton (300 000 lb thrust) design study was sound, although they had yet to operate a rocket using liquid hydrogen and oxygen or to cool a hydrogen-oxygen rocket with hydrogen rather than water.

Switch in Emphasis from Military to Science

While the contracts for industrial research were producing satisfactory results, the Navy's change in tactics—emphasizing scientific purposes rather than purely military ones—required closer liaison with civilian scientists. This, in turn, implied a shedding of the secrecy that envelops military projects. Admiral Leslie Stevens of the Bureau of

---

*The average rate was 13 J/s·m²; in the section just before the nozzle, a peak of 29 was measured. Pressure was 20 atm at the injector end and the mixture was fuel-rich (oxidizer-to-fuel mass ratio of 5). The average heat transfer rate was about 6 times greater than Ohio State's values when the latter used a plenum chamber at about the same operating conditions and performance (fn., p. 24). Some of the difference can be attributed to the much greater gas velocities in the flared tube as well as the different types of propellant injection.

Aeronautics recommended in January 1947 that the Joint Research and Development Board remove the satellite project from the jurisdiction of the Aeronautical Board and "establish an agency for the coordination, study, evaluation, justification, and allocation of all phases of the Earth Satellite Vehicle Program. . . ."[29] The need for something like the National Aeronautics and Space Administration was envisioned a decade before it became a reality.

Stevens's recommendation meant the voluntary relinquishment of control over satellites by the Joint Research and Development Board. Not surprisingly, the recommendation was referred to the Aeronautical Board where it was studied for a couple of months with the not unanticipated conclusion that jurisdiction should remain where it had been. By then it was mid-1947 and although the reports of Martin and Aerojet were in, satellite considerations were becalmed in a sea of changing organizations.

On 26 July, President Truman signed the Armed Forces Unification Act. The Departments of War and Navy were abolished and the National Military Establishment was created, headed by the Secretary of Defense. The Army Air Forces became the Department of the Air Force, equal in status with the Departments of the Army and Navy. By the end of September, the old Joint Research and Development Board was replaced by the Research and Development Board under the same chairman, Vannevar Bush. Reorganization had little effect on the board and its subgroups, but there was much additional work to be done, especially in defining the role of the Air Force with respect to missiles. The Aeronautical Board and the subcommittee on earth satellite vehicles continued to function. In November, the Office of Naval Research asked to be designated the coordinating agency for the "High Altitude Research and Earth Satellite Program." Before the subcommittee reached a decision, the parent Research and Development Board gave responsibility for earth satellites to the Committee on Guided Missiles, which formed a Technical Evaluation Group under the chairmanship of Professor Clark Milliken of the California Institute of Technology.[30]

The Canright Report

During changes in government R&D organizations and objectives in 1947, rocket analysts were looking beyond the merits of exhaust velocity in comparing propellants and focusing on the importance of propellant density and its influence on vehicle design and performance. Not satisfied with an analysis by von Braun, Hager, and Tschinkel in 1946 that placed considerable emphasis on propellant density, Richard Canright of JPL developed a method of comparing propellants for rockets of the V-2 class and larger with propellant masses 70 to 90 percent of initial vehicle mass. Equal total impulse (thrust · time) was assumed; tank volume was adjusted to provide the necessary propellant in each case; and total vehicle mass was calculated. The vertical height attained by the rocket was the comparison criterion, which was almost the same as comparing initial masses.[31]

For large vehicles, Canright found that the exhaust velocity of propellant combinations was decidedly more important than propellant density and that emphasis on high energy propellants was justified. Although his analysis showed that hydrogen was superior to any other fuel using the same oxidizer, Canright favored hydrazine, finding it favorable under all the conditions assumed.*

Aerojet's Third Series of Experiments, 1947–1949

When the Navy renewed Aerojet's contract in mid-1947, the central task was to develop a liquid hydrogen–oxygen rocket engine suitable for a small-scale version of the earth satellite vehicle. The engine was to be in the thrust range of 9–13 kilonewtons (2000–3000 lb), have a minimum exhaust velocity of 2972 meters per second, and be capable of operating for 60 seconds. Maximum mass was specified as 34 kilograms. Propellants were to be supplied to the thrust chamber by a turbopump. Other tasks, which were concerned with drawings and operating instructions, indicated that the Navy intended to be prepared for development of a small-scale experimental vehicle. The contract also called for several analyses and a design study of a rocket engine of 37.8 kilonewtons (85 000 lb thrust), apparently for the Martin minimum-sized vehicle. Although there was little reason for optimism, the Bureau of Aeronautics was keeping its options open.

The Aerojet work with hydrogen from mid-1947 to mid-1949 was the climax of five years of effort along three major lines: (1) the supplying of liquid hydrogen, (2) turbopump development, and (3) thrust chamber development.[32] These will be described separately.

Supply of Liquid Hydrogen

From the first tests in 1945 through the second series of rocket experiments in 1947, Aerojet had to use gaseous hydrogen because liquid hydrogen was not available. Starting in early 1946, Aerojet enlarged its facilities to handle gaseous hydrogen and oxygen. Gaseous hydrogen under a pressure of 136 atmospheres was available directly from a trailer of high pressure tubes with a capacity of 800 cubic meters (at atmospheric pressure) and from a stationary bank of high pressure tubes of about the same capacity. Gaseous oxygen at pressures up to 163 atmospheres was supplied from two trailers with a capacity of 560 cubic meters. The total quantity of gases from these sources allowed only a few minutes of operation—a situation conducive to continued frustrations, as the following incident illustrates. One day the test crew was ready to run the rocket and waiting impatiently for a commercial firm to deliver some needed gas. When it came, the crew quickly connected the trailer to the pipes leading to the test cell and ran the test. Meanwhile, the truck driver had gone to the office to get the delivery ticket validated. On his return he was told the trailer was empty and could be taken back. Used to leaving such trailers for a considerable time at other places, the

---

*On the basis of an altitude index of 100 for alcohol-oxygen and a tank pressure of 20 atm, hydrogen-oxygen was 153, 21 units higher than hydrazine-oxygen; the advantage of hydrogen increased if a lower tank pressure was assumed. In his initial calculations, Canright considered hydrazine-fluorine, which he found superior to hydrogen-oxygen. Later, however, Canright indicated that hydrogen-fluorine should give the maximum range obtainable from chemical reactions.

driver simply would not believe the crew until it was explained rather forcibly to him. He departed with the trailer, shaking his head.[33]

By early 1947, the Aerojet group was planning ahead to the next phase of hydrogen-oxygen experimentation and acutely felt the handicap of not having a supply of liquid hydrogen. Envying their former associate, Marvin Stary at Ohio State University, with his assured supply of liquid hydrogen from the Johnston liquefier, they decided to attack the problem directly. They discussed liquid hydrogen with several possible users on the West Coast and the idea blossomed into a proposed cooperative venture among several government agencies, universities, and industrial firms. Confident that they could get liquid hydrogen—and having gone to as high a thrust as was reasonable with gaseous hydrogen—the Aerojet engineers proposed to use liquid hydrogen in their third series of experiments starting in July 1947. They went even further and proposed to build a flyable rocket engine, complete with its own controls and turbine-driven pumps. They also recommended that the government build a medium-scale hydrogen liquefier on the West Coast.

Aerojet got its new contract in July 1947, but immediately faced a problem: the cooperative venture to get liquid hydrogen failed to materialize. Aerojet decided to try to interest private industry in supplying liquid hydrogen, and if that failed, to get authority and funding from the Navy to build a liquefier. The first step was to get an estimate of the amount of liquid hydrogen needed. The Jet Propulsion Laboratory agreed to participate and estimated a need for 600–900 kilograms a year. Aerojet added their needs and settled on a 3600-kilogram total requirement for two years. Three possible commercial sources were then queried. The Shell Development Corporation could not supply liquid hydrogen, but had a surplus of high-purity gaseous hydrogen for sale. The National Cylinder Gas Company believed that the sale of liquid hydrogen was neither economical nor safe and recommended liquefaction at the point of consumption. The Linde Air Products Company submitted an oral bid for $62 per kilogram at their plant in Los Angeles, but later lowered the price to $55 per kilogram for the first 1800 kilograms and $44 thereafter.

While soliciting industry, Aerojet began investigating the possibility of building a liquefier modeled after Johnston's and estimated that it would cost $100 000, including the cost of the liquefier, materials, and labor for producing 3630 kilograms of liquid hydrogen. This was half the revised Linde estimate and had the added advantage of being under Aerojet control and located near the rocket test stand. Aerojet officials became enthusiastic over the prospect and set about convincing the Navy. By late September they received oral approval, which was formalized on 16 December 1947. Aerojet engaged Johnston as a design consultant; he was also to supply parts of the liquefier. Herman L. Coplen was the principal Aerojet engineer for design, construction, and operation.

Aerojet expected to have the liquefier in operation by late spring or early summer. As so often happens, the optimistic schedule fell victim to late equipment deliveries. However, the liquefier produced its first liquid hydrogen—12 liters—on 3 September 1948. The initial operation turned up the usual number of bugs; the second operation on 21–23 September produced 120 liters. Of this, 75 liters were shipped to the Jet Propulsion Laboratory for rocket tests there.

Aerojet was pleasantly surprised to find that the actual capacity of the liquefier was 30 liters per hour instead of the design value of 25. The increased capacity came from a larger hydrogen compressor; the Johnston-built heat exchangers were oversized. This led Aerojet to propose, in early 1949, the doubling of the liquefaction capacity by installing additional hydrogen compressors.

At first, the liquefier was operated intermittently. Beginning on 8 November, a two-shift operation was begun to meet the needs of the rocket test engineers, and from 27 December three shifts were employed. By the end of 1948, 7500 liters (535 kg) of liquid hydrogen had been produced, over 90 percent of it in November and December. Only about 30 percent of the hydrogen liquefied was used in test operations; the bulk was lost during storage and test delays.

In the first three months of operation, the liquefier was shut down twice, but the troubles were quickly fixed; the time lost was four days. Overall, the liquefier was highly successful and made possible the testing of pumps and thrust chambers.

By the end of March, Coplen had added two more compressors and the liquefaction rate rose to 80 liters (5.67 kilograms) per hour. But early March had brought catastrophic news to the liquid hydrogen producers. On 2 March 1949, the Bureau of Aeronautics directed Aerojet to change fuels from liquid hydrogen to anhydrous hydrazine, which is a liquid at room temperature and pressure.* The directive allowed Aerojet to continue liquid hydrogen testing until the end of June, but the irony was that the switch came just as the producers of liquid hydrogen were finally prepared to meet rocket test needs.

In its operations through June 1949, the Aerojet liquefier produced 47000 liters (3357 kilograms) of liquid hydrogen at an estimated cost of $29.72 per kilogram. The cost of commercial gaseous hydrogen and liquid nitrogen were major expenses.

Sometime after the contract ended in mid-1949, Aerojet received a government directive to dismantle and prepare the liquefier for shipment. Very few at Aerojet knew, but the liquefier was destined for reassembly on a remote Pacific isle for use in the first test of a thermonuclear device, the predecessor of the hydrogen bomb.

Turbopump Development, 1947–1949

The principal engineer for turbopump development was George Bosco. This was a new field for Aerojet, and during the second half of 1947, Bosco and his group learned about the pump work of others and made preliminary design studies. Aerojet representatives visited Ohio State University where Florant was working on hydrogen pumps, and consulted Dietrich Singelmann, a German pump expert at Wright Field.

---

*The author has been unable to pin down the reason for this sudden change, but it is not surprising. Hydrazine is storable and considerably easier to handle than liquid hydrogen, its performance is high, and interest in it during the 1940s and 1950s was high. For example, Canright, in his analysis of relative importance of exhaust velocity and density, preferred hydrazine to hydrogen even though hydrogen gave higher performance (pp. 47–48).

Bosco subsequently used Singelmann's data in designing Aerojet's first hydrogen pump.*

By mid-1948, Aerojet had selected centrifugal pumps for both liquid hydrogen and liquid oxygen. They obtained some German radial-vane pumps from the Navy and tested them during the second half of the year.†

By the end of 1948, Aerojet had designed, built, and tested a liquid-hydrogen pump (15 cm diameter). Initially, it used ball bearings that were run clean and dry, because the low temperature made conventional lubrication impractical. The pump was first operated at low speeds to allow its parts to cool down to operating temperature. When temperature gauges showed that liquid hydrogen had reached the pump, an attempt was made to accelerate from 5000 to 35000 revolutions per minute. The pump failed and examination of the pieces pointed to a failure of the bearing, as well as the impeller. After some testing, super-precision bearings, lubricated by oil that was atomized and directed by a stream of gaseous nitrogen, were used. On the next run, the bearings worked satisfactorily but the stresses were too great for the brazed impeller and it flew apart. A new one was made by milling from a solid block of aluminum. Time was running out, as the contract had less than six months to go. The next two runs with the new pump were a great disappointment; the instruments showed no significant flow or pressure rise. The problem was traced to the exit diffuser of the pump, which was too small and insufficiently cooled during the cool-down cycle so that it limited the flow. This was corrected by adding vent holes in the pump housing; the vents were opened during cool down and closed when the pump was cold. With this fix, two additional runs were made in March 1949 and both were successful. Flow rate and pressure were found to be in approximate agreement with theoretical predictions. The maximum pressure was 26 atmospheres and the flow was 0.25 kilogram per second.

Thrust Chamber Experiments, 1947–1949

From their previous work, Young and Gordon were confident that the flared tube configuration, with its very small combustion chamber, was the best design for the thrust chamber of 13.3 kilonewtons (3000 lb thrust). They intended to use a porous inner wall but were still undecided about the coolant. They decided to determine the relative merits of both water and liquid hydrogen as transpiration coolants. They also planned to study injection methods for liquid hydrogen. Stary was studying the same things at Ohio State University and had just made his first run using liquid hydrogen (p. 20).

From mid-1947 to mid-1948, the Aerojet men made few thrust chamber tests. None was made with liquid hydrogen, for the liquefier was not yet in operation. The major experimental work was an investigation of the performance loss at sea level in operating a nozzle designed for maximum performance at altitude.

---

*The initial design provided for pumps for hydrogen, oxygen, and water (coolant), each with inlet and discharge pressures of 2.4 and 51 atm, respectively. The liquid hydrogen flow rate was 0.39 kg/s; oxygen, 2.1; and water, 0.54. An estimated 9.7 kW (130 horsepower) turbine was needed to drive the three pumps.

† The pumps, made by the Bayerische Motoren Werke, were from the BMW 109-718 booster rocket engine used on the ME-262 aircraft.

The force produced by a nozzle from expanding exhaust gases is the result of a momentum force and two pressure forces. One of the pressure forces aids the momentum force and the other opposes it. An ideal nozzle is one that expands the exhaust gases from the pressure in the combustion chamber to the outside ambient pressure. The nozzle thereby maximizes the momentum force and the two pressure forces cancel each other. Since a rocket nozzle is a fixed design, the designer must choose a single ambient pressure for his design. If he chooses sea-level pressure, he gets less than optimum performance at altitude; if he chooses a lower pressure corresponding to some altitude, he theoretically loses performance at sea level. Since much of the operation occurs at reduced ambient pressure, the designer usually wishes to make the nozzle as large as mass and size restrictions permit. The question at Aerojet was: What penalty would result from sea-level operation of a nozzle designed for best operation at altitude? In experiments with a small rocket chamber they found, to their great joy, that the actual performance loss was much less than theoretically predicted—their nozzle designed for altitude had only a 10 percent loss at sea level.*

Aerojet was still committed to transpiration cooling but had encountered a series of new and worrisome material problems. It was difficult to obtain porous materials of uniform permeability—but worse yet, the porous structure became clogged in unpredictable and nonuniform ways. These problems began to raise doubts about using the flared tube configuration as well as transpiration cooling. When the project received new funding and directions in mid-1948, Aerojet planned to use a group of thrust chambers of various sizes and shapes, as well as a variety of injection methods. The engineers believed regenerative cooling would be possible with either oxygen or hydrogen, or both. Preparations were made to study the heat transfer properties of oxygen and hydrogen by means of an electrically heated tube. All of these activities signaled a major change in direction by Aerojet, from emphasis on their flared tube design using transpiration cooling to a conventional plenum thrust chamber with regenerative cooling. It was about this time, mid-1948, that George H. Osborn became the chief test engineer.

The first Aerojet test with liquid hydrogen and oxygen was made on 20 January 1949 with a 1780-newton (400 lb thrust) chamber. By the end of March, 10 runs had been made with disappointingly low exhaust velocities—about 2920 meters per second or 82 percent of theoretical. Of equal concern was the unsteady operation, or "chugging," which indicated unstable combustion. The injector, designed by Osborn, used a diverging cone of liquid oxygen intersecting a converging sheet of liquid hydrogen. The only good news was a low heat transfer rate, which was attributed to incomplete combustion.

In the midst of all the bad experimental results came the worst news of all. On 2 March 1949, as previously mentioned, the Bureau of Aeronautics directed Aerojet to change the fuel from liquid hydrogen to anhydrous hydrazine, but allowed the experiments with liquid hydrogen to continue for the three months remaining in the contract. No evidence has been found that Aerojet protested this change—perhaps it

------

*The exhaust gases did not overexpand as much as theory implied, but separated from the nozzle walls at a shock front. The exhaust gases filled the nozzle up to a certain point and then separated from the wall and flowed as though the rest of the nozzle were not there.

was welcomed after the first series of experiments with liquid hydrogen. However, the Aerojet designers were determined to do a creditable job with liquid hydrogen in the time remaining and the record shows that they did. The key was injector design.

Osborn was designing new injectors even before all the dismal results with the spray type were in. The second design was a "showerhead" type with 115 fuel and oxidizer holes across the face and 30 fuel holes around the circumference for film cooling. The film, or layer, of fuel-rich gas next to the chamber and nozzle walls kept them cool. The design gave low performance and failed structurally on 4 April, three months before the end of the contract.

The pressure on the team to succeed must have been great. Fortunately, Osborn had designed a third injector, called a multitube concentric orifice, in March and it proved to be highly successful. Liquid hydrogen was injected through a number of thin-walled tubes surrounded by an annular flow of liquid oxygen, as illustrated by figure 10. For

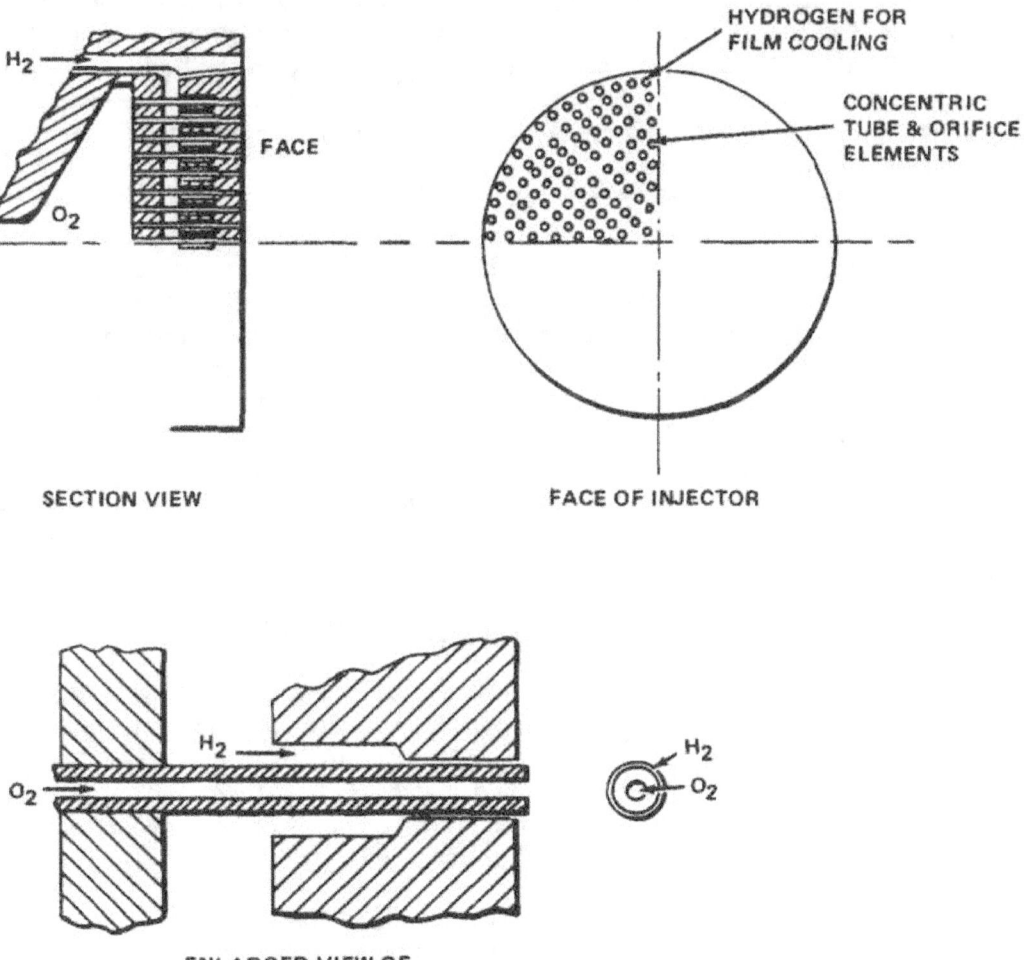

Fig. 10. Aerojet's multitube concentric orifice injector. One design had 489 concentric tube orifice elements for the 13.3-kN (3000-lb-thrust) experimental rocket.

the 1780-newton (400-lb-thrust) chamber, 61 of these "tubes within tubes" provided a very fine degree of mixing. As in the previous design, axial orifices were spaced around the circumference for hydrogen film cooling. Two runs with this injector gave an exhaust velocity of 3590 meters per second, or virtually 100 percent of theoretical. The propellants mixed so well that combustion occurred very close to the injector face and burned it. Osborn sought to correct this with design changes, but the fix did not work as well as the original design. However, he knew how he wanted to design the 13-kilonewton (3000-lb-thrust) injector. When he signed the drawing for it on 5 May, there were less than two months left to complete the work. The injector had 489 sets of circular oxygen orifices surrounding hydrogen tubes, plus 60 hydrogen orifices for a fuel-rich layer at the walls. The thrust chamber, which had been designed and fabricated earlier, was a conventional plenum chamber, water cooled, with an inner liner of copper. The copper was machined from a solid billet and its size limited the nozzle design so that it was not ideal.* Starting on 27 May three successful runs were made with this engine at pressures from 24 to 31 atmospheres. Exhaust velocities of 3380 to 3520 meters per second were obtained, approaching 95 percent of theoretical performance. On 16 June, with two weeks to go before the contract expired, they attempted to make a fourth run, but an explosion occurred in the liquid hydrogen propellant system—the second in that system. Aerojet attributed the cause to contamination of the liquid hydrogen with solid oxygen. That ended Aerojet's rocket experiments with liquid hydrogen.

In reporting the results, Osborn and Wayne D. Stinnett included experiments by Gordon on heat transfer and injectors using a smaller, water-cooled engine where the multitube, concentric injector had initially proved successful. Heat transfer rates were reported as excessive for both engines, leading the authors to conclude that additional film cooling over that used in the larger engine would be necessary. Although they had not fulfilled the objective of a self-cooled, lightweight rocket engine using liquid hydrogen–oxygen, the investigators believed that their results were highly encouraging, and no fundamental difficulties were encountered. From their rapid progress during the last four months of the contract, there is little doubt that Aerojet was on the right track in thrust chamber design and with additional work would have been able to perfect self-cooling. Concurrent with their work, Dwight I. Baker at nearby Jet Propulsion Laboratory was doing just that.

JPL Experiments with Hydrogen-Oxygen, 1948–1949

It is ironical that Young's experimental team at Aerojet, early in getting started with hydrogen-oxygen in 1945—even building a liquefier to get a supply of liquid hydrogen—was not the first to experiment with liquid hydrogen in a rocket on the West Coast. Baker, using Aerojet-furnished liquid hydrogen, beat them by four months. JPL had been interested in hydrogen-oxygen as a high-energy propellant combination since starting a study for the Bureau of Aeronautics in 1945.†

---

*The nozzle ratio of exit-to-throat area was 4, a ratio that theory indicates would underexpand the exhaust gases; hence the momentum force was not a maximum.

†JPL was also interested in the possible use of nuclear energy to heat hydrogen. In 1947, Walter B. Powell of JPL attempted to measure the performance of gaseous hydrogen heated electrically in a tube, but found that the thrust and flow rates were so low that accurate measurement was impractical.

When Aerojet queried JPL in 1947 for interest in using liquid hydrogen, JPL responded with an estimated need for 600 to 900 kilograms for a year of experimentation. While Aerojet's liquefier was under construction, a 100-liter dewar was built for use in transporting liquid hydrogen from the Aerojet plant to the JPL test cell. When Aerojet produced liquid hydrogen on 21 September 1948, Baker was ready and waiting. Aerojet provided 75 liters of liquid hydrogen to JPL and Baker used it in a rocket run the same day. The results were first reported in the JPL Combined Monthly Summary No. 8 for the period 20 August–20 October 1948:

The first motor test with liquid hydrogen and liquid oxygen was made during the past period on a 100 lb thrust [445 N] motor at a nominal chamber pressure of 300 psia [20.4 atm]. . . . Three points . . . were obtained at mixture ratios [oxygen to hydrogen by weight] of 6.27, 5.46, and 4.99 . . . during a single test having a duration of 105 seconds.

With these words, JPL became the second U.S. laboratory to report rocket experiments using liquid hydrogen, a little over a year after Ohio State University's first test.

The performance obtained in the first JPL test with liquid hydrogen–oxygen was 2717 meters per second, within 15 percent of theoretical—not bad for the first attempt. The average heat transfer rate was 3.6 joules per second per square meter, much lower than measured by Aerojet but in agreement with the data from Ohio State University.

Baker was appalled at how little liquid hydrogen he was able to use in the rocket firing. Only 37 percent was burned in the rocket chamber. An estimated 21 percent was lost in cooling the transport dewar, 16 percent evaporated during transit from Azusa to Pasadena, and 26 percent was lost in cooling the propellant tank of the test rocket. If Baker had not already precooled the hydrogen containers and system with liquid nitrogen, the liquid hydrogen loss would have been much greater. This experience led JPL to use gaseous hydrogen for injector testing while reserving liquid hydrogen for heat transfer and cooling tests. They were already conducting some experiments with gaseous hydrogen which also were reported in Monthly Summary No. 8.

The gaseous hydrogen–liquid oxygen rocket experiments were conducted with a 445-newton (100-lb-thrust) chamber and the results indicated that liquid oxygen above its critical pressure cooled two-thirds of the combustion chamber, with water cooling the rest. At that time, cooling with liquid hydrogen was a big unknown, for fundamental heat transfer data on hydrogen above its critical pressure were missing. Walter B. Powell, who had built an electrically heated tube for heat transfer research, agreed to obtain the missing data. This was given first call on the next available supply of liquid hydrogen while injector testing continued with gaseous hydrogen–liquid oxygen at a higher thrust (2.2 kN or 500 lb). Baker was to use the data Powell obtained to design a regeneratively cooled thrust chamber, possibly using both liquid hydrogen and liquid oxygen as coolants.

Early in 1949, Baker succumbed to enthusiasm, confidence, or impatience and decided to go ahead with designing and testing a hydrogen-cooled thrust chamber without waiting for Powell's results. He had already calculated that liquid hydrogen had twice the heat absorbing capacity of liquid oxygen at their relative flow rates and therefore would be a better coolant. He designed a rocket engine of 445 newtons (100-lb thrust) for operation at 20 atmospheres chamber pressure. On 15 April 1949, Baker became the first person in the United States, if not the world, to operate a liquid hydrogen–liquid oxygen rocket thrust chamber that was cooled with liquid hydrogen. The test ran for 77 seconds and performance was relatively low (2630 meters per second); succeeding runs, however, established beyond any doubt that high performance and regenerative cooling with liquid hydrogen were realizable. Sixteen runs were made through 10 June 1949 over a range of hydrogen-oxygen mixture ratios, with an average running time of 69 seconds for the series. Three runs were made at a combustion pressure of 33 atmospheres and three sizes of combustion chambers were used during the series. Maximum performance was an exhaust velocity of 3420 meters per second at 33 atmospheres combustion pressure and an oxygen-to-hydrogen mass ratio of 4. Baker encountered no serious difficulties and concluded that large size, regeneratively-cooled rocket thrust chambers using liquid hydrogen–liquid oxygen were practical.[34]

Although Baker had no serious problems with burning hydrogen or cooling with it, he was still concerned over the supply of liquid hydrogen. The cost was about $45 per kilogram and he was able to burn half or less of the amount purchased, with the rest lost in transit and cooling. The hydrogen delivered was about half orthohydrogen and half parahydrogen. Baker was aware that the spontaneous conversion of orthohydrogen into parahydrogen released heat, and suggested that savings could be made if all the hydrogen were converted to parahydrogen by means of a catalyst at the liquefier. With this sensible suggestion, he anticipated developments during the 1950s.

Fading Interest in Hydrogen-Oxygen

The successful results at Ohio State University, Aerojet General Corporation, and the Jet Propulsion Laboratory with liquid hydrogen–liquid oxygen for rocket engines in the late 1940s had little effect on the higher levels of the Air Force and Navy. In late 1948, Harvey Hall and his colleagues at the Bureau of Aeronautics attempted to maintain the Navy satellite program by proposing a reconfigured HATV as a super-performance sounding rocket to obtain information on the upper atmosphere. The proposal, backed by a detailed engineering report by the Glenn L. Martin Company, was made to the NACA Subcommittee on the Upper Atmosphere and to the Geophysical Sciences Committee of the Research and Development Board. The NACA subcommittee endorsed it—but it was only moral support, for the NACA had no funds for such work. The Geophysical Sciences Committee simply listened and took no formal action. This last-ditch effort was essentially the end of the Bureau of Aeronautics struggle for a high altitude test vehicle.[35]

In 1949, the Air Force again considered satellites for military operations and directed RAND to resume satellite studies. By the end of the year, Ohio State University was the only laboratory engaged in experimental investigations of liquid hydrogen for rockets, and there William Doyle had switched emphasis from hydrogen-oxygen to hydrogen-fluorine. The Ohio State hydrogen investigations in rockets ended in 1951.

SUMMARY, PART I

From 1945 to 1950, liquid hydrogen received considerable attention in analytical and design studies and in experimentation. The Jet Propulsion Laboratory of the California Institute of Technology and Project RAND at Douglas Aircraft Company compared rocket vehicle performance using hydrogen with the performance from other fuels. The superiority of liquid hydrogen was clearly indicated, but the biggest uncertainty related to the mass of vehicles using liquid hydrogen. North American Aviation and the Glenn L. Martin Company both made detailed designs of rocket vehicles using liquid hydrogen to obtain better vehicle mass values. Both incorporated thin-wall, pressure-stabilized, lightweight tanks as Oberth had proposed in 1923. Although not yet proven, this later became a key concept in the successful use of liquid hydrogen. Both the North American and Martin designs indicated superior vehicle performance with liquid hydrogen.

Concurrent with analytical and design studies were experiments on using liquid hydrogen–liquid oxygen in rocket engines. The Air Force sponsored experiments at Ohio State University on rockets, as well as scientific investigations of hydrogen's properties. At the same time, the Navy sponsored work at the Aerojet Engineering Corporation on liquid hydrogen–liquid oxygen rockets to determine the feasibility of launching a satellite with a single-stage-to-orbit vehicle. JPL, supported by the Army, also investigated the experimental performance of liquid hydrogen–liquid oxygen rockets and regenerative cooling.

All three laboratories conducting experiments had little difficulty obtaining efficient combustion and high exhaust velocities. Aerojet concluded that efficient combustion could be obtained with as little as 1/10 the volume normally used for rocket combustion. This, plus measurements indicating very high heat transfer, led them to propose and investigate an unusual thrust chamber design featuring a very small combustion volume and porous walls for transpiration (sweat) cooling, but difficulties with materials and cooling led to abandonment of the concept in favor of a more conventional design. Ohio State and JPL both used the more conventional thrust-chamber design and obtained much lower heat transfer values than Aerojet. This led to the successful use of liquid hydrogen as a regenerative coolant, a major contribution to liquid hydrogen technology.

In the investigation of injection techniques for efficient combustion, it was found that a concentric tube design, where an annulus of hydrogen surrounds an oxygen stream, was superior to the conventional impinging stream concepts, and an injector with many such concentric tube elements gave good performance. This concept, verified by Aerojet, was another major contribution to liquid hydrogen technology for rocket engines.

Both Ohio State and Aerojet investigated the pumping of liquid hydrogen and both found it feasible with a centrifugal pump. Ohio State also found that ball bearings for the pump could be operated without lubrication when immersed in liquid hydrogen, a very important finding for simplifying hydrogen pump design. The two investigations indicated liquid hydrogen could be successfully pumped.

Aerojet, using Herrick L. Johnston's design, built a hydrogen liquefier of 80 liters per hour, over three times greater than previous liquefiers. This showed that greater hydrogen liquefaction capability could be achieved through relatively straightforward engineering design. Dwight I. Baker of JPL found, however, that losses of liquid hydrogen prior to experimentation were too high to be tolerated and suggested that orthohydrogen be converted to parahydrogen at the liquefier by means of a catalyst—a key concept for practical use of large quantities of liquid hydrogen.

All the foregoing technical developments indicated that the basic technology for successful development of a rocket vehicle using liquid hydrogen–liquid oxygen was at hand, yet interest in using liquid hydrogen waned near the end of the 1940s. There are several explanations for this lack of interest. One is technical, for in spite of their successes, the experimenters encountered more than the usual number of difficulties in using liquid hydrogen, largely because of its lack of availability, very low temperature, explosive hazard, losses from orthohydrogen to parahydrogen conversion, and above all, the very low density. These were formidable obstacles for designer and experimenter alike, indicating that development of a hydrogen-fueled vehicle would be a long and costly development.

A second reason for lack of interest in hydrogen was the absence of a clear-cut need for its high performance. There were many other candidate fuels to be investigated including the boron compounds, hydrazine, and ammonia; and none had as many handicaps as liquid hydrogen. Of these, hydrazine looked particularly attractive.

A third reason was political. High Navy officials did not strongly support satellites. The Air Force made a major policy decision near the end of the 1940s to emphasize air-breathing propulsion rather than rocket propulsion.

Taking these reasons together, it is not surprising that interest in liquid hydrogen as a propulsion fuel receded in all but a few places where research-minded people remained interested in all high-energy rocket propellants. One such place was the Lewis Flight Propulsion Laboratory of the National Advisory Committee for Aeronautics at Cleveland, Ohio. The Lewis group was planning to conduct research with liquid hydrogen in 1950, but faced the same problem as Aerojet—the lack of liquid hydrogen. As they struggled with this problem, another development involving liquid hydrogen was begun on a crash basis and greatly advanced liquid hydrogen technology—thermonuclear research leading to the hydrogen bomb. These two contrasting activities—propulsion and explosives research—would renew interest in liquid hydrogen during the early 1950s.

Part II
1950–1957

PART II

1950–1957

During 1949–1950, changes in international relationships led to accelerated research in weaponry and aeronautics, both of which involved liquid hydrogen technology.

In the early postwar years, the United States was supremely confident of its superiority in atomic weaponry and did little to advance the technology. In September 1949, President Truman announced that Russia had exploded an atomic bomb; with it, went U.S. complacency. Relations between the two countries had been steadily deteriorating. Late in 1948, the Russians announced the withdrawal of occupation forces in Korea north of the 38th parallel and the establishment of a North Korean communist government. The North Koreans soon added to the tension by conducting raids south of the parallel. In June 1950, after massive invasion by North Korea, Truman authorized U.S. armed forces to assist the South Koreans.

Unlike the stagnation in weapons technology, U.S. progress in aeronautics during the postwar years had been significant. Effort concentrated on exploring transonic and supersonic flight regimes. The Air Force's Bell X-1 was flying at supersonic speeds in 1948, and a year later so was the Navy's Douglas D-558-II. Both were part of a military-industry-NACA flight research program which, by 1949, included more than a half dozen advanced experimental aircraft. In NACA's 1949 *Annual Report*, the chairman, Jerome C. Hunsaker, reported that this program had "given aeronautics perhaps the greatest impetus in its history." The same year, Congress passed the unitary wind-tunnel bill to coordinate and expand the nation's aerodynamic research.

In this environment of international tensions and greater emphasis on weaponry and aeronautics appeared three different research and development activities that involved liquid hydrogen. Each drew upon the technology developed by Ohio State University, Aerojet Corporation, and the Jet Propulsion Laboratory of the California Institute of Technology during the second half of the 1940s. And in the next seven years each added significant contributions to hydrogen technology. One, beginning in 1950, was the crash effort to develop a thermonuclear weapon, the hydrogen bomb. The second was research on high-energy rocket propellants by the National Advisory Committee for Aeronautics, which began to focus on liquid hydrogen in 1950. The third, started in 1952 or perhaps earlier, was an escalation of interest in high-altitude aircraft by the Air Force, which led to considerations of liquid hydrogen as an aviation fuel by both the Air Force and the NACA. The first and third activities dwarfed the second in terms of funding and manpower, but all three provided the basis for later development of launch vehicles using liquid hydrogen. The three activities will be described in the five chapters of this part.

4

Hydrogen Technology from Thermonuclear Research

Thermonuclear research began in the 1930s with the hypothesis that thermonuclear reactions are the energy sources of the Sun and stars. The nuclei of deuterium (heavy hydrogen) react more easily than the nuclei of normal hydrogen, and after Harold Urey separated deuterium in 1931, interest in the possibility of reacting deuterium increased. In 1942, Edward Teller began working on the possibility of initiating such reactions by means of an atomic explosion, but his initial conclusion was negative. Later the same year, he attended a conference on thermonuclear reactions where the group agreed that tritium (isotope of hydrogen) should be studied as well as deuterium and concluded that a thermonuclear explosion could be accomplished. The following year, plans for research on thermonuclear reactions were put aside at the newly formed Los Alamos laboratory to concentrate on uranium fission. Teller and a few others, however, continued their research.[1]

Until 1948, thermonuclear research received little support. Robert Oppenheimer, an early supporter, became opposed to further research on thermonuclear reactions after Hiroshima. Following the announcement that Russia had exploded an atomic bomb, the general advisory committee of the Atomic Energy Commission, chaired by Oppenheimer, recommended against proceeding with the development of a hydrogen bomb on technical, political, and moral grounds. The committee felt that the H-bomb was not yet technically feasible or economical and that lack of restrictions would mean high danger to civilization. This position was unpopular with many scientists at Los Alamos, where work continued, and with politicians, who recommended to President Truman that H-bomb development be initiated. The final catalyst appears to have been Klaus Fuchs's treachery; four days after he confessed to having given U.S. atomic secrets to the Russians, Truman directed that development of the H-bomb start.

With the presidential go-ahead in January 1950, Teller and associates at Los Alamos intensified their efforts to design a practical bomb and began preparations for some critical tests. Hydrogen liquefiers were needed for these, and Herrick L. Johnston of Ohio State University became a key figure in setting up and operating the equipment.

63

Johnston's New Career

The announcement that the U.S. would proceed with the H-bomb had special interest for Professor Johnston; he saw it as a golden opportunity to capitalize on his position as an authority on large-scale hydrogen liquefaction and associated equipment.[2]

Whether Johnston realized that this new opportunity would eventually take him completely away from his academic career is a matter of conjecture. His colleagues knew that he harbored a long-time disappointment over what he considered a lack of sufficient recognition in the scientific community.[3] Perhaps some of this feeling was associated with his earlier work on deuterium. His preceptor, William Giauque, had been awarded the 1949 Nobel chemistry prize for his achievements in low-temperature physics, and Giauque had generously credited Johnston with significant contributions in the description of the prize-winning work.[4] Whatever his reasoning, Johnston resolved to seek greater compensation for his expert knowledge of cryogenics. The hydrogen bomb development provided this opportunity and he seized it. From then on, Johnston gave less attention to science and education and more to developing a business in cryogenics equipment. By 1954 the metamorphosis was complete, but during the crucial 1950–1954 period, he simultaneously pursued three careers—scientist, educator, and businessman—all involved liquid hydrogen.

Johnston was colorful, unconventional, controversial; it was difficult for those who came in contact with him to remain neutral about him. To aspiring undergraduates, he was a person who made or broke them, for a good grade in his tough thermodynamics course was required for continuing a career in chemistry.[5] To graduate students, post-doctoral fellows, and his peers, Johnston was a first-class preceptor and scholar, a man of great inspiration and integrity.[6] To university officials, he was a mixed blessing; his contracts brought equipment, staff, and prestige; but his utter disregard for normal operating procedures brought endless problems.[7] To employees, he was a paternalistic and high-handed autocrat, impatient and demanding, who would, and did, fire a person at the slightest provocation.[8] To business associates, he was a formidable competitor, capable of quick responses, low bids, and early delivery of his products.[9]

To Johnston, there was no problem that could not be solved and solved quickly. He demanded and got the best in equipment and services for himself and his people. He disdained normal administrative procedures and anything resembling bureaucracy infuriated him. He was often at odds with one official or another and never hesitated to go over their heads to appeal to higher authority. An unsung hero and loyal supporter of Johnston was Edward Mack, Jr., chairman of the department of chemistry at Ohio State from 1941 to 1955, who spent long hours sorting out and solving the endless problems that always seemed to surround the fast-moving Johnston and his activities.[10] In spite of the problems, Johnston's work was internationally recognized and many of his graduate students and assistants became prominent in the scientific community.*

---

*One graduate student was Clyde Allen Hutchinson, professor of chemistry at the University of Chicago; a key assistant was David A. White, now chairman of the department of chemistry of the University of Pennsylvania; another assistant was Thor Rubin, professor of chemistry at OSU. The old War Research Building at OSU is now named the Johnston Building in his honor.

Fig. 11. Herrick Lee Johnston (1898–1965), scientist, educator, entrepreneur, and a pioneer in the science and technology of liquid hydrogen. (Courtesy of the Photo Archives, Ohio State University.).

In the 1950–1951 period, Johnston supplied the Los Alamos Scientific Laboratory with two hydrogen liquefiers. When the decision was made to conduct thermonuclear tests at Eniwetok, he was given the contract to reassemble the old Aerojet liquefier and add a second one. He chose Gwynne Wright to head the team to do this and the next two years became a period of swiftly moving activity for all of them. In May 1951, the first thermonuclear test, Operation Greenhouse, was successful and not long after it, preparations began for the next test. By late 1951, Johnston was so involved that he wrote to the president of Ohio State University requesting that: (1) his services to the university be reduced to 25 percent of full load, effective 1 January 1952; (2) selected members of the cryogenics laboratory be given leaves of absence; (3) OSU shop facilities be allowed to continue their work for Los Alamos; and (4) air and hydrogen liquefiers be made available for an essential training program. Johnston ended by assuring President Bevis that the university would be recompensed for its expenses and services. He was off and running again, doing very high priority work for the government, and the university had little choice but to go along with his wishes.[11]

In 1952, Johnston set himself up in business as the H. L. Johnston Company, Inc., and lured some of the key people from the OSU cryogenics laboratory. In May, the graduate school notified Mack that Johnston's name would be removed from the list of faculty approved to advise graduate students for masters and doctoral degrees. Mack protested vigorously, and in December he was joined by seven students who petitioned to retain Johnston as their preceptor. During this period, and working against odds, Johnston and his men delivered on their promise to produce deuterium on Eniwetok for the Mike Event of Project Ivy (fig. 12). On 1 November 1952, the event took place and was the first test of a thermonuclear "device"—a device that wiped out the islet where it had been set up. It was the most powerful explosion man had devised up to that time.[12]

National Cryogenic Engineering Laboratory

The development of the hydrogen bomb gave the National Bureau of Standards the opportunity to establish itself as the leader in cryogenic engineering research during the 1950s. The bureau had been involved with liquid hydrogen and cryogenic research since purchasing its first liquefier from the British Oxygen Company in 1904.[13] In 1925, Frederick G. Brickwedde became head of the cryogenic laboratory in the heat and power division, a post he held until 1957. He distilled liquid hydrogen to obtain a sample of deuterium for Harold Urey in 1931 (appendix A-3). In 1947, Brickwedde and William Gifford began a cryogenic engineering project. Some years earlier, Professor Samuel C. Collins of the Massachusetts Institute of Technology had designed a helium cryostat, which was being produced and marketed by the Arthur D. Little Company as the ADL-Collins cryostat. It had a capacity of about 4 liters of liquid helium per hour, which was ample for most university research needs, but Brickwedde wanted a cryostat of greater capacity. He and Gifford, working with the Arthur D. Little Company, designed one with a capacity five times greater than the ADL-Collins cryostat. It was placed into operation in 1952. From the summer of 1948, Brickwedde visited the Los Alamos Scientific Laboratory as a consultant. There he worked with Edward F. Hammel, head of the cryogenic laboratory, and E. R. Grilly. In 1949, Hammel began

Fig. 12. Dr. Herrick L. Johnston and staff who installed the cryogenic facilities at Eniwetok for Operation Ivy. Front row, L to R: Lester Cox, Ross Justus, Man Treziaki, Gwynne Wright, Herrick Johnston, Al Sesonski, Don McGill, Richard Reo, James Pierce; middle row: George Koch, Mike Nicata, Raymond Ward, Everett Allen, Paul Hansel, Nip Nunimaker, William Johnston, Richard White, Dewey Sandell; top row: Paul Camky, Charles Ames, Dave Wood, Howard Altman, Nate Hallet, Leon Wagner, Clarence Cunningham, ------ Groves, Douglas Chapin, Art Kurz, Jay Alberti. (Photo courtesy of James Pierce, identifications by Gwynne Wright.)

suggesting that the country needed a single large national laboratory for cryogenic engineering. He formed a committee of advisors on cryogenic engineering that included Manson Benedick, Brickwedde, Samuel Collins, Herrick Johnston, Earl Long, and Darrell Osborne. This group discussed Hammel's idea for a laboratory and supported it.[14]

The Bureau of Standards had become pinched for space in Washington and had decided to locate its expanding radio facility elsewhere. In 1949, the citizens of Boulder, Colorado, donated a 0.9-square-kilometer tract at the foothills of the Rockies for the facility. All these events might have remained unconnected except that Truman's decision to go forward with the hydrogen bomb put considerable support behind Hammel's suggestion for a cryogenic laboratory. The Atomic Energy Commission selected the Bureau of Standards to build and operate a cryogenic engineering laboratory at the Boulder site. In the summer of 1951, the Stearns-Roger Manufacturing Company began construction and within a year, two buildings were completed—one for the hydrogen and nitrogen liquefiers and another for research. Brickwedde and Gifford became the first members of the staff. By March 1952 liquid hydrogen was being produced and by August the laboratories were open.[15] The NBS-AEC Cryogenic Engineering Laboratory, with Dr. Russell B. Scott as its first chief, was in full swing in the fast-moving preparations for the hydrogen bomb development.

The gas liquefaction capacity was 350 liters per hour of liquid normal hydrogen (or 240 liters per hour of liquid parahydrogen) and 480 liters per hour of liquid nitrogen; storage capacity was 4500 liters of liquid hydrogen and 22000 liters of liquid nitrogen. It was the largest liquid hydrogen plant in the country and started operation less than three years after the Aerojet liquefier, built for hydrogen rocket experiments, closed down. By 1954, the Cryogenic Engineering Laboratory had an extensive program that included: (1) precise measurement of the thermal conductivities of metals and dielectrics, (2) mechanical properties of materials at low temperatures, (3) superinsulations, (4) high vacuum techniques, (5) transfer of liquefied gases, (6) development of vessels for storage and transport of liquid hydrogen, (7) ortho- to parahydrogen conversion, (8) hydrogen liquefiers and pilot plant evaluation, and (9) cryogenic testing, particularly with respect to vibration.[16]

Mobile Liquid Hydrogen Equipment, 1952–1954

The Air Force worked closely with the Atomic Energy Commission in hydrogen bomb development and as part of its responsibility, contracted for the development of mobile equipment. This work was cancelled in 1954, but not before some remarkable equipment had been built, which later became available for rocket research by the National Advisory Committee for Aeronautics.

One piece of equipment the Air Force developed was the air-transportable dewar for carrying liquid hydrogen or deuterium in the B-36 or B-47. The National Bureau of Standards and H. L. Johnston, Inc., both developed tactical dewars for the Air Force, and the vessels were described in a 1954 cryogenic engineering conference.[17] Essentially they utilized the same thermal insulation method as the familiar dewar vacuum flask, but the design was much more elaborate and complex in order to store liquid hydrogen at 20 K. The design minimized heat transfer by its three modes: conduction, convection, and radiation. The liquid hydrogen was held in an inner tank. Surrounding it was a space evacuated to a high vacuum to minimize heat transfer by conduction and convection. The wall on the other side of the space was maintained at the temperature of liquid nitrogen, 77 K, which minimized heat transfer by radiation. The radiation shield—the walls of a liquid nitrogen container—was itself insulated from the outer shell of the dewar by another vacuum space.

The air tactical dewars held 750 liters of liquid hydrogen. The heat flow to the liquid hydrogen shell was slight, amounting only to about 4 watts; liquid hydrogen boil-off was about 7.5 liters per day, or 1 percent of rated capacity. The dewars were equipped with an array of valves, instruments, and a vacuum pump.

The tactical dewars (fig. 13) had to be built for rough treatment. The Johnston design employed hardened stainless steel rods to suspend the inner tanks and minimize conduction. These rods had to be tuned to an exact frequency to meet vibration and shock load specifications. Howard Altman solved this problem in an ingenious fashion typical of Johnston's operation. He calculated the required frequency and pitch and brought his violin to work one night to tune the rods; the Johnston dewar then passed its Air Force test. Several years later it passed another—unplanned—test. Altman was helping to unload it from a truck at the NACA Lewis Laboratory in Cleveland when it slipped and dropped four feet to the ground. He heard the rods vibrating sweetly, and the dewar survived undamaged.[18]

Another type of transportable dewar was fabricated by the Cambridge Corporation for the Air Force, using a design by the Arthur D. Little Company. Called the

Fig. 13. Air-transportable dewar for 750 liters of liquid hydrogen developed by H. L. Johnston, ca: 1952. Howard Altman is on the left. (Courtesy of H. A. Altman.)

Refrigerated Transport Dewar, it used a closed cycle helium refrigerator, and the 2000 liters of liquid hydrogen could be stored or transported indefinitely with no loss as long as the refrigerator was operated. It was a large piece of equipment, occupying all of a 10.7-meter semi-trailer, and weighed 18.1 metric tons, including a diesel generator for refrigerator operation away from electric power lines.[19]

The third type of mobile equipment developed for the Air Force in the 1952–1954 period was a mobile hydrogen liquefier, again built by H. L. Johnston, Inc. It was mounted on three semi-trailers and was capable of producing 100 liters per hour of 45 percent liquid parahydrogen. Two of the trailers housed huge Norwalk horizontal compressors. The trailers also contained a gas holder and auxiliary equipment for the compressors. The third trailer housed the complete hydrogen purification and liquefaction equipment (fig. 14). All three trailers were capable of highway transport at 89 kilometers per hour. Gross weight was about 25 metric tons, and they required 105 kilowatts of electric power for operation.[20] The author remembers inspecting these trailers when they were loaned by the Air Force to the NACA (about 1956) and marvelling at how much equipment had been packed into such a small space. Particularly impressive were the big compressors with their large flywheels. Johnston's students had designed the layout of these trailers using cardboard cutouts to arrange the equipment. In their first operation at Kirkland Air Force Base, the whole 25-

Fig. 14. Mobile hydrogen liquefier developed for the Air Force by H. L. Johnston, Inc., in 1952–1953. The lower trailer contains a gas holder and large compressor; the upper, hydrogen purifier and liquefier. (Courtesy of W. V. Johnston.)

metric-ton trailer began to bounce and "walk," moving 8 to 10 centimeters forward with each bounce—quite an awesome sight. The problem was solved by raising the trailer off its tires on large jacks.[21]

Cryogenic Information Exchange

With the fast-paced research and development in cryogenics that began in 1950, there was a need for exchange of information among the engineers and scientists engaged in the program. To that end, the NBS-AEC Cryogenic Engineering Laboratory sponsored an engineering conference at Boulder, 8–10 September 1954.[22] Sixty papers were presented on cryogenic equipment, instrumentation, insulation, and materials. A second conference, held in 1956, had fifty papers grouped into four categories—cryogenic processes, equipment, properties, and applications—and one special application, bubble chambers for research on the physics of particles. There were papers on the fundamentals of hydrogen liquefaction, ortho-to-para catalysts, distillation of hydrogen-deuterium mixtures, and safety.[23]

Among the papers in the third conference in 1957 was one by three Bureau of Standards men on the design of orthohydrogen-to-parahydrogen converters (the necessary step seen by Dwight I. Baker in 1949, p. 56.) The investigators reported that the 240-liter-per-hour hydrogen liquefier at the NBS cryogenic laboratory used 1.5 liters of 30–100 mesh granules of hydrous ferric oxide catalyst, and this converted the 240-liter-per-hour hydrogen output to about 94 percent parahydrogen. Another paper described liquid oxygen transfer equipment capable of 3800 liters per minute, developed by the Cambridge Corporation; while another described a 6000-liter liquid hydrogen dewar made by Beech Aircraft Company.[24] These papers illustrated the level of cryogenic and liquid hydrogen technology in 1957; a quantum jump had been made since the beginning of the decade. A fourth conference was held in Cambridge, Massachusetts, in 1958; the fifth at the University of California, Berkeley, in 1959; and the sixth back at Boulder in 1960.[25]

Summary

Under the stimulus of hydrogen bomb development, liquid hydrogen technology advanced rapidly in the first part of the 1950s. Hydrogen liquefier capacity had risen from the 80 liters per hour of the Aerojet plant in 1949 to the 350 liters per hour of the NBS-AEC Cryogenic Engineering Laboratory. The new national laboratory and the increased number of contractors who entered cryogenic engineering brought many new developments. Dewars were built that allowed as much as 6000 liters of liquid hydrogen to be stored indefinitely or transported cross-country. Applications of this cryogenic technology began to increase. Among them was the use of liquid hydrogen as a working fluid for nuclear rockets that began in 1955. The use of hydrogen in a nuclear rocket is not as a fuel, however; the energy comes from the reactor, and hydrogen is the ideal working fluid because of its low molecular weight. For this reason, the nuclear rocket development of the 1950s will not be discussed further except as it relates to technology used in the application of liquid hydrogen as a fuel.[26]

5

NACA Research on High-Energy Propellants

The National Advisory Committee for Aeronautics (NACA), established in 1915 to. develop practical solutions for the problems of flight, showed interest in liquid hydrogen as a fuel in 1939 but did nothing about it for over a decade. The early interest came as a surprise to Robert Goddard when he visited NACA's director, George W. Lewis, in March 1939. He learned that "the NACA is considering liquid H [hydrogen] as a fuel (!) possibly used with air for rocket propulsion."[1] Four days after his visit, still amazed, Goddard wrote to a friend:

> On talking with Dr. Lewis of the NACA I found that they are contemplating using liquid hydrogen, because of its low weight and high heat value, as a fuel with atmospheric air. I mention this because liquid hydrogen is expensive and difficult to transport and store . . . and also because tanks of it have to be surrounded by liquid oxygen or liquid nitrogen. It makes my use and advocation of liquid oxygen seem really conservative by comparison. The main point is that even with the extreme difficulty of liquid hydrogen, its use is being considered by a body as serious as the NACA.[2]

What did Lewis have in mind? The use of atmospheric air rather than readily available liquid oxygen suggests that he may not have been thinking of a simple rocket for propulsion but a rocket as a component in an air-breathing engine, possibly applied to a turbine engine. He may have heard about the early work of Hans von Ohain, employed in April 1936 by Ernst Heinkel to develop a turbojet engine. Pressed for time, von Ohain turned to gaseous hydrogen as a fuel for convenience in tests beginning in early 1937 and found that his turbojet engine worked well using hydrogen.*

---

*Von Ohain's work made hydrogen one of the first fuels to be used in turbojet engines. Lewis visited Germany in September–October 1936, returning on the hydrogen-filled *Hindenburg* which impressed him so much he wrote a report on it. Lewis may have learned about von Ohain's work during his visit or through later reports of John J. Ide, NACA representative stationed in Paris, or through intelligence reports which he received about aeronautical developments. Hans von Ohain to author, 21 May 1974; George W. Lewis file, NASA History Office; telephone interview with Robert E. Littell, former NACA aide at headquarters. 20 Aug. 1973.

On the other hand, Lewis had long been thinking about rocket research, for 18 months earlier he had asked NACA member Charles A. Lindbergh, then in England, "for recommendations with reference to any rocket research for the National Advisory Committee for Aeronautics to carry on."[3] Lindbergh, in turn, sought the advice of Robert Goddard who suggested "several lines of research: for example, liquid propulsion rockets for gliders; application of rockets to turbines; rockets for accelerating and decelerating planes; development of combustion chambers of large thrust."[4] In 1938 Lewis wrote to Goddard expressing interest in his high-speed work and Goddard asked for NACA wind-tunnel tests to determine the flight stability of his rockets. Was Lewis thinking of Goddard's suggestion of applying rockets to turbines, a concept appearing later as a "turborocket"? Whatever Lewis had in mind remains a mystery for, characteristically, he kept his planning informal and shared it with few others. The incident, however, illustrates the dual nature of NACA during that period—receptive to new ideas but conservative and slow in entering new fields of research. It also indicates the ease with which liquid hydrogen comes to mind when engineers think of high-energy fuels.

In 1944, seven years after asking Lindbergh about recommendations for rocket research and apparently after some prodding by Wright Field, Lewis authorized the construction of four simple rocket test cells at the Aircraft Engine Research Laboratory in Cleveland.[5]

The information on German jet propulsion and rocket developments, which increased from a trickle in 1943–1944 to a flood of captured documents in 1945, made NACA officials realize how far behind they had fallen in these new propulsion systems. In the fall of 1945, a sweeping reorganization of the Cleveland engine laboratory caught all but senior officials by surprise. Overnight, research emphasis shifted from piston engines to jet engines (turbojet and ramjet) with some work on rockets. The rocket research was kept small because of the conservative nature of NACA and the influence of its chairman, Jerome C. Hunsaker, who shared with many the belief that rockets were more applicable to artillery than aircraft and had no place in an aeronautical research laboratory. The word "rocket" was avoided in the organizational name in favor of "high-pressure combustion."

The rocket group at the Cleveland laboratory concentrated on high-energy, liquid-propellant rocket engines with teams working on propellant performance (theoretical and experimental), combustion, and cooling.* The propellant work followed the logical path of computing the theoretical performance of several fuel-oxidizer combinations over a range of operating conditions and selecting the most promising for experimental investigation. By 1948, Riley Miller and Paul Ordin reported calculations of a number of propellant combinations containing hydrogen, nitrogen, and oxygen atoms, with liquid hydrogen giving the highest exhaust velocity and having the lowest propellant density.[6] The same year, Vearl Huff and his associates made a major contribution to theoretical performance techniques by developing a convergent, successive approximation method that saved considerable time over other methods.[7]

---

*The rocket section was part of a combustion branch headed by Walter T. Olson in a division headed by Benjamin Pinkel. Joseph R. Dietrich was the first head of the rocket section, followed by Everett R. Bernardo, and the author in 1948.

High-energy rocket propellants were difficult to obtain, for most were available only in small quantities. The NACA researchers passed by liquid hydrogen in favor of hydrazine and diborane as fuels and 100 percent hydrogen peroxide, chlorine trifluoride, liquid oxygen, and liquid fluorine as oxidizers.[8]

Calculated risks were taken to transport comparatively rare samples to the laboratory. Louis Gibbons, chief of fuels research, brought a gallon of pure hydrogen peroxide from Buffalo clamped between his knees in an all-night train ride. Paul Ordin used much the same method in bringing a sample of hydrazine from St. Louis. The first diborane, nested in dry ice, was delivered by private automobile from Buffalo. The first liquid fluorine was obtained from downtown Cleveland and transported in a special laboratory-built trailer escorted by police.[9]

During the 1947–1949 period, diborane was of great interest as a rocket fuel, but experiments soon revealed that it had great disadvantages and its theoretical promise could not be realized. When used with liquid oxygen or hydrogen peroxide, diborane formed boron oxides which deposited in the rocket nozzle and degraded performance.[10] When used with liquid fluorine, the combustion products were volatile, but the absence of deposits was replaced with a greater problem—difficulty in cooling. The theoretical flame temperature of diborane-fluorine under rocket operating conditions is about 5400 K, far higher than many other propellants. Moreover, neither diborane nor fluorine is suitable as a regenerative coolant, which means a third fluid is required for cooling, seriously degrading performance. Although experimental performance of diborane-fluorine was measured, it became apparent by the early 1950s that diborane was not a good rocket fuel.[11] The experience with diborane showed not only the limitations of theoretical considerations in selecting propellants but also the value of experiments in revealing practical problems.

In 1949, the acceleration in aeronautical research brought another major reorganization to the Lewis Flight Propulsion Laboratory.* Its director remained Edward R. Sharp, a gregarious and able administrator who had started as an apprentice at the Langley laboratory. Technical management was strengthened by elevating Abe Silverstein to chief of research.

The reorganization brought a pleasant surprise to the small rocket group. It was moved up one level in the organizational hierarchy, named for what it was, and given more personnel. Silverstein was the highest NACA official to show significant interest in rocket research, although much of it was new to him. One of the things he wanted to understand better was the propellant selection process and, particularly, how candidates for research were chosen.

During the same period, organizational changes were occurring in related military research and development. In 1949, a USAF advisory committee headed by Louis N. Ridenour recommended that the Air Force research and development activities be consolidated into a single command. In January 1950, the Air Force established the Air Research and Development Command which included the facilities at Wright Field and the Air Engineering Development Center at Tullahoma, Tennessee—the latter renamed in honor of General Arnold the following month.

---

*The NACA Aircraft Engine Research Laboratory at Cleveland was named the Lewis Flight Propulsion Laboratory in 1948 in honor of NACA's first director of research.

The reassessment of research plans that followed the organizational changes had special significance for the Lewis rocket group in March 1950 when a group of Wright Field officials visited the Lewis laboratory.* The visitors showed considerable interest in rocket research in general and propellant selection in particular. Also discussed were the merits of forming a NACA subcommittee on rockets, a need recognized by the Durand committee nearly a decade earlier.

Conference on Propellant Selection

Apparently as a direct result of the visit by Wright Field officials, the NACA called a meeting of rocket experts at the Lewis laboratory on 19 May 1950 to discuss the selection of rocket propellants for long-range missiles.[12] A secondary purpose was to use the meeting as a "test run" to determine if a NACA subcommittee on rockets was desirable and feasible.

Propellant selection for any mission is always a compromise between performance and other desired characteristics such as density, cooling capacity, storability, handling, and availability (appendix A-4). The selection process had advanced through several levels of sophistication. The simplest method used exhaust velocity as the criterion, since range varies approximately with the square of exhaust velocity. This ignores the effect of propellant density, which affects tank and vehicle size and mass. In 1947, Richard Canright of the Jet Propulsion Laboratory developed a method of relating exhaust velocity and propellant density for large rockets and found that exhaust velocity was the more important of the two. Combinations using liquid hydrogen ranked the highest, although Canright favored hydrazine for its overall characteristics (p. 47–48). Later studies involved more complex considerations of missile design and flight than Canright's, but all suffered from lack of data that could be obtained only when rockets were designed, built, and flown.

In addition to the major flight parameters, the military was very interested in the logistics problems of propellants—such characteristics as vapor pressure, freezing point, stability during storage, corrosiveness, toxicity, availability, and cost.

At the Lewis meeting, the military representatives and their contractors presented their views and research results. The NACA-Lewis recommendation for propellants, presented by the author, consisted of a primary selection and alternatives. The primary fuel was liquid hydrogen and the primary oxidizer was liquid fluorine. If propellant density proved too great an obstacle for liquid hydrogen in a practical application, the alternate fuels selected were hydrazine, ammonia, or a mixture of the two. The Lewis choice of alternative oxidizer was oxygen.

The NACA recommendation, its first firm choice of liquid hydrogen as a rocket fuel, was not opposed by anyone at the meeting. After all, the selection was for research

---

*They were: Col. Marvin Demler, chief of the power plant laboratory; Lt. Col. J. M. Silk; Opie Chenoweth, chief scientist; C. W. Schnare, in charge of rockets; E. C. Simpson, in charge of turbojet engines; W. A. Wolfe; E. Brown; and R. E. Ray. NACA attendees were: Edward R. Sharp, Abe Silverstein, E. J. Manganiello, Walter T. Olson, Willson Hunter, John H. Collins, Jr., and the author.

purposes, not for a development. The selection of alternatives reflected the uncertainty over the effect of fuel density on long-range missile design and performance. The NACA position satisfied both those who believed in the potential of liquid hydrogen and those who did not.

Views about using liquid fluorine, however, varied considerably. During the morning session, William Doyle, of Ohio State University, listened with growing impatience to presentations by Rocketdyne and Aerojet on their fluorine experiments. As a strong advocate of both liquid hydrogen and liquid fluorine, he felt that the presentations were too pessimistic. He could hardly wait to rebut them, but lunch intervened. After lunch, the meeting chairman, Abe Silverstein, noted the meeting was behind schedule and cancelled discussion of the morning papers. This was too much for the peppery Doyle who jumped to his feet, announced that he knew more about fluorine than anyone else present, and proceeded "to lambast the hell out of the two fluorine papers" for their pessimism. Silverstein allowed Doyle to make his point before clamping down.[13]

Following the May propellant selection meeting, the NACA rocket group planned experiments with liquid hydrogen but faced the familiar problem: how to get a supply of it. Obtaining dewars of liquid hydrogen from Herrick Johnston at Ohio State University and transporting them to Cleveland was rejected as impractical. Since it was not available commercially, the only course open was to build a liquefier at the laboratory. Since the money needed was too much to come from operating funds, the NACA, for the first time, went to Congress in 1951 with a request specifically for rocket research. The fiscal year 1952 budget for construction of facilities contained an item of $150000 to buy a hydrogen liquefier and a building to house it. The justification stated in part:

> Of the chemical combinations that are available as propellants for rocket engines for maximum range, liquid hydrogen offers great potentialities. With certain oxidizers liquid hydrogen has the greatest thrust-per-pound propellant flow [exhaust velocity] of any of the chemical combinations, an important factor for long flight. Insufficient experimental research has been done in this Nation on the use of liquid hydrogen with suitable oxides [sic]. . . . Although there are no commercial cuppliers of liquid hydrogen, simple liquefaction equipment developed during the war, is available commercially.[14]

Congress approved the request and NACA contracted with the Arthur D. Little Company of Cambridge, Massachusetts, for a hydrogen liquefier scaled up from a Collins cryostat. The company ran into some difficulties which delayed delivery, and the NACA spent the interim period investigating other high-energy rocket propellants.

NACA Rocket Subcommittee

NACA officials were pleased with the outcome of the May 1950 meeting of rocket experts and the following January established the Special Subcommittee on Rocket Engines under the Power Plants Committee. The chairman was Professor Maurice J.

Zucrow of Purdue University, a well known and respected authority on jet propulsion.*

Establishment of the rocket subcommittee represented a significant milestone in NACA recognition of the importance of rocket research. In addition to its great value for coordinating and exchanging information on rocket research and development, the subcommittee was a political force for assuring a fair share of attention to rocket propulsion. Although the number of research personnel assigned to rocket research at the NACA Lewis laboratory was still small—less than 3 percent—the group had the strong support of both Silverstein, an associate director of the laboratory, and a body of national experts on rockets whose advice and recommendations would carry weight.

Research Conference on Supersonic Missiles

On 13 March 1952, the NACA held a research conference at the Lewis laboratory to present the latest results of research pertaining to supersonic missile propulsion. Papers on turbojet and ramjet propulsion dominated the meeting, but there was one paper on the status of liquid-propellant rocket engines by Gerald Morrell and Vearl Huff. Their paper covered propellants, combustion, and cooling—the three subjects of NACA research. Experimental performance data for rocket engines using ammonia and ammonia-hydrazine mixtures as fuels and liquid fluorine as the oxidizer were presented. With respect to high-energy propellants in general, the authors stated:

> The high specific impulse [exhaust velocity] propellant systems are of greatest promise for application in long-range missile propulsion; recommendations for propellant systems which require development include hydrogen-oxygen, hydrogen-fluorine, and ammonia-fluorine. Experience with these systems is still in the early experimental stages, but the performance obtained to date is encouraging. With the hydrogen-oxygen system, other laboratories (JPL, Aerojet and Ohio State) have obtained 96 to 97 percent of the theoretical specific impulse calculated for equilibrium expansion; that is, maintenance of chemical equilibrium is assumed during the expansion process. Experiments with the hydrogen-fluorine system in a 100-pound-thrust [445 newton] engine at JPL have yielded equally good performance.[15]

Boost from the Subcommittee

Since the start of the Korean conflict in 1950, the NACA had submitted larger budget requests for aeronautical research each year, only to have the requests cut

---

*Other members: Richard B. Canright, JPL-CIT; Comdr. K. C. Childers, USN, Bureau of Aeronautics; R. Bruce Foster, Bell Aircraft; Stanley L. Gendler, Rand Corp.; Joseph L. Gray, Office of Chief of Army Ordnance; Paul R. Hill, NACA-Langley; G. E. Moore, General Electric; Thomas E. Meyers, North American Aviation; C. W. Schnare, Wright Air Development Center; Jack H. Sheets, Curtiss-Wright; Capt. Levering Smith, USN, Naval Ordnance Test Station; R. J. Thompson, Jr., M. W. Kellog; Paul Winternitz, Reaction Motors; David A. Young, Aerojet; the author; and Henry E. Alquist, NACA, secretary. The next year, Lt. Col. Langdon Ayers, USAF, replaced Schnare; Eugene Miller of Redstone Arsenal replaced Gray; and Benson E. Gammon, NACA, replaced Alquist.

sharply in final appropriations by an economy-minded Congress. Within the NACA, the rocket subcommittee, aided and abetted by the NACA rocket group, became convinced the NACA was not doing enough rocket research. To support this view, a comprehensive review of the NACA rocket program was conducted at the 26–27 June 1952 meeting of the subcommittee. By this time, theoretical work on propellant performance, carried out with the aid of computers, was far ranging and included hydrogen with oxygen and fluorine. In addition, the relationship of propellants and propulsion systems to missions was being studied. Experimental work centered around ammonia and ammonia-hydrazine mixtures as fuels and fluorine as oxidizer, using small engines. Research with liquid hydrogen was still in preparation.[16]

After reviewing the program, the rocket subcommittee passed a resolution that was to have far-reaching consequences:

WHEREAS, The rocket propulsion research effort of the NACA is highly commendable and of good quality, and

WHEREAS, The NACA rocket propulsion research activity is at much too low a level to be consistent with the importance of rocket propulsion to military services, and

WHEREAS, The rocket propulsion research at the NACA is, in general, being conducted on equipment which is of such small scale that the results obtained are only of limited value to the rocket engine contractors, and

WHEREAS, A function of the NACA is to serve the rocket propulsion industry as an advanced research agency,

BE IT RESOLVED, That the Special Subcommittee on Rocket Engines recommends to the NACA that the research activity on rocket propulsion be expanded and emphasis placed on a broader and more advanced approach to the solution of rocket propulsion problems.[17]

The subcommittee then listed nine problem areas that should be added to the NACA program, but none mentioned hydrogen or other high-energy propellants.*

The rocket subcommittee resolution was presented to the parent Power Plant Committee by Zucrow; it was approved and passed to the NACA Executive Committee, which also approved it. Word passed from Washington to Cleveland to intensify the planning of rocket facilities.

Plans for Rocket Facilities

By 1952, the Lewis rocket facilities consisted of four original test cells and four newer and larger cells built with operating funds by the "Hurry-Up Construction

---

*They included scaling factors for designing large-thrust rockets, causes and remedies of combustion oscillations, composite and multiple-stage missiles (in cooperation with structural and aerodynamics research teams), nitrogen oxides as oxidizers, rocket propulsion for fighter aircraft, problems of using nitric acid as an oxidizer, variable-expansion nozzles, and altitude performance of rockets.

Company"—the self-styled, in-house construction group. With the word from Washington, the imaginative and ambitious rocket group began turning out a series of grand plans for rocket testing that required a huge site in a remote area of the West. These went far beyond the intent of NACA officials and the bubbles burst one by one, until planning narrowed down to what could be built at the Cleveland laboratory site. By the time the NACA executive committee approved the rocket subcommittee's resolution, NACA had decided to request an $8.5 million rocket-engine facility at the Lewis laboratory. It was described at the November 1952 meeting of the rocket subcommittee by Walter T. Olson and the author.[18]

The proposed facility provided complete engine systems research using two major classes of propellants: high-energy propellants for long-range missiles, and high-availability, low-cost propellants for boosters, superperformance aircraft, and medium-range missiles.* The facility was unique in four features: (1) high thrust and long durations (89 kN and 3 min. for high-energy propellants; up to 445 kN and 3 min. for high-availability propellants); (2) hydrogen liquefaction and fluorine generation and liquefaction in quantity at the test site; (3) exhaust-gas scrubbers, designed from data provided by NACA research, to remove hydrogen fluoride from the exhaust; and (4) silencing equipment to muffle the rocket's roar.

The subcommittee endorsed the proposed facility and its chairman, Maurice Zucrow, added his hearty endorsement.[19] This support was crucial but despite it, the attrition process that had befallen earlier plans reappeared at the Bureau of the Budget. The rocket group began to get telephone calls about cutting various features to reduce the cost. One of the first items to go was the fluorine plant, but this was not too serious as Allied Chemical was becoming interested in supplying fluorine for rocket applications.† Hans Neumark of Allied Chemical, under contract with the Air Force, was developing an over-the-highway trailer for transporting liquid fluorine. The next item to go from the proposed facility was the hydrogen liquefaction plant, followed by the large-scale facilities for engines of 445 kilonewtons. The rocket group became depressed as they watched their dreams melting away. One day a call came from Washington: What can you do for $2.5 million? The answer: the high-energy propellant features with exhaust-gas scrubber and silencer. This was accepted and the facility was authorized and funded by Congress. Construction began in 1953 with scheduled completion in 1956. During this period, an existing rocket test-cell was modified to handle high-energy propellants in engines of 22 kilonewtons and the Arthur D. Little hydrogen liquefier was installed. The research program remained essentially the same, but four years after selecting liquid hydrogen as its first choice, the NACA had yet to experiment with it.

In spite of the increased NACA support, rocket research remained comparatively small during the construction of the new facility. Disappointed, the rocket subcommittee at its October 1954 meeting noted that the NACA was spending twice as

---

*The high-energy propellants were liquid hydrogen, hydrazine, and ammonia as fuels, with fluorine and oxygen as oxidizers. The high-availability propellants were hydrocarbons as fuels with nitric acid and liquid oxygen as oxidizers.

†Six years later (October 1958), the author participated in opening ceremonies of Allied Chemical's fluorine plant at Metropolis, Illinois.

much on ramjet research as on rocket research,* whereas the military services were emphasizing rockets, not ramjets—a clear signal that the NACA was about to miss the boat again as it had earlier with its late start in jet propulsion.

Switch from Air-Breathing to Rocket Engines

Up to 1952, military concepts for long-range missiles emphasized rocket-boosted, winged missiles powered by air-breathing engines.† Beginning that year, a series of events brought great changes in military thinking about strategic missiles. These events, according to Herbert York, a participant, were "the invention and demonstration of the hydrogen bomb, the election of Eisenhower and the concomitant extensive personnel changes throughout the executive branch, . . . and the growing accumulation of intelligence reports . . . that the Soviet Union had already launched a major program for the development of long-range rockets."[20]

In June 1952, the Department of Defense established a study group on guided missiles which led to the Strategic Missiles Evaluation Committee, chaired by John von Neumman, the famed mathematician.‡ The von Neumman committee studied the Air Force's strategic missile program and reported in February 1954 that both the missile systems and their specifications were out of date and unsatisfactory. An urgent need for greater strategic missile capability was seen because of improved Soviet defenses against manned bombers as well as rapid development of Soviet strategic missiles. The committee pointed out that progress in weaponry research allowed reduction of warhead mass as well as a relaxation of accuracy requirements.[21] The von Neumman committee recommendations had great influence and when adopted and implemented, long-range missile development swung from winged missiles using air-breathing engines to ballistic rockets—the beginning of the accelerated ballistic missile development of the 1950s. In the first series of liquid-propellant missiles, the propellants were a kerosene-like hydrocarbon and liquid oxygen.

The choice of a single propellant combination for development of long and intermediate range liquid-propellant ballistic missiles did not stop research on high-energy propellants, which became candidates for a second generation of improved missiles.

The NACA was not oblivious to the changes in military emphasis from air-breathing to rocket engines, but took no strong steps to realign its research emphasis until about 1956. Meanwhile, the small rocket group at the Lewis laboratory was steady on its course and late in 1954 was ready, at long last, to experiment with liquid hydrogen.

---

*In October 1954, NACA rocket research was $1.2 million or 2.4 percent of the budget; ramjet research was $2.5 million or 5.1 percent of the budget. Minutes of meeting, NASA History Office.

† Three strategic missiles in development were: Snark, a winged cruise missile, developed by Northrop and powered by a turbojet engine after assisted take-off with solid-propellant rockets; the Matador, developed by Glenn L. Martin, similar to but smaller than the Snark; and the largest, Navaho, developed by North American Aviation, powered by a ramjet engine after boost to supersonic speeds by three liquid-propellant rocket engines.

‡Other members: George Kistiakowsky, Charles A. Lindbergh, Simon Ramo, Jerome Wiesner, and Dean Wooldridge.

First Attempt to Use Liquid Hydrogen

When the rocket subcommittee met in October 1954, preparations for the first experiment with liquid hydrogen were almost complete. The liquefier was producing liquid hydrogen. One of the larger test cells had been equipped to use liquid hydrogen with either of two oxidizers—liquid oxygen or liquid fluorine. Edward Rothenberg headed the team using hydrogen-oxygen and was ready first.* On 23 November 1954, the first successful run with liquid hydrogen was made; thrust and chamber pressure were at design values and exhaust velocity was 90 percent of theoretical. Ten days later, two more successful runs were made, but performance data were incomplete. A fourth successful run on 6 January 1955 yielded lower performance than the previous runs.[22]

After the four successful runs with liquid hydrogen-oxygen in 22-kilonewton engines, no more experiments with liquid hydrogen were undertaken for almost a year. The reasons were several. One was a need to reassess injector design. On the first three runs the oxidizer injection rings had burned, and the low performance of the fourth was a clear signal of poor injection. Another reason was a need for improved start and shutdown techniques. Although a satisfactory method had been worked out, it depended a good deal on the reaction time and skill of the operator. In starting, a low hydrogen flow was ignited outside the engine and flashed back into the engine when oxygen flow began. When flashback occurred, full hydrogen and oxygen flows were established. After the first run, the operators discovered that a fire had started during the ignition phase, which ignited hydrogen escaping from the supply tank. The problem was solved by opening the hydrogen valve and burning the escaping hydrogen until the tank was exhausted. The experience was somewhat similar to the leaks encountered by Walter Thiel in Germany about 1937 (p. 269).

Two other factors contributed to the delay in hydrogen testing. The Air Force loaned the laboratory the mobile hydrogen liquefaction equipment developed by Herrick L. Johnston (fig. 14) which would produce almost twice as much liquid hydrogen as the installed equipment and help keep pace with growing demands for hydrogen in other laboratory work. Glenn Hennings of the rocket staff was placed in charge of getting the mobile equipment into operation. The other factor was increasing interest in the possibility of upgrading the performance of existing missiles using JP fuel–oxygen by adding a small quantity of fluorine to the oxygen.† The fluorine not only increased performance but made the combination self-igniting. By 1955, three reports had been written on investigations of JP-4 fuel with mixtures of oxygen and fluorine.[23] This concept was not tried in a missile, however, because of concern over the toxicity of fluorine.

---

*Paul Ordin headed the team working with hydrogen-fluorine in 1954 and planned performance and regenerative cooling experiments. However, Silverstein picked Ordin to head a special hydrogen flight project (described later), and Howard Douglass took his place. For various reasons, the hydrogen-fluorine experiments were delayed until 1957.

†JP (jet propulsion) fuel was the designation for the petroleum blend similar to kerosene, used at the time. Later rockets used RP (rocket propulsion) fuel.

The rocket subcommittee was still concerned over the low level of NACA rocket research when it met in November 1955.* A resolution was passed detailing the importance of rocket research, the concern over the low level of NACA activity, and the problems needing attention; it ended by recommending the NACA effort "be considerably increased so that significant progress can be made at the pace keyed to the swiftly moving national defense effort in rocket propulsion."[24] Development of the Atlas and Titan ICBMs and Thor and Jupiter IRBMs was accelerating and subcommittee members from propulsion and airframe manufacturers as well as the military were feeling the pressure. They believed that the NACA ought to help solve their development problems.

NACA officials recognized the increased emphasis on missiles but continued research on advanced air-breathing engines as well. In the 1955 NACA annual report, Chairman Hunsaker stated: "Today, problems associated with a nuclear engine for aircraft propulsion and with an intercontinental ballistic missile are perhaps the most pressing."[25] The nuclear engine for aircraft was soon to fade into oblivion, but the intercontinental as well as the intermediate-range ballistic missiles became key elements in U.S. military preparedness.

Second Attempt to Use Liquid Hydrogen

In December 1955, the Lewis rocket team resumed experiments with liquid hydrogen but with slight success. Seven runs were made on the 8th and 10th of December; two engines burned out. A successful run was made on 16 February 1956, followed by three more two weeks later. Preparations were then made to operate with liquid hydrogen–fluorine. On 9 March, the first attempt was made, but the engine burned out in four seconds. This was long enough, however, to measure performance: thrust and pressure were near design values and exhaust velocity was 3510 meters per second, or 93 percent of theoretical.[26] This was the highest rocket performance value obtained at the Lewis laboratory up to that time. Cooling, however, remained an obvious problem, and emphasis was placed on it.

NACA Reconsiders Missiles

Although the NACA had always maintained an interest in the problems of high-speed flight and had made some significant contributions,† actual effort remained relatively small until about 1956. Meanwhile, the military ballistic missile effort had risen rapidly since early 1955; in FY 1956, it passed the half-billion-dollar mark and was nearly three times larger the next year.[27] Interest in extending ballistic missile

---

*Thomas E. Myers of North American Aviation replaced Maurice Zucrow as chairman in 1954, with Zucrow continuing as a member. Other members: Lt. Col. Langdon F. Ayers, USAF-ARDC; R. B. Canright, Douglas Aircraft; B. F. Coffman, Bu. Aer., Navy; H. F. Dunholter, General Dynamics; R. B. Foster, Bell Aircraft; W. P. Munger, Reaction Motors; J. R. Patton, Office of Naval Research; C. C. Ross, Aerojet-General; C. N. Satterfield, M.I.T.; F. E. Schultz, General Electric; A. J. Stosick, JPL; R. C. Swann, Redstone Arsenal; F. I. Tanczos, Bu. Aer., Navy; the author, NACA-Lewis and B. E. Gammon, secretary.

†For example: the blunt-body theory for warhead reentry into the atmosphere from a ballistic trajectory conceived by Harvey Allen of NACA-Ames in 1951 and published in the open literature in 1958.

capabilities to launching satellites was also growing. In NACA's 1956 annual report, Chairman Hunsaker recognized the need for more missile research and added, "we are striving for the knowledge that will make possible satellites probing the regions beyond the earth's atmosphere . . . ."[28]

Following up on the need for greater effort in both aircraft and missile research, the NACA established a panel to determine the type of facilities required for the coming years. Hugh L. Dryden, NACA's director of research, wrote to Thomas Myers, the rocket subcommittee chairman, acknowledging the three-year attempt by the subcommittee to increase rocket research and informed him that the facilities panel "now has under consideration . . . a proposal for a rocket systems research facility to provide the necessary space and equipment to implement the Subcommittee's recommendation."[29] Dryden was fully aware of the new high-energy propellant facility nearing completion at Lewis and had in mind "basic research leading to improved turbopump designs for rocket engines and to improvements in propellant systems generally." Dryden had pinpointed a deficiency in NACA research that could be swiftly remedied. Since its formation in 1945, the Lewis rocket group had been limited to problems associated with the thrust chamber of a rocket engine system. The laboratory had a large division specializing in compressor and turbine research for jet engines; a technical field closely related to turbopumps of rocket engines. By building suitable facilities for turbopumps and complete engine systems, NACA could tap this pool of technical talent and accelerate its contributions to missile problems.

In his letter to Myers, Dryden asked for comments and answers to specific questions such as, "Why should NACA enter the field of rocket propellant systems?" His choice of words was unfortunate for apparently Myers thought he meant rocket propellant combinations. Myers replied with an eloquent plea for NACA research on propellants, giving three reasons: high-energy propellants can increase missile range by an order of magnitude or for satellites, permit increases in payload; improved propellant systems have multiple applications and research data from NACA spread throughout the country have a beneficial effect on rocket developments; and the NACA's achievements in propellant research are widely recognized and used by the rocket industry.[30]

Advanced Propulsion Concepts

Dryden's request to Myers was typical of NACA's conservative approach in entering a new field—solicit opinions and build a broad base of national support so that it would appear the agency was practically pushed into the new work. This process continued when Myers, at the May 1956 meeting of the rocket subcommittee, asked the members for suggestions for rocket research for the next 10 to 15 years. By the fall of 1956, the responses were grouped into five discussion topics, one of which was high-energy propellants, but without specific reference to liquid hydrogen.* In discussions of these topics in meetings the following year, evaluation of high-energy propellants

---

*The others: nonconventional rocket propulsion, such as solar energy, ions, electrons, charged particles, and free radicals; nuclear energy; comparison of nuclear and non-nuclear propulsion; and summary of the other topics from the viewpoint of applications and military requirements.

was of first-order importance, but again no specific propellant combinations were singled out.[31] The NACA-Lewis rocket group, however, was still greatly interested in liquid hydrogen and believed that they had the support of the rocket subcommittee.

In mid-1957, at the invitation of the chairman of the NACA subcommittee on aircraft fuels, Richard Canright (who had left JPL for Douglas Aircraft in 1953) submitted a paper on rocket propellants as viewed by an airframe manufacturer. Canright had a long interest in rocket propellants and his employer, Douglas Aircraft, had development contracts involving both solid and liquid rocket engines. From this viewpoint, he dismissed the application of air-breathing engines for missiles as "extremely limited if not completely non-existent." After discussing the relative advantages of solid and liquid rocket propellants, Canright gave his views on liquid hydrogen:

> This is, of course, the non-carbonaceous fuel that offers the highest performance of any fuel. However, because of its low density, it is useful only in certain extreme applications. Hydrogen offers excellent combustion characteristics, both in the gas generator and in the main motors, and good heat transfer characteristics in the supercritical regime; on the other hand, it is hazardous to handle and there is no large engine experience with this fuel to date.[32]

Except for the mention of combustion and heat transfer characteristics, Canright offered little more on liquid hydrogen than Tsiolkovskiy 30 years earlier, an indication of the continuing gap between what was known about liquid hydrogen and its practical application.

When asked to summarize the contributions of the rocket subcommittee in its five years of existence, Canright replied:

> We have constantly spurred the NACA on to tests on a larger scale. We have urged them to become familiar with complete engines rather than work only on component R&D. We have tried to emphasize the importance of rocket technology to this country's defense effort and urged that the NACA devote a greater portion of its personnel and funding to this important field.[33]

He added, however, that NACA interest in high-energy propellants was praiseworthy and that the subcommittee supported it.

High-Energy Propellant Facility

The new $2.5 million rocket facility for high-energy propellants, requested in 1952, was completed in the fall of 1957 (fig. 15). It comprised a test cell, propellant supply system, and a unique combination exhaust-gas treatment and silencing system. A service building and high-pressure helium bottles were adjacent to the test cell. Farther away were storage areas for fuels (hydrocarbons, ammonia, hydrazine), liquid oxygen, and water for the scrubber, all piped to the test cell under gravity. Fluorine was loaded into propellant pressure tanks from trailers and a similar provision was made for liquid hydrogen. The oxidizer tanks were in a pit behind the cell, suspended on a weighing system within another tank into which liquid nitrogen was placed as a coolant when

Fig. 15. The NACA-Lewis high-energy rocket propellant test facility, placed into operation in 1957. The building with the slanted roof is the test cell. The large horizontal duct with vertical stack is the exhaust-gas scrubber and silencer. The two men are in front of racks of high-pressure cylinders of helium used for pressurization. On the far right is the water storage tank and on the far left is the water detention tank and treatment system.

fluorine was used (fig. 16). The tanks and system were sized to allow three minutes of operation at a thrust of 89 kilonewtons, a considerable increase in size over other high-energy rocket facilities in the country. The exhaust duct at the rocket nozzle exit was 3.7 meters in diameter; the horizontal section, 7.6 meters, and the vertical stack, 6 meters. During a run, water from the 1500-cubic-meter (400 000-gallon) tank on the upper level flowed to the exhaust scrubber at a rate of 190 cubic meters (50 000 gallons) per minute. This was well over a hundred times the mass flow of exhaust gases. The hot gases, emerging from the nozzle at velocities of 3000–4000 meters per second and temperatures of about 2300 K, were met with a drenching spray of water and quickly cooled to steam temperature and slowed to a velocity of about 8 meters per second. Additional water sprays condensed the steam, and the non-condensable exhaust gas emerged from the stack at about 340 K and a velocity of 3 meters per second. The hydrogen fluoride of a fluorine-hydrogen rocket is highly soluble in water; the water containing it was collected in the detention tank. After the run, calcium hydroxide was introduced into the detention tank and the hydrogen fluoride converted to calcium fluoride—which is inert and harmless—and a slurry of it was pumped into tank trucks and hauled to a dump. The facility was equipped with barrels containing ordinary charcoal and connected to the fluorine system. Harold Schmidt of the rocket research group had found that fluorine reacts readily with charcoal and is converted into inert carbon fluorides—an excellent way to dispose of unwanted fluorine. Monitors to sniff

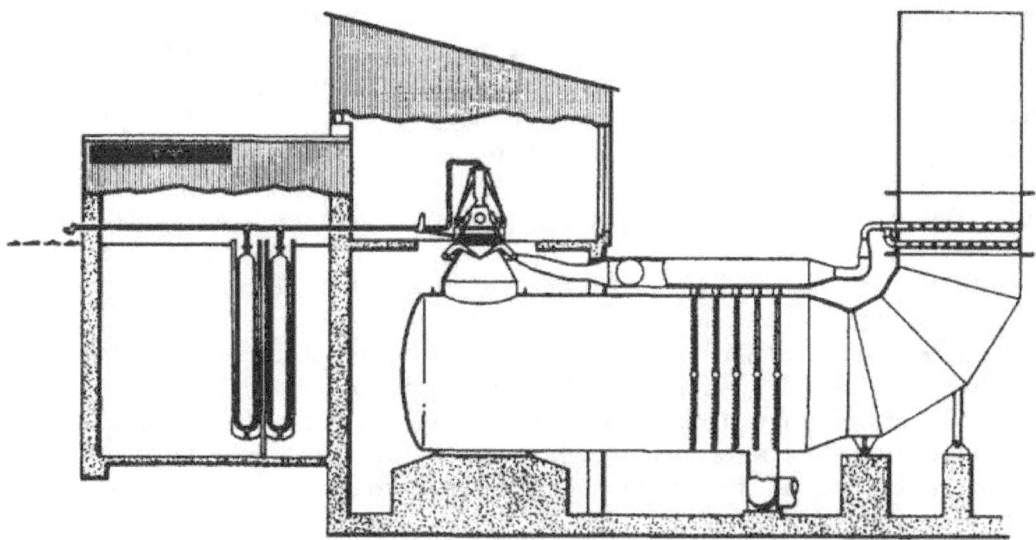

Fig. 16. Sectional view of the NACA-Lewis rocket facility. On the left are high-pressure propellant tanks in pits. The rocket engine and test stand, for vertical downward firing, are dwarfed by the large ducting for exhaust gas scrubbing and silencing.

a variety of gases and combustible mixtures were located at strategic places about the test area. The test cell was controlled by a well-equipped center in the rocket operations building about a kilometer away.

Thrust Chamber Design and Fabrication

The initial failure of the Lewis experiments with liquid hydrogen was primarily one of thrust chamber design. The key to a successful design lies in the injector which can mean high or low performance, durable operation or quick burnout. The function of the injector is to mix the fuel and oxidizer thoroughly and uniformly for complete combustion, while the propellants also cool the injector face. With a good injector, combustion chamber design becomes a matter of providing sufficient volume for the reaction to go to completion, sufficient wall strength to contain the pressure, and sufficient cooling to keep wall temperature within its working limits. The design of the nozzle involves a compromise between providing the optimum contour for complete gas expansion and size, the latter limited by vehicle design.

Rocket experimenters exploring the performance of various propellant combinations usually used either a water-cooled thrust chamber and nozzle or uncooled types that could withstand high temperatures for a few seconds. Most effort was concentrated on obtaining an injector yielding high performance. Following this, the next step was to cool the chamber and nozzle with the fuel. We have already seen that both the Jet Propulsion Laboratory and Ohio State University succeeded in operating regeneratively-cooled hydrogen-oxygen thrust chambers during the 1940s. The Lewis experimenters were trying the same but with larger engines (22 and 89 kN), combining regenerative cooling with thrust chambers of light weight to approach a practical flight design. These objectives had been spelled out in 1952 and reaffirmed each year.

The first injector used for liquid hydrogen–liquid oxygen at Lewis in 1954 was a like-on-like impingement where jets of the same fluid impinge, breaking the streams into droplets. Mixing is obtained by locating the impinging streams of fuel and oxidizer near each other so that the resulting droplets mix well. Ohio State University used this type and it was popular among rocket experimenters. JPL used an injector with impinging hydrogen jets and an oxygen spray. Aerojet's best injector was a multiple-tube, concentric type where each jet of hydrogen was surrounded by a sheath of oxygen (fig. 10). The three successful runs by the Lewis group in February 1956 used a "tube bundle" injector where a large number of small tubes carried the hydrogen and oxygen into the chamber with a fine degree of mixing.

By all experience and design principles, the hydrogen-fluorine injector used in the first Lewis laboratory run in March 1956 should have worked well. It consisted of four rings of hydrogen holes producing streams parallel to the combustion chamber axis, alternating with four rings of similar holes for fluorine. The holes were small and mixing was good; but when tried, the operator summed the results in four cryptic sentences in his log:

H_2-F_2 was run on "B" stand, Cell 22. Made only 1 run. Injector burned out causing chamber to go. Run time=4 sec.[34]

Parallel to these experiments, more detail studies were under way at the Lewis laboratory on fundamentals of injector design. Such work had been in progress since the early 1950s, but it was not until 1956 that experiments in this basic work focused on hydrogen. Carmen M. Auble studied six types of injection methods for hydrogen-oxygen in a small (900 N) thrust chamber.[35] Gaseous hydrogen, chilled to the temperature of liquid nitrogen (77 K), simulated the physical characteristics of hydrogen after it served as a coolant prior to injection. Not surprisingly, Auble found correlation between the mixing and spreading of the propellants and performance over a range of propellant mixture ratios, all his designs doing well at fuel-rich ratios. Increasing the temperature of the hydrogen to room temperature was beneficial. Compared with hydrocarbon fuels, hydrogen needed a fifth as much volume for comparable combustion efficiency. Separate and parallel jets, as used in the hydrogen-fluorine run, did as well as injectors that promoted mixing. Auble found, however, that combustion efficiency was controlled more by the degree of oxygen vaporization than by hydrogen dispersion and mixing.

Late in 1957, Marcus F. Heidmann and Louis Baker, Jr., extended Auble's investigation, combining it with earlier analyses of propellant vaporization as a rate-controlling step in combustion. They investigated fourteen injectors for hydrogen-oxygen in an engine of the size that Auble had used. Their investigation confirmed that the degree of oxygen atomization was the primary factor affecting combustion efficiency.[36]

Concurrent with injector and performance studies were several investigations of fabrication techniques for lightweight and cooled combustion chambers and nozzles. In 1953, John E. Dalgleish, a fabrication expert, and A. O. Tischler, a rocket researcher, worked together on lightweight thrust chambers using an electroforming technique.[37] In 1954, Tischler placed orders in the shop for two other types. One used tubes formed according to the contour of the combustion chamber and nozzle and brazed together—a method used by several rocket manufacturers starting with Reaction Motors. The other type was similar except that, instead of tubes, channels were formed and then brazed or welded together with a closure over the channels to complete the coolant passage and strengthen the whole assembly. Both of these experimental types were still in the shop two years later as they had been given a low priority.

Until 1956, the primary responsibility for designing thrust chambers rested with an engineering service group headed by William A. Anderson. He developed a fabrication technique consisting of an inner shell of spun metal, wire spacers to form spiral coolant passages on the outside of the shell, and a welded "clam-shell" outer wall to enclose the coolant passages. A variation of this method was to form the outer shell of square wire brazed together. The Anderson design was successfully used on engines of 4.5 kilonewtons and was the prime design for larger engines until 1956–1957.

Obtaining experimental engines was hampered by increasing congestion in the fabrication shop. The NACA shops were unexcelled in advanced fabrication techniques and willingly accepted all challenging work, but delivery was sometimes delayed by an avalanche of orders or work of higher priority. In 1956, the shops had orders for over a dozen thrust chambers of various designs and delivery was delaying experimentation. Steps were taken to reduce the number of designs, and Silverstein assigned Edward Baehr, a gifted design and fabrication engineer, to assist the rocket group. Baehr made a major contribution to the rocket effort by choosing a design something like Tischler's channel-wire wound type and successfully fabricating it. It consisted of a number of longitudinal channels of varying depth according to the coolant velocity required. These were bonded together to make up the chamber and bound by stainless steel wire wrapping which was brazed to make a fluid-tight and strong outer skin (fig. 17). This design was used in 1957 and subsequently.[38]

Since the early 1950s, Lewis associate director Abe Silverstein had been interested in liquid hydrogen as a fuel for both jet engines and rockets. In the spring of 1957, he decided that it was time to hold a research conference on results of the laboratory's investigations. That conference, plus additional emphasis on rocket research at the laboratory, meant unprecedented support for the rocket group, and they made the most of it. Silverstein became more involved in rocket problems and on 9 August 1957 held one of his famous after-hours staff conferences on the subject of injector design. The informal session was held in the control room of the new facility—complete with beer and pretzels, compliments of Silverstein—and ran past midnight. Everyone contributed his views.* The author remembers stressing the concepts of mixing on a very fine scale coupled with uniform mixing except for a fuel-rich cooling region at the chamber walls. These concepts, not new, were adopted along with other design ideas such as selecting angles of jet impingement well away from the injector face, avoidance of recirculation of reactants across the injector face, and fuel-cooling of the injector face. In September, Silverstein held another meeting on injectors as well as other rocket design problems for experiments intended to be reported at the coming conference.[39]

---

*Attendees were: Silverstein, W. T. Olson, Edward Baehr, Vearl Huff, M. F. Heidmann, A. O. Tischler, Howard Douglass, George Kinney, William Anderson, and the author.

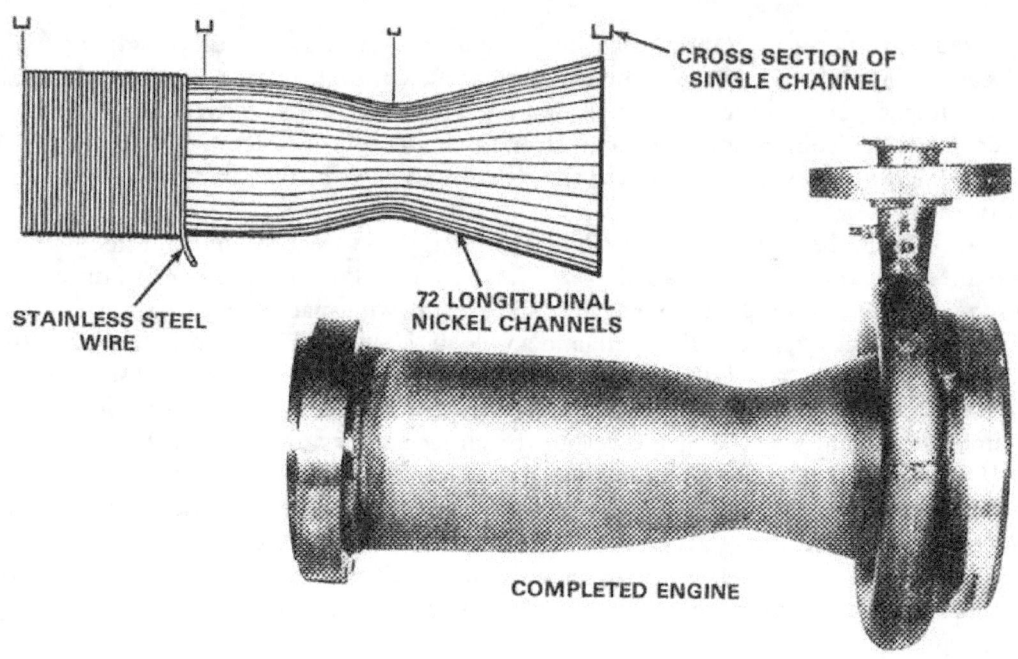

CROSS SECTION OF
SINGLE CHANNEL

STAINLESS STEEL
WIRE

72 LONGITUDINAL
NICKEL CHANNELS

COMPLETED ENGINE

Fig. 17. Experimental rocket chamber of 22 kN, regeneratively cooled. Fabricated by a method developed by Edward Baehr. NACA-Lewis, 1957.

Space Becomes an Acceptable Word

The fall of 1957 was one of fast-paced activity. The new rocket facility at the Lewis laboratory was completed and system checkouts were under way. Plans were made to include it as an exhibit stop in the laboratory's triennial inspection, scheduled for October. These inspections were NACA's way of showing its facilities and latest research progress to congressmen, government officials, professors, engineers from industry, reporters, families, and friends. The affairs were exhaustively planned and rehearsed and executed with split-second precision.

For its part in the inspection, the rocket group showed the great advantages of high-energy propellants, including hydrogen-oxygen and hydrogen-fluorine. As an example, they illustrated the case of a manned satellite in an 1850-kilometer orbit with a winged glider for returning to earth. The use of high-energy propellants would reduce the required booster size and weight by half. The exhibit also demonstrated the powerful oxidizing property of fluorine. A steel bar, chemically cleaned, was exposed to a small jet of gaseous fluorine and nothing happened. The bar was then contaminated by a slightly greasy thumbprint and again exposed to the fluorine jet. The fluorine then reacted with the contamination, the reaction heating the steel bar until it burned—a spectacular and impressive demonstration of fluorine's potency.

Among the many rehearsals was a review by officials from NACA headquarters. The climate in Washington in the fall of 1957 was very negative towards space. It was all right to talk about the slow-paced scientific Vanguard satellite, part of the International Geophysical Year, but anything beyond it was considered "space-cadet" enthusiasm. When John Victory, NACA executive secretary and one of its original employees, heard the word "space," he ordered that it not be used for fear of offending some of the visitors, particularly congressmen and other government officials. Before the inspection, however, the Soviet Union's Sputnik I put the word "space" in the headlines of every American newspaper, and guests heard the word in many of the laboratory's presentations.

Emphasis on Hydrogen

When the NACA 1957 Flight Propulsion Conference was held at the Lewis laboratory on 21–22 November 1957, it could have been called, as one member of the audience remarked, a conference on liquid hydrogen as a fuel. The primary emphasis was on air-breathing engines, but the rocket group had a sizeable part of the program—the last three of eight presentations.[40] Silverstein had decided that rather than having individual papers, each subject would be handled by a panel taking turns presenting the subject and discussing it. The subjects were broad. The three on rockets were propellants, turbopumps for high-energy propellants, and performance and missions. The last two were firmly on the subject of high-energy propellants, but somehow the one on propellants got out of line. It covered the spectrum of propellants, with high-energy propellants receiving attention only at the beginning and at the end, and even then the emphasis was on cooling rather than performance. This emphasis on cooling was due to circumstances. Of the four panelists on propellants, only Howard Douglass was experienced in investigating high-energy propellants. Two of the panelists were newcomers to rockets; one was studying solid propellants, and the other was a controls specialist. It was characteristic of Silverstein to stimulate research by obligating the staff to a research conference with tight deadlines and by assigning individuals the responsibility of discussing subjects broader than their immediate research. Douglass had intended to cover high-energy propellants more thoroughly and add experimental results, but a series of delays in experimental operations almost caused his discussion to be all theoretical.

First Regeneratively-Cooled Hydrogen-Fluorine Rocket

For many months—since the first run of hydrogen-fluorine in March 1955—Howard Douglass, Harold Price, and Glen Hennings* had worked to design, build, and operate a rocket engine of 22 kilonewtons using liquid hydrogen and liquid fluorine, with the liquid hydrogen serving as a regenerative coolant. Edward Baehr worked with them in designing and fabricating the engine. Two kinds of injectors were designed and fabricated: a showerhead type and an impinging-jet type with two jets of fuel impinging on one of oxidizer (fig. 18). The face of each injector was fuel-cooled.

Hennings was operations chief. A perfectionist ideally suited to cope with the hazards of handling fluorine, he made many equipment changes. Operations with

---

*Hennings had been in charge of liquefying hydrogen until a liquid hydrogen plant in Painesville, Ohio, built by the Air Force, made liquid hydrogen available to the laboratory in quantity.

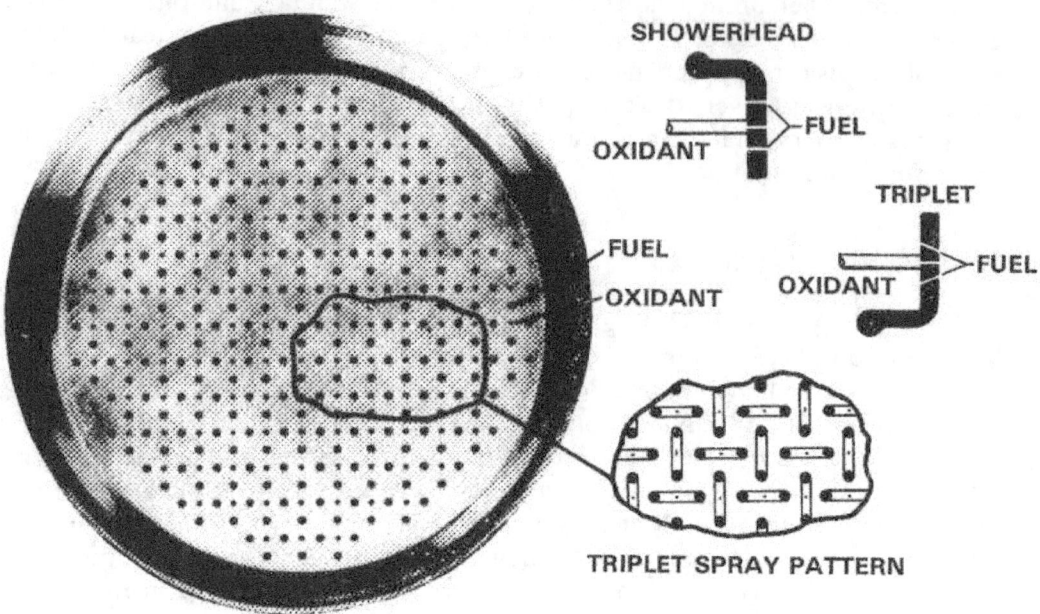

Fig. 18. Showerhead and triplet injectors for rocket engines at 22 kilonewtons using hydrogen-fluorine at NACA-Lewis, 1957–1959. The same general concept was also used for hydrogen-oxygen engines of the same thrust and also for larger engines (89 kN).

fluorine were scheduled for weekends to minimize possible hazards for laboratory personnel working nearby. In a series of attempted runs, an incredible number of problems arose which Hennings doggedly attacked and solved one by one in his careful, methodical fashion. Time was running out, however, and with the conference only a few days away, Douglass drafted his part of the propellants paper around the cooling aspects of hydrogen-fueled engines and hoped for the best.

An attempted run on the weekend before the conference was aborted and prospects appeared grim. Hennings cleared up several vexing problems, determined to operate the engine as soon as possible. On Wednesday evening, 20 November—the day before the conference began—a run was attempted but again problems halted operations. The crew worked all night to solve the problems and through the day Thursday, aiming for another try Thursday night. Douglass was to speak on Friday. The author, fearing that fatigue could cause a misjudgment and an accident, urged the crew to give up, take a rest, and try another day. They continued, however, and worked all through Thursday evening. Finally, at 5:59 Friday morning, they succeeded with a beautiful run that lasted eight seconds, with no sign of overheating. The exhaust velocity measured was 3400 meters per second, 96 percent of theoretical performance. But these values were not known to the tired crew when Douglass went home for a short rest and freshening before his appearance with the panel at 9:30. Harold Price remained to work up the data as fast as possible and bring it to Douglass at the conference.

When the propellants panel began its discussion, Price had not appeared; he was having trouble with security guards because his name was not on the list of those permitted to attend the classified conference. Finally he managed to convince them, hurried to the projection room, marked a data point on Douglass's slide, hurried downstairs and up the aisle to the stage where he handed Douglass a note with the data and the engine itself, which had been dismounted to display at the conference. He was in the nick of time, for Douglass was the next to speak. Some in the audience thought the entrance was staged, but it was the real thing and a great moment of triumph for the NACA rocket group.[41]

Two other panels at the conference also made a persuasive case for high-energy propellants, particularly liquid hydrogen for rockets. The turbopump panel found no great obstacles in developing turbopumps for hydrogen-oxygen or hydrogen-fluorine combinations and estimated that the mass of such a turbopump would be comparable to one for conventional propellants. The panel on performance and missions found that the greater the energy requirements for a mission, the greater the need for high-energy propellants. For the case of a moon landing and return, the difference in initial mass between vehicles using kerosene-oxygen and those using hydrogen-fluorine or hydrogen-oxygen could be a factor as high as 8 to 1. Silverstein and the Lewis rocket group were convinced that liquid hydrogen was an extremely attractive fuel.

Significance

Although NACA started late in rocket research, kept its effort relatively small, and was but one of many organizations investigating high-energy propellants, its technology contributions were significant to later vehicle developments. NACA was the only government laboratory conducting in-house experiments on high-energy propellants, and NACA data were quickly available to and influenced the work of all other groups. The strong NACA-Lewis preference for liquid hydrogen, which began in 1950 and persisted in spite of delays in securing a supply, was instrumental in keeping others interested in hydrogen. Abe Silverstein, the Lewis associate director, strongly supported liquid hydrogen, and he later occupied a key position in the nation's space program. The NACA rocket subcommittee, a unique body of rocket experts from government, industry, and universities, exchanged information between all interested groups and assisted in national planning of rocket research and development.

6

NACA Research on Hydrogen for High-Altitude Aircraft

Hydrogen was considered as an aviation fuel by P. Meyer in 1918 (p. 12); Tsiolkovskiy considered and rejected it for a rocket-powered airplane in 1935 (p. 256). In 1939, George W. Lewis, director of research for the National Advisory Committee for Aeronautics (NACA) was talking about using liquid hydrogen with atmospheric air, presumably for aircraft propulsion (p. 73). During World War II, F. Simon, a respected physicist in England, nearly confounded the practical fuel experts in the United States by suggesting that liquid hydrogen be used to increase aircraft range (pp. 11–12). Opie Chenoweth, Robert Kerley, John Duckworth, and their associates at Wright Field's power plant laboratory contracted with Ohio State University in 1945 to investigate the application of liquid hydrogen to aircraft and rockets (p. 18). None of these, however, got very far, principally because hydrogen's very low density made its application in volume-limited airplanes appear totally impractical. If this was not enough, opponents to hydrogen clinched their case by citing its very low availability as a liquid and its handling hazards.

Beginning in the 1950s, however, several factors combined to make liquid hydrogen appear exceedingly attractive as an aviation fuel. Among them: incentives to operate airplanes at very high altitudes, advances in liquid-hydrogen technology, and experiments showing that hydrogen burned readily at low pressures.

One of the places where an intense interest in hydrogen for aircraft developed during the 1950s was the NACA Lewis Flight Propulsion Laboratory in Cleveland, where it was pushed hard by the associate director, Abe Silverstein. NACA involvement with hydrogen for this application, however, had its roots in earlier work in fuels and combustion.

One of the initial facilities built at the NACA Cleveland laboratory in the early 1940s was a well equipped chemical laboratory for fuels and lubricants. New fuels or blends for piston engines could be synthesized, and during the war, for example, the laboratory studied alternate high-octane fuels such as the aromatic amines. With the switch from piston to jet engines after the war, the type and characteristics of desired fuels also shifted. The amount of heat obtainable per unit mass and volume became of great importance. Research involved not only the theoretical energy content of fuels, but how to release and harness that energy over a range of operating conditions. How well did the fuel mix in an air stream? Would the fuel ignite and propagate over a range of combustible mixtures? How efficient was the combustion process over a range of operating conditions. particularly at the reduced pressures of high altitude? Such questions became important for research to answer.

In 1948, the Lewis laboratory presented its research on fuels at a conference; six of the nine papers were on fuels for turbojets and ramjets.[1] Melvin Gerstein discussed powdered metallic fuels such as aluminum and beryllium, which had heats of combustion per unit volume up to four times greater than gasoline. Gerstein also discussed diborane, reporting that its flame speed was fifty times greater than that of hydrocarbons. It was part of the great love affair with diborane and pentaborane by the laboratory and others which extended beyond the mid-1950s.

In 1950, the uneasy international situation, and especially the outbreak of the Korean war, led to an acceleration of aeronautical research and development. One goal was aircraft capable of operating at very high altitudes, and one obstacle in doing this was described by Walter T. Olson, J. Howard Childs, and Edmund R. Jonash of the Lewis laboratory in 1950:

> Experience has shown that, as operating altitudes are progressively increased beyond 25 000 feet [7600 m], the effects of altitude on combustion efficiency ultimately result in severe penalties in thrust and specific fuel consumption. The problem of maintaining high combustion efficiency is one of the most important problems of altitude operation.[2]

The investigators found that combustion efficiency increased with fuel volatility, with greater hydrocarbon content as compared to aromatics, and with more straight-chain and fewer branched-chain hydrocarbons.

The following year, Olson and Louis Gibbons surveyed fuels suitable for ramjets and summarized results achieved by several organizations, including the experiments on liquid hydrogen at Ohio State University. Although Olson and Gibbons included liquid hydrogen among the fuels of interest, they were more interested in investigating diborane. pentaborane, and slurries of magnesium and aluminum.[3] The same year, Benson E. Gammon examined the performance of liquid hydrogen and two other fuels for ramjets. finding hydrogen superior per unit mass but inferior per unit volume.[4] Another Lewis laboratory analyst, Hugh M. Henneberry, considered fuels for aircraft during 1951 and concluded that:

> neither the very high nor very low fuel densities have any advantage for long-range flight . . . the practical difficulties associated with the use of liquid hydrogen cannot be justified on a range basis, but *if tactical considerations predicate flight at extremely high altitudes, liquid hydrogen must be considered as a possible fuel* [emphasis added].[5]

There it was—the advantage of hydrogen for attaining extremely high altitudes—but Henneberry, like others at Lewis, was impressed by the potential of another high-energy fuel, diborane, and consideration of hydrogen went no further at that time.*

---

*For flight at an altitude of 21 000 m and speed of Mach 3.6. Henneberry concluded that diborane had a 59 percent greater range than hydrocarbon fuels.

The military services and their advisors also showed little or no interest in hydrogen for aircraft prior to 1954. The Navy had embarked on a massive investigation of boron-hydride fuels for jet engines and was joined in this effort by the Air Force and NACA.[6] The fuels and propulsion panel of the USAF Scientific Advisory Board, in considering high-energy fuels at its April 1952 meeting,* noted that rockets favored fuels with combustion products of minimum molecular mass but that "this condition is irrelevant in a turbojet."[7] This indifferent attitude towards hydrogen appeared to prevail generally for two more years until a series of events, starting in 1954, swept it aside like fog before the wind.

New Interest in Hydrogen

Beginning in February 1954 and extending through March, the fuels and propulsion panel of the Scientific Advisory Board met three times, in an exhaustive survey of the major aspects of the propulsion program of the Air Force.[8] Although no mention was made of hydrogen in the minutes, the panel was greatly interested in high-energy fuels and the Air Force program on them. On the same day as the last meeting (24 March 1954), Randolph S. Rae visited Wright Field with a proposal to use hydrogen in a high-altitude aircraft powered by a unique engine called Rex I. By all indications his visit touched off a strong renewal of interest in liquid hydrogen for aircraft, which will be described in the following chapter.

The origin of interest within NACA to use hydrogen as an aviation fuel has not been fully established, but experiments began in 1954. Several events apparently contributed to the NACA interest. In Washington, A. M. Rothrock, chief of propulsion research, completed a comprehensive survey and analysis of turbojet propulsion and its effect on airplane performance in August; it was published seven months later. Rothrock discussed seven major propulsion factors and the state-of-the-art concerning them. One was the heat of combustion of the fuel—where, of course, hydrogen excels. Rothrock's favorite way of beginning such a discussion was to show a plot of heat of combustion as a function of atomic number, and hydrogen was higher than the upper limit of his scale. Despite this, Rothrock's discussion of hydrogen revolved more around hydrogen as an element in fuel molecules than as a fuel per se. He acknowledged interest in hydrogen mainly in focusing on the current favorite fuels, the boron hydrides, and did not mention hydrogen in his conclusions. A month after completing his report, Rothrock attended a meeting of the fuels and propulsion panel of the Scientific Advisory Board when Rae's Rex engine using hydrogen was discussed.[9] Apparently Rothrock was not sufficiently impressed with the idea of using low-density hydrogen in volume-limited aircraft to change his report, which was still in the process of publication; but he may have passed word on the hydrogen proposal to the Lewis laboratory.

In 1954, current turbojet engines could operate at altitudes of 13 700 meters without serious loss of combustion efficiency. Under the direction of Olson and Childs, a group

---

*The chairman was Prof. C. Richard Soderberg, M.I.T. Other members: Louis G. Dunn, William M. Holladay, Andrew Kaletensky, and W. D. Rannie. A. M. Rothrock of NACA attended the meeting as a guest and was later a member, 1953–1955.

of researchers at the Lewis laboratory was engaged in a series of experiments to relate the effect of fuel characteristics, combustor design, and altitude operation on combustion efficiency. From this research, the altitude limits for good combustion had been extended to 21 000 meters, but the goal was 30 500 meters. As part of this research, Jonash, Arthur Smith, and Vincent Hlavin turned to gaseous hydrogen in 1954 and were not disappointed.* In their report completed two days before Christmas, 1954 (published five months later), they indicated that hydrogen burned well in a single turbojet combustor at pressures as low as 1/10 atmosphere; at 1/4 atmosphere, combustion efficiency was above 90 percent. These results were within the combustion pressure range for turbojet engines operating at 30 500 meters altitude. The authors believed that they could attain 100 percent combustion efficiency with better mixing of hydrogen and air. Propane was investigated briefly and found to be greatly inferior to hydrogen, with the difference attributed to hydrogen's higher flame speed and wider flammability limits.[10]

Sometime during 1954 or early 1955, Abe Silverstein, the associate director of the Lewis laboratory, was struck with an idea concerning hydrogen. Well aware of high-altitude flight objectives and well versed in aircraft design principles, he suddenly saw a way of using hydrogen's superior combustion characteristics and coping with its principal disadvantage, low density. At high altitudes and low speeds, large wings are needed and these call for a proportionately large fuselage. Under these flight conditions, the drag of the airframe is low. The large volumes available in the wings and fuselage favored the use of low-density liquid hydrogen, provided lightweight hydrogen tanks proved feasible.

As was his custom when struck with a new idea, Silverstein made some approximate or "back of the envelope" calculations. He became so enthused over the results that he went to Washington to discuss them with Hugh Dryden, NACA director of research. Dryden, too, was impressed, and the two discussed the idea with Air Force officials. Silverstein was convinced that he had something good, but needed more detailed calculations to back it up.[11]

Silverstein-Hall Report

When Silverstein returned from Washington, he asked Eldon Hall, one of the laboratory's top analysts, to assist him in refining his analysis on using hydrogen for high-altitude aircraft. While this was under way, the fuels and propulsion panel met in March 1955 to discuss high-energy fuels.† The panel was very impressed with the potential of liquid hydrogen and boron hydrides.[12] The work of Jonash, Smith, and Hlavin was described, as well as current work by Thaine W. Reynolds of the Lewis laboratory. Reynolds, who was assisting Hall in the analysis for Silverstein, was studying lightweight tanks for hydrogen and was convinced that they were feasible.

---

*Whether the Lewis group thought of using hydrogen independently or as a result of a suggestion by the Air Force or by Rothrock has not been established.

†Mark M. Mills had succeeded Soderberg as chairman. Other members: W. D. Rannie, E. S. Taylor, Gale Young, and A. M. Rothrock.

The fuels and propulsion panel suggested that the Air Force begin work on hydrogen fuel systems, hydrogen-fueled engines, and preliminary designs of hydrogen-carrying aircraft. This meeting apparently spurred the Lewis analysts to faster action, for Silverstein and Hall completed their report on 1 April 1955 and published it two weeks later—a near record for fast NACA publication and an indication of the importance Silverstein attached to the subject.

In their introduction, Silverstein and Hall noted that despite hydrogen's high heating value and good combustion characteristics, it had received only casual attention. They acknowledged the deterrents of low density, low availability, and difficult handling, but made a case for considering hydrogen based on four points: a military need that could not be met in any other way, advantages of hydrogen for high-altitude flight, improvements in jet engines that indicated their mass could be halved for the same power, and large wing and fuselage requirements for high-altitude flight. The first two points were based on hydrogen's unique properties. The third favored light weight, and the fourth high volumes, to overcome hydrogen's disadvantage of low density. As for availability and handling, Silverstein and Hall cited past experiences, implying that if the flight problems could be solved, so could those on the ground.[13]

Of the flight problems, the authors singled out hydrogen tankage as a major problem. They drew on the technology of long-range missiles, particularly the Atlas, and suggested that liquid hydrogen tanks be constructed as cylindrical balloons of light-gage metal, depending upon internal pressure to maintain shape (fig. 19). This, of course, was the same idea proposed by Oberth in the 1920s and Martin and North American engineers in the 1940s, and being used for the first time on the Atlas ICBM amid some skepticism.

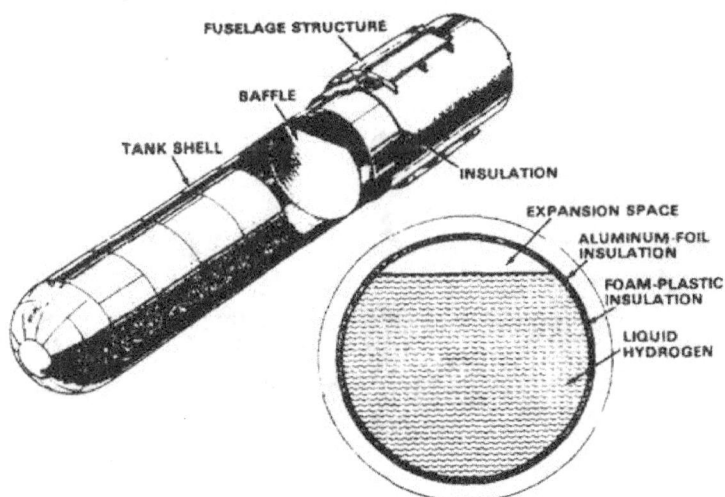

Fig. 19. Liquid-hydrogen tank suitable for aircraft as envisioned by Abe Silverstein and Eldon Hall, "Liquid Hydrogen as a Jet Fuel for High Altitude Aircraft," NACA RM E55C28a, 15 Apr. 1955. Of light-gage metal that depended on internal pressure to maintain its shape, the tank was 25 m long, 3 m in diameter with a volume of 175 m³. Liquid-hydrogen capacity was 11 300 kg. The estimated tank mass was 10 percent of the fuel mass.

Using the basic hydrogen-tank design, Silverstein and Hall analyzed the use of liquid hydrogen for a subsonic bomber, subsonic reconnaissance airplane, and supersonic fighter. Of these, the reconnaissance type will be described as a typical example and for its relationship to later events.

The subsonic reconnaissance airplane had a gross mass of 40000 kilograms and carried hydrogen tanks in wings and fuselage, as well as optional drop tanks for additional range (fig. 20). It operated at an altitude of 24000 meters and could make observations 13500 kilometers from its base. A supersonic version was about 1/4 lighter, operated at the same altitude at a speed 3 times faster, but had a range less than 1/5 the subsonic type.

The subsonic version was powered by advanced turbojet engines weighing about half those in current use. The supersonic type also used an advanced turbojet that was equipped with an afterburner. Additional data on the airplanes and engines are given in table 2.

Silverstein and Hall concluded that "within the state of the art and the progress anticipated, aircraft designed for liquid-hydrogen fuel may perform several important missions that comparable aircraft using hydrocarbon (JP-4) fuel cannot accomplish." They also concluded that "substantial applied research and development effort will be required in many technical fields to achieve the goals outlined."[14] It was a convincing case for hydrogen if the assumptions were accepted. Silverstein, as the chief research executive of the Lewis laboratory, thereupon initiated a massive research program on hydrogen to give substance to his assumptions.

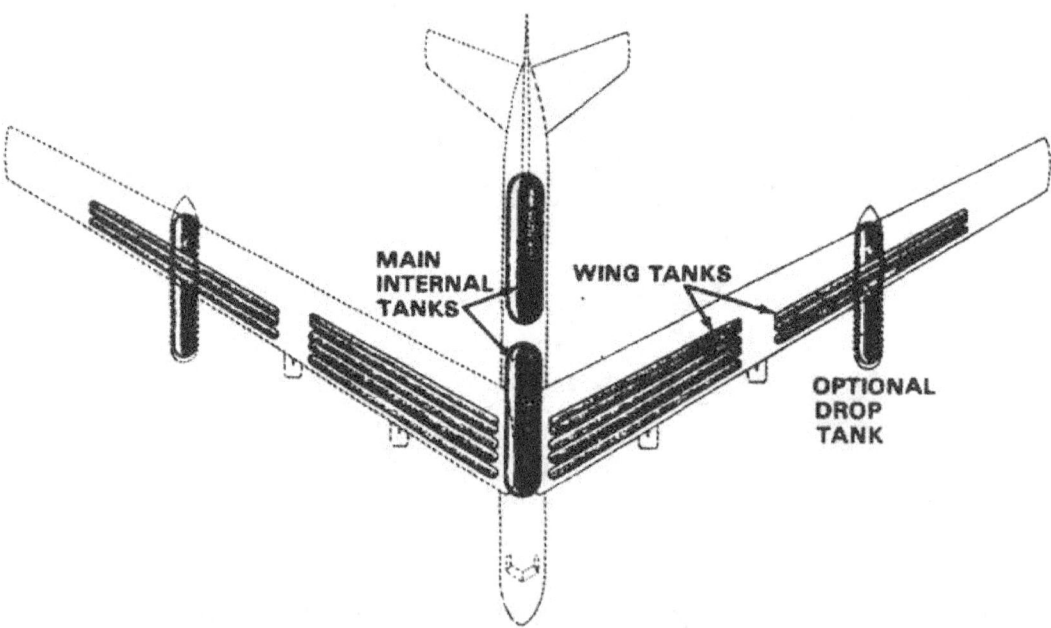

Fig. 20. High-altitude, subsonic reconnaissance airplane using liquid hydrogen as fuel. The liquid hydrogen tanks are in both fuselage and wings. Flight Mach number, 0.75; altitude 24400 m. From Silverstein and Hall, "Liquid Hydrogen as a Jet Fuel," 1955.

TABLE 2.--*Characteristics and Performance of Reconnaissance Airplane and Engine Designs*

Item	Reconnaissance Airplane	
	Subsonic	Supersonic
Cruise Mach No	0.75	2.5
Cruise altitude, m	20000	22000
Target altitude, m	24000	24000
Gross mass, kg	39800	34000
Fixed (instruments, cameras, controls)	2268	2268
Structures	13000	13000
Engine	6328	6169
Fuel tank	2372	1610
Fuel	15760	10730
Wing		
Area, m^2	348	107
Sweep angle, deg	31	0
Aspect ratio	13	3
Average section thickness ratio	0.12	0.03
Taper ratio	2	2
Empennage area, m^2	87	32
Fuselage		
Length, m	45	52
Diameter, m	3.5	3.7
Lift coefficient, initial cruise	0.54	0.14
Lift/drag ratio (airplane less engine nacelles, initial cruise)	25.4	4.33
Radius, km	13500	2490
Engines (turbojet)		
Number	4	4
Compressor diameter, m	0.87	0.84
Sea-level thrust N	64050	72500
(lb)	(14400)	(16300)
Cruise specific fuel consumption, kg/hr/N	0.040	0.072

From Abe Silverstein and Eldon Hall, "Liquid Hydrogen as a Jet Fuel for High-Altitude Aircraft," NACA RM E55C28?
15 Apr 1955, p 21.

A key assumption of the Silverstein-Hall analysis was the feasibility of lightweight, insulated flight tanks suitable for liquid hydrogen. Reynolds continued his investigation and reported the results in August 1955. Table 3, taken from the report, summarizes the results. Reynolds concluded that it was feasible to design a tank that had a mass less than 15 percent of the liquid hydrogen it contained. Estimated hydrogen vaporization rates were less than 30 percent of hydrogen consumption during cruise, and prior to flight, the tank could be held in stand-by condition and readied for flight in a short time.[15]

Following the completion of the report with Hall on flying aircraft fueled with hydrogen, Silverstein again visited the Air Force with missionary zeal. He also set in motion a great wave of research related to hydrogen's use in aircraft at the Lewis laboratory. This included properties, combustion, materials, tankage, bearings, pumpings, controls, and complete engines. In 30 months, the investigations led to three dozen reports and were climaxed by a research conference in November 1957.

In September 1955, Jerrold D. Wear and Arthur L. Smith completed an investigation of six types of injectors for burning gaseous hydrogen in a turbojet

TABLE 3.— Flight-Type Liquid-Hydrogen Tank Design

Size:		
Diameter, m		3.05
Length, m		24.9
Volume, m^3		174.2[a]
Surface area, m^2		238.2
Working pressure, atm		2
Styrofoam[b] insulation:		
Thickness, cm		5.7
Density, kg m^3		20.8
Mass of tank:		
Shell, kg		1163
Insulation, kg		284
Covering, kg		29
Allowance for baffles and stiffeners, kg		112
Approximate total mass, kg		1588[c]
Estimated performance with ambient temperature at:	300 K	218 K
Outer surface temperature, K	285	154
Heat-leak rate, W	25 770	14 500
Hydrogen-vaporization rate, kg hr	206	116
No-loss time on ground, min	165[d]	

[a] Holds 11 340 kg liquid hydrogen with 9 percent expansion volume

[b] Covered with layer of Mylar-aluminum foil

[c] About 14 percent of fuel mass

[d] For a tank with 5.7 cm insulation, precooled with liquid nitrogen. No-loss time is the time for heat leaking into the tank to vaporize enough hydrogen to raise the pressure to the working pressure (2 atm)

From: T W Reynolds, "Aircraft-Fuel-Tank Design for Liquid Hydrogen," NACA E55F22 9 Aug. 1955, p. 9.

combustor.[16] They found that at conditions simulating full power, all six injectors gave high combustion efficiency—an indication of the ease of burning hydrogen. Some relatively low combustion efficiencies were obtained, but these were at conditions where ordinary jet fuel would not burn. These experiments were followed by others as the laboratory probed deeper and deeper into the combustion of hydrogen under a variety of conditions.

Bee Project

The component and engine testing of hydrogen in the laboratory, essential as they were, did not answer an important question: Was it practical to use liquid hydrogen in an aircraft? Silverstein had been interested in finding this out from the beginning and his big opportunity came from a parallel interest by the Air Force.

In the fall of 1955, the power plant laboratory of Wright Field, headed by Col. Norman C. Appold, planned an experiment to determine the feasibility of flying an airplane fueled with liquid hydrogen. The bids for a contract—about $4 million a year for 3 years—were higher and longer than anticipated. Lt. Col. Harold Robbins, ARDC headquarters and former Air Force liaison at Lewis, suggested that the NACA be approached to do the work. Silverstein jumped at the opportunity. He promised to do the job in 12 months and with $1 million for special equipment. The agreement was reached in December 1955, and Silverstein lost no time in getting started. He chose Paul Ordin to be the project manager, assisted by Donald Mulholland. The project staff was quickly selected and put to work on their new assignment.[17]

Although Silverstein was technical head of a laboratory with a complement close to 3000, it was characteristic of him to direct the project personally. He had a room in the basement of the administration building cleared for use by the project group. It was directly below his office and convenient for his close supervision. The project was classified secret and known as Project Bee.

The airplane selected for the project was the B-57B twin-engine bomber powered by Curtiss Wright J-65 turbojet engines. The basic plan was to equip the airplane with a hydrogen fuel system, independent of its regular fuel system, and modify one engine to operate on hydrogen as well as its regular fuel, which was JP-4 (kerosene). The airplane was to take off and climb on its regular fuel. After reaching level flight at about 16 400 meters, the fuel on one engine was to be switched from JP-4 to hydrogen. When the hydrogen experiment was complete, the fuel flow would be switched back to JP-4 and the airplane would return to base under its normal operating conditions.

The project team, aided by others in the laboratory, began to design and test the various components for the flight system. A liquid hydrogen tank was designed for mounting beneath the tip of a wing. Two methods for pumping liquid hydrogen were selected. The first was to pressurize the hydrogen tank with helium, a simple and fast method but requiring a fairly heavy tank to withstand the pressure. The second was to employ a liquid-hydrogen pump, but this required time for development. Consequently the first tests were made with the pressurization system.

Earlier combustion experiments showed that gaseous hydrogen burned easily in the turbojet engine. To feed gaseous hydrogen to the airplane engine required some means for gasifying the liquid. A heat exchanger was designed and tested for this purpose. Ram air passed through it during flight to heat and gasify the liquid hydrogen.[18]

The dual fuel system and transition between the two fuels, JP-4 and gaseous hydrogen, called for an integrated control system, the key component of which was a flow regulator for the gaseous hydrogen. The speed of the engine was controlled by coupling the hydrogen flow regulator to the engine's JP-4 fuel control.[19]

The flights were the province of the laboratory's test pilots headed by William V. (Eb) Gough, Jr., the fourth Navy pilot to qualify in helicopters and the thirtieth in jets; he joined the NACA as a test pilot after the war. By early May, Gough had checked out on the B-57 at the Glenn L. Martin plant in Baltimore and the Air Force had ferried a B-57 to Cleveland for the experiments.[20]

Assisting Gough was Joseph S. Algranti, another test pilot, who would fly in the rear seat and operate the special controls of the hydrogen fuel system. He participated in the ground testing of the system from the beginning of the project. A third test pilot served as back-up and was in charge of the ground control station.

The testing of the flight components required a considerable amount of liquid hydrogen—the problem that had plagued the rocket group at Lewis for a long time. The Air Force made available mobile hydrogen liquefaction equipment and tanks from the hydrogen bomb program. Glenn Hennings got the equipment in good working order and was soon producing liquid hydrogen for the various laboratory needs.[21] In the first half of 1956, as part of another program, the Air Force let a contract to build at Painesville, Ohio, a hydrogen liquefaction plant with a capacity of 680 kilograms per day. When this plant began production late in 1956, it supplied all of Lewis's hydrogen needs.

Concurrent with the development of the flight system for supply and controlling hydrogen to the engine, a number of experiments were conducted with single turbojet combustors and full-scale engines using gaseous hydrogen as a fuel. The engine performance was high and insensitive to initial hydrogen temperature.[22]

In other research, hydrogen in a combustor 2/3 as long as a standard one, outperformed JP-4 and also operated at an altitude of 26 000 meters—6000 higher than the limit for JP-4.[23] This meant that a shorter engine was possible with hydrogen, with accompanying substantial savings in mass. In another investigation, a team led by William A. Fleming compared the altitude performance of two turbojet engines, one burning hydrogen and the other JP-4. The engines were single-spool, axial-flow types, developing 33–45 kilonewtons (7500–10 000 lb thrust). Hydrogen provided stable operation to the limits of the test facility—about 27 400 meters and Mach 0.8. In comparison, the same engine using JP-4 flamed out at altitudes 3000 to 4500 meters lower. Further, the specific fuel consumption (mass flow of fuel per hour divided by thrust) of hydrogen was 40 percent that of JP-4 fuel.[24]

Silverstein wanted a thorough check of the engine and control system, using both JP-4 and hydrogen fuels in the altitude wind tunnel before attempting flight. This was carried out by Harold R. Kaufman and associates, including test pilot Algranti. The hydrogen system consisted of a stainless steel, wing-tip fuel tank, a heat exchanger that utilized air passing through it to vaporize the liquid hydrogen, and a regulator to control the flow of hydrogen to the engine. The J-65 turbojet engine was modified by the addition of a hydrogen manifold and injection tubes. The modification did not change the engine's regular fuel system using JP-4. Kaufman reported that with JP-4 the maximum altitude for stable combustion was about 20 000 meters and flame-out occurred at 23 000 meters. In contrast, hydrogen was stable to the limit of the facility at 27 000 meters at flight-rated speed and temperature. The thrust was 2 to 4 percent higher, and specific fuel consumption was 60 to 70 percent lower, than with JP-4 fuel.[25]

In the simulated flight tests, 38 transitions were made from JP-4 fuel to hydrogen. Over three-fourths of these were satisfactory. The others had some engine speed variations, but they were so small and short in duration that the engineers believed there would have been no detrimental effect on aircraft performance. These satisfactory results in the altitude chamber cleared the way for testing the hydrogen system in the B-57.

The hydrogen fuel tank on the left wing of the airplane (figs. 21 and 22) was 6.2 meters long with a volume of 1.7 cubic meters. The stainless steel tank was designed for a pressure of 3.4 atmospheres and insulated by a 5-centimeter coat of plastic foam, covered by aluminum foil and encased in a fiberglass covering. On the opposite wing was the helium supply consisting of 24 fiberglass spheres charged to 200 atmospheres. The helium was used for pressurizing the hydrogen tank and for purging. A heat exchanger for vaporizing the liquid hydrogen, a flow regulator, and a manifold for feeding gaseous hydrogen to the engine comprised the rest of the hydrogen system.

As Christmas neared, pilots Gough and Algranti made a series of checkout flights without hydrogen, and finally the big day came. On 23 December 1956, Scotty

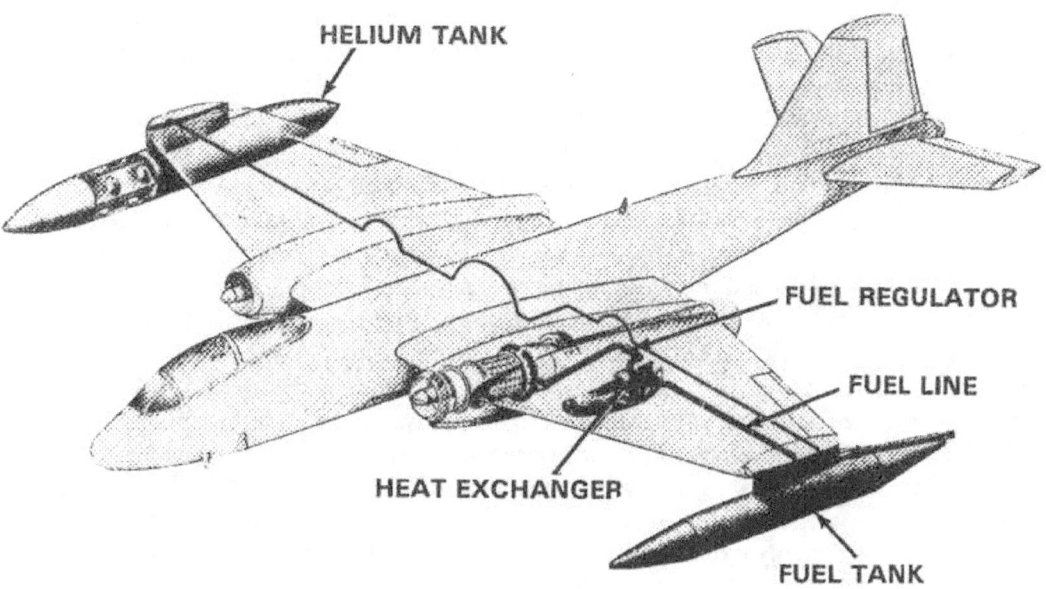

Fig. 21. Liquid-hydrogen fuel system for one engine of a B-57 airplane installed by the NACA Lewis laboratory.

Fig. 22. B-57 airplane modified by the NACA Lewis laboratory to use liquid hydrogen in one engine. The wing-tip pod on the right (the airplane's left wing) is the hydrogen tank; the opposite pod contains helium for pressurization and purge. The dense smoke is normal in starting this engine on conventional fuel.

Simpkinson made the final check of instruments and the B-57 was fueled with JP-4. It was then towed to a remote site for loading liquid hydrogen. The vent of the tank was connected by pipe to a discharge area well away from the airplane and the system purged with helium. After countdown, 94 kilograms of liquid hydrogen were loaded into the wingtip tank. The ground crew left the vent-pipe system connected until Gough started the plane's engines on JP-4. At that time, Algranti closed the vent valve, the ground crew disconnected the vent line, and Gough began to taxi. He was accompanied by an Air Force chase plane equipped with a camera.[26]

As the B-57 taxied into position for take-off, Algranti was maintaining the pressure in the liquid hydrogen tank. With the vent valve closed, the vaporization of a small amount of hydrogen caused the pressure in the gas pocket above the liquid hydrogen to rise. The vaporization was caused by heat leakage through the insulation, which is unavoidable in a practical installation. From ground testing, Algranti knew that the pressure would rise from 1 to 3.5 atmospheres in about five minutes, and he had to manually vent the tank when the pressure began to rise above 3.5 atmospheres. While taxiing, he noticed that the rate of pressure rise was considerably slower than in ground tests; the instrument records indicated that sloshing and agitation of the hydrogen during taxiing slowed the pressure rise by a factor of two. During takeoff, the tank pressure dropped sharply from agitation. Once airborne, however, the agitation ceased and the pressure began to rise at about the same rate as in the stationary tests. This phenomenon was caused by thermal gradients and stratification of liquid hydrogen and its vapor and was the subject of detailed investigation later.

The takeoff and climb to the cruising altitude of 15 200 meters took almost an hour, and during that time, Algranti vented the tank 8 times to keep the pressure within limits. This resulted in a loss of about 16 percent of the hydrogen. On signal, Algranti made the transition from JP-4 to hydrogen. The engine responded by overspeeding and vibrating hard. The startled pilots quickly shut it down, purged the lines, and jettisoned the liquid hydrogen in the wing tank. The B-57 was difficult to fly on one engine, but Gough's training included this contingency. The experiment had taken place over Lake Erie and the weather had deteriorated. Gough dismissed the chase plane, but the pilot elected to accompany him back to the Cleveland airport. The two landed side by side on dual runways in a light rain.

Although the first flight was unsuccessful in operating the engine with hydrogen for an extended period, it was successful in showing that hydrogen could be handled and jettisoned safely. In addition, data were obtained on the phenomenon of hydrogen thermal stratification in the tanks.

The second flight was also only partially successful. The transition from JP-4 to hydrogen was made successfully, but insufficient hydrogen flow prevented satisfactory high-speed engine operation. Again, the bulk of the hydrogen was jettisoned without incident. The jettisoning took less than 3 minutes, with the hydrogen forming a dense plume which vanished about 6 meters aft of the tank.

On 13 February 1957, the first of three successful flights was made and the fuel system worked well.[27] The transition to hydrogen was made in two steps. The hydrogen lines were first purged, then the engine was operated on JP-4 and gaseous hydrogen simultaneously. After two minutes of operations on the mixture, Algranti switched to hydrogen alone. The transition was relatively smooth and there was no appreciable change in engine speed or tailpipe temperature. The engine ran for about 20 minutes on hydrogen. The pilots found that the engine responded well to throttle changes when using hydrogen. When the supply was almost exhausted, the speed began to drop. As this became apparent, Algranti switched back to JP-4 and the engine accelerated smoothly to its operating speed. The engine burning hydrogen had produced a dense and persistent condensation trail, while the other engine operating on JP-4 left no trail.

On 26 April, Silverstein held a special conference to report what had been learned by the Bee project using hydrogen in flight. The 175 attendees heard 7 papers by 19 members of the project team. They covered hydrogen consumption, fueling problems, airplane tankage, airplane fuel system, and the flight experiments. The results were also given in a series of research reports published later.[28]

The first series of flights of the hydrogen-fueled B-57 was made with a helium pressurization system to force the liquid hydrogen from the wing-tip tank to the engines. This required a fairly heavy tank to withstand the pressure. Later, a liquid-hydrogen pump was developed which permitted a reduction in tank weight that more than offset the weight of the pump. Arnold Bierman and Robert Kohl developed the five-cylinder piston pump, driven by a hydraulic motor, for installation in the wing-tip liquid-hydrogen tank.[29]

Flight experiments with the pump extended into 1959. Three successful flights were made. Although the pump speed and discharge pressure varied, the hydrogen regulator maintained a constant engine speed during operation with hydrogen. All the transitions from JP-4 to hydrogen, burning hydrogen, and transition back to JP-4 were made without incident. The feasibility of using liquid hydrogen in flight had been thoroughly demonstrated.[30]

Flight Propulsion Conference

The Bee project of flying an airplane fueled with hydrogen was part of a broader investigation of advanced engines for airplanes and missiles at the NACA Lewis laboratory. The broader vein was presented at a second research conference held on 21–22 November 1957, with 300 attendees. Hydrogen was the chief fuel discussed. The papers were presented by a series of eight panels, five of which were on air-breathing engines. The other three were on rockets (p. 91). Edgar M. Cortright, J. Howard Childs, DeMarquis D. Wyatt, and David S. Gabriel led, describing the air-breathing engine concepts. They pictured military planning as being at a critical stage. The choice of deterrent weapons included the manned bomber, unmanned missile, glide bomber utilizing aerodynamic lift, intercontinental ballistic missile, and satellite bomber for flight beyond the atmosphere. Development of each was expensive and time consuming; the purpose of the first five panels of the conference was to present "an appraisal of the ultimate performance capabilities of aircraft and missiles powered by air-breathing engines"—range, speed, weight, and payload were used as criteria of merit. Flight at very high speed heats aircraft surfaces and requires cooling for sustained flights. Cortright's panel found that only liquefied methane and hydrogen had significant cooling capacity at flight speeds above Mach 5. Hydrogen was the best fuel for cooling, primarily because it was thermally stable and useful up to the maximum allowable temperatures of the vehicle surfaces.[31]

Fuel heating value was also examined and not surprisingly, the panel singled out the superiority of hydrogen, noting that it was 70 percent better than diborane. Hydrogen's high heating value, combined with its greatly superior cooling capacity, made it extremely interesting as a fuel for long-range hypersonic flight.

The first panel noted the disadvantages of hydrogen's low density—a problem considered by a later panel. Also noted was possible dissociation loss that might limit the realization of full heating value of the fuels considered. These and other considerations provided the basis for detailed discussions of two applications: a manned bomber flying at a speed of Mach 4, to be powered by a new engine; and an unmanned ramjet missile—with all surfaces glowing red hot from air friction at its flight speed of Mach 7—cooled and fueled by hydrogen.

Air-Breathing Engines for High-Speed Flight

After panel discussions of inlets, exits, and cooling, an engine panel headed by H. M. Henneberry analyzed four types of engines for the Mach 4 manned mission and two types of ramjet for the Mach 7 unmanned missile.* The Mach 4 engines were: turbojet, fuel-rich turbofan, hydrogen expansion, and air-turborocket. All four had common elements of air inlet, fan or compressor for increasing the pressure of the incoming air, afterburner where additional fuel was burned, and nozzle. The fan or compressor was driven by a gas turbine, but the turbine and its driving gas differed among the four engines. In the turbojet the driving gas was primarily air, in the fuel-rich turbofan and air turborocket the gas was hydrogen rich, and for the hydrogen-expansion engine the gas was all hydrogen.

Of the four engines, the hydrogen-expansion type is of particular interest because it was under development in a super-secret Air Force project to be discussed later. The hydrogen-expansion engine described by Henneberry had a complex flow system which will be described with the aid of figure 23. Air entering the engine was compressed by a two-stage fan driven, through suitable gearing, by a high-speed turbine. A small amount of air was directed to the primary combustor; the main stream flowed directly to the afterburner. The liquid hydrogen was raised to a high pressure by a pump and served as a coolant for various purposes (such as cooling hot vehicle surfaces) prior to entering the engine. The heat absorbed during these cooling functions converted the hydrogen to a gas. In the engine, the hydrogen flow was split, with one part flowing directly to the afterburner. The other part flowed through a heat exchanger where its temperature was increased substantially. The hot hydrogen was used to drive a 3-stage turbine which, in turn, powered the fan for compressing the air. After leaving the turbine the hydrogen entered the primary combustor where it burned hydrogen-rich with air. The hot, hydrogen-rich combustion gas entered the other side of the heat exchanger where it provided the heat for the separate flow of hydrogen gas

---

*With Henneberry on the engine panel were A. V. Zimmerman, J. F. Dugan, W. B. Schram, R. Breitwieser, and J. H. Povolny.

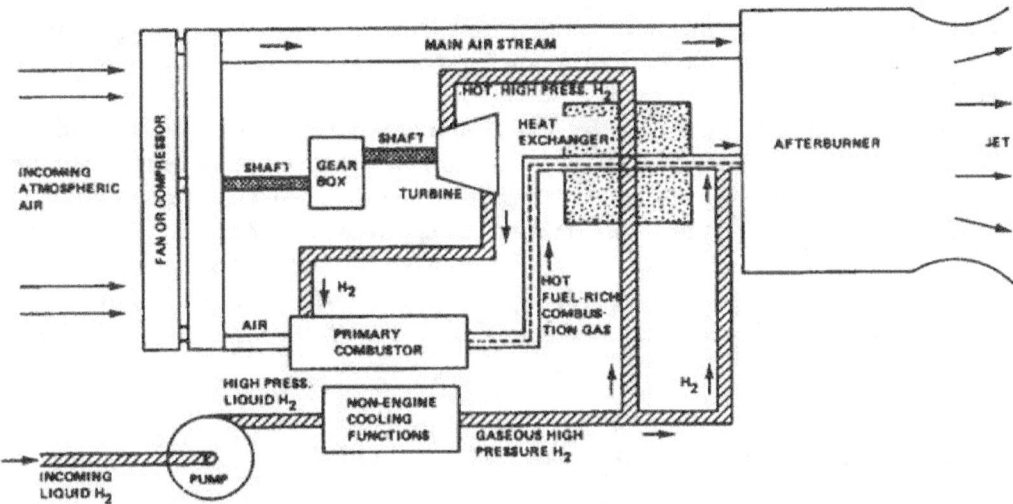

Fig. 23. Schematic of hydrogen expansion engine as described by Henneberry at NACA 1957 Flight Propulsion Conference, 21–22 Nov. 1957.

for the turbine previously mentioned. After leaving the heat exchanger, the hydrogen-rich combustion gas flowed to the afterburner where it and the other part of the hydrogen flow burned to completion, after which the hot gases expanded through the nozzle to provide thrust. Henneberry and his panelists estimated the weight of the hydrogen-expansion engine to be 10 percent heavier than a turbojet, and to have many difficult development problems.*

The turborocket, being pushed by W. C. House of Aerojet-General Corporation, used a small rocket to provide the hot gases for driving a turbine, with the turbine driving the air compressor or fan. The rocket used either a monopropellant or bipropellants—the latter being fuel-rich with additional burning in the afterburner, as in the other engines. The panel described a turborocket using liquid hydrogen as the fuel. After being compressed, part of the incoming air was diverted through a heat exchanger, chilled on its other side by liquid hydrogen. The air was liquefied and pumped at a high pressure to the rocket chamber. The main air-stream flowed directly to the afterburner. Liquid hydrogen, after serving to liquefy the air in the heat exchanger, went directly to the rocket chamber where it mixed and burned fuel-rich with the air. The combustion products drove a turbine (which drove the air fan or compressor). After leaving the turbine, the hydrogen-rich gas flowed to the afterburner where it mixed and burned in the main air-stream. A potential problem in this type of engine was icing from moisture in the incoming air.

---

*As will be discussed later, Pratt & Whitney built a hydrogen-expander engine and tested it for the first time two months before the NACA conference. The engine development was part of a highly secret Air Force project, and it is very doubtful that the Henneberry panel was aware of it. Another, somewhat similar, type of cycle—the Rex engine—was known to the Lewis laboratory and the Henneberry panel may have drawn on this knowledge. Interview with A. V. Zimmerman, Roger Luidens, and Richard Weber, NASA Lewis Research Center, 30 May 1974.

After comparing the four types of engines, the Henneberry panel concluded that the turbojet was the simplest, would require the least development, and was adaptable to a wide variety of fuels.

Of the two ramjets considered by the Henneberry panel, one was described as conventional, the other fuel-rich. In the former, enough fuel was provided for complete combustion with the oxygen available, while the latter used an excess of fuel. The excess fuel helped to offset decrease in thrust at speeds above Mach 10. The panelists found that hydrogen was superior to diborane and methane for cooling and for performance at high supersonic speeds. They concluded that a fuel-rich ramjet using liquid hydrogen could extend the usefulness of air-breathing engines to speeds up to Mach 18.

Mach 4 Configurations and Missions

Another panel at the NACA 1957 conference on propulsion, headed by Roger W. Luidens, bridged the papers on propulsion with an analysis of the range capabilities of airplanes using the advanced propulsion concepts; it was followed by another panel led by S. C. Himmel that tied all the previous discussions of air-breathing engines, airplanes, and missiles together.* One mission selected for discussion was a speed of Mach 4, altitude of 30 500 meters, a payload of 4500 kilograms, with airplanes using turbojet engines. When designed for hydrogen, the airplane was 91 meters long and had a gross mass of 136 000 kilograms of which about 1/3 was hydrogen. The airplane using JP-4 fuel was half as long, but had a mass 40 percent greater than the hydrogen configuration, with the JP-4 making up 60 percent of the mass. The hydrogen-fueled airplane had a range of just over 5000 kilometers compared to 3050 for the JP-4 airplane (fig. 24). While the hydrogen airplane had the greater range, it was short of the goal of 10 200 kilometers. Use of an air-turborocket increased the range 13 percent, but this was not enough to warrant the cost and time of development. Even with additional engine improvements and by using advanced airframe design, the range of the hydrogen-fueled airplane could be increased to only 7600 kilometers, still short of the goal.

The airplane designs using turbojets were outclassed by a hydrogen-fueled ramjet missile. With a mass of 17 400 kilograms and boosted to its cruising speed, it carried the same payload (4500 kg) at Mach 7 for a distance of 16 700 kilometers. Liquid methane and diborane were both inferior to liquid hydrogen.

The November 1957 propulsion conference at NACA-Lewis proved to be the climax of efforts to promote air-breathing hydrogen-fueled engines as competitors to rocket-powered intercontinental ballistic missiles. Strangely enough, the rich amount of experimental data on hydrogen from Lewis ground and flight experiments was not apparent to members of the audience. What came across strongly from the papers were concepts and trends of what the future could be like with hydrogen in advanced turbojet and ramjet engines. These potentials, however, came too late to catch up with

---

*With Luidens were J. H. Disher, Murray Dryer, and T. W. Reynolds; with Himmel were E. W. Conrad, R. J. Weber, R. R. Ziemer, and W. E. Scull.

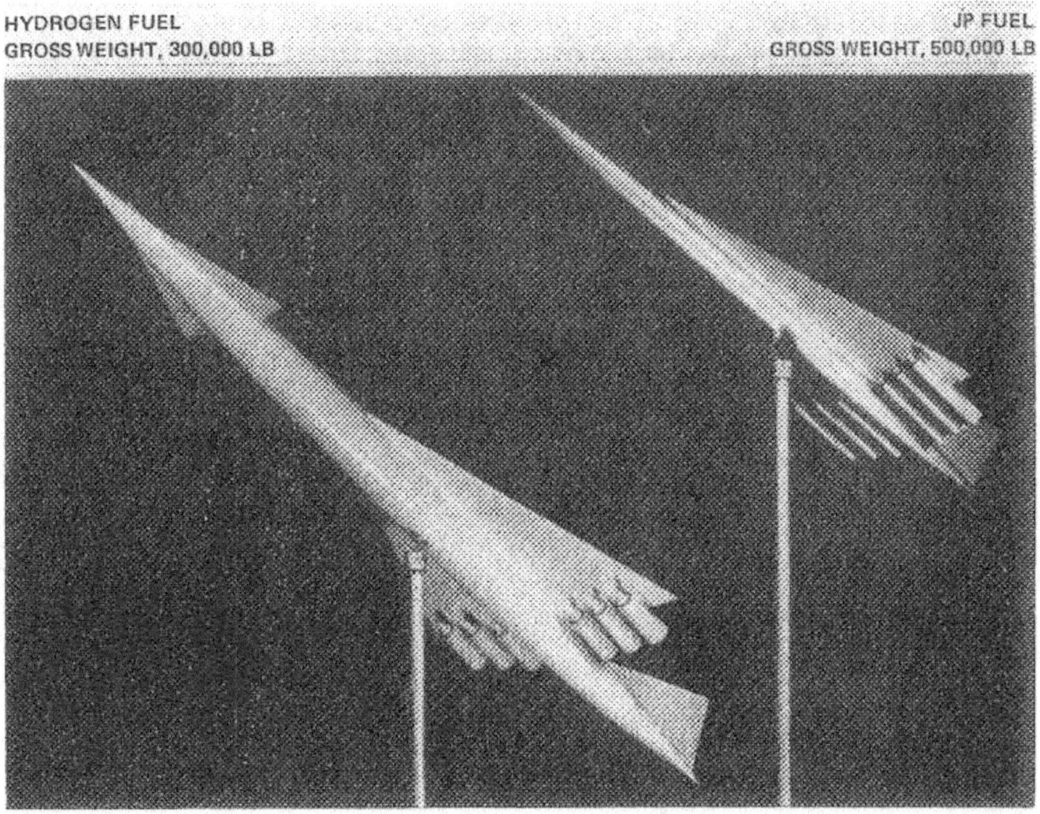

HYDROGEN FUEL
GROSS WEIGHT, 300,000 LB

JP FUEL
GROSS WEIGHT, 500,000 LB

Fig. 24. The effect of fuel type is shown by these models of high-altitude supersonic aircraft. The gross weight of the larger, hydrogen-fueled aircraft (left model) is only 60 percent of that of the smaller, JP-fueled aircraft (model on right). NACA 1957 Flight Propulsion Conference.

ballistic missile development. As a final clincher, Sputnik had ushered in the space age seven weeks earlier and turned attention to space, where the rocket was the undisputed propulsion system.

The many research scientists at the Lewis laboratory who worked on hydrogen as a fuel for high-flying aircraft were completely unaware that a huge and highly secret effort on hydrogen for high-altitude flight had been started in the Air Force the previous year. That work was managed by Col. Norman C. Appold, who attended the NACA conference. The Air Force project will be described in a later chapter.

Summary

During the period 1954–1957, the NACA-Lewis Flight Propulsion Laboratory at Cleveland investigated liquid hydrogen as a fuel for high-altitude aircraft and missiles. The experiments began in 1954 with an investigation of low-pressure combustion in a single turbojet combustor, extended to other components (tanks, pumps, heat exchangers, controls) and complete turbojet engine systems, and culminated in the first (and only) flight experiments. Among the many contributions:

(1) Gaseous hydrogen burns well at low pressures in a turbojet combustor.

(2) Promotion of hydrogen as a turbojet fuel; especially the concept that high-altitude, low-speed flight using turbojet engines demands efficient combustion at low pressure, best provided by hydrogen; and, at the same time, aircraft configurations for that flight regime favor large-volume aircraft which alleviates the disadvantage of hydrogen's low density.

(3) Lightweight, low-loss liquid hydrogen tanks are feasible.

(4) Liquid hydrogen can be pumped satisfactorily for turbojet engine conditions.

(5) Hydrogen requires less combustion volume than hydrocarbons, making possible shorter and lighter engines.

(6) A complete turbojet engine for subsonic flight can be operated with hydrogen at higher altitudes and with less fuel consumption (mass basis) than the same engine using hydrocarbon fuels.

(7) Existing turbojet engines can be easily adapted to use hydrogen.

(8) Flight demonstrations that liquid hydrogen can be handled safely in ground operations and in flight.

(9) Liquid hydrogen is an excellent heat-sink for very high-speed flight where air friction heats the vehicle surfaces.

(10) Turbojets using hydrogen give good performance at flight speeds of Mach 4 and ramjets for flight speeds of Mach 7, with much higher speeds feasible with the latter.

All these advantages made hydrogen appear to be the fuel of the future for advanced air-breathing engines; but, in fact, its prospects were already being tested, as we will see.

7

New Initiatives in High-Altitude Aircraft

In 1953, military aviation was in transition from subsonic to supersonic flight. Chance-Vought delivered the last propeller-driven fighter, an F4U Corsair, to the Navy in February. Three months later, the YF-100A, produced by North American Aviation for the Air Force, became the first service supersonic fighter—the start of the Century series. These were made possible by more powerful turbojet engines such as the Pratt & Whitney J-57, which went into production in 1953. Speed, however, is but one of the familiar trinity of major military aviation goals—higher, faster, farther. Higher altitudes meant less vulnerability for bombers and reconnaissance aircraft. The altitude goal frequently mentioned during the period was 30 500 meters. Greater range was not neglected as a goal, but global bases and in-flight refueling sometimes made it possible to compromise range in favor of other goals. In addition, military aviation planners during the 1950s felt the keen competition of guided missiles, which were in rapid ascendancy. The rivalry between aviation and missile men was strong.

From late 1952 to early 1954, three men of diverse backgrounds initiated proposals for achieving flight at very high altitudes. One was an Air Force major stationed at Wright Field, John D. Seaberg; another was a famous airplane designer, Clarence L. (Kelly) Johnson of Lockheed Aircraft; and the third was a lone British inventor with a novel idea, Randolph Rae. These initiatives and the activities they generated proceeded concurrently with, but largely independent of, the NACA research described in the previous chapter. The initiative of Seaberg led to the new altitude capability of the B-57; Johnson's led to the extraordinary U-2 high-altitude reconnaissance airplane; and Rae's led to his personal disillusionment, but new interest within the Air Force for using liquid hydrogen in aircraft.

Origins of Very-High-Altitude Aircraft at Wright Field

At the outbreak of the Korean war, John D. Seaberg, an aeronautical engineer at Chance-Vought, was called back to active duty as an Air Force major. Seaberg, who had served as an engineering and base executive during World War II, was assigned to the new development office for bombardment aircraft at Wright Field. Late in 1952, he went to his boss, William E. Lamar, with some new ideas about achieving flight at very high altitudes. Seaberg saw in the new generation of turbojet engines, with their inherent high altitude potential, the opportunity of matching engine and airfoil to achieve an airplane of low wing-loading capable of higher altitude operation than anything yet conceived. The ideal application for such an airplane was reconnaissance; the high altitude would make detection very difficult and provide protection until effective countermeasures were developed.[1]

By March 1953, Seaberg's idea had jelled into a set of specifications for preliminary design studies by aircraft manufacturers. Operating conditions selected were an altitude of 21 340 meters or higher, a range of 2800 kilometers, and subsonic speeds. Propulsion was to be by turbojet or turboprop suitably modified for the high altitude operation. The airplane would carry a crew of one and photographic equipment weighing between 45 and 318 kilograms. No armament or ejection equipment was provided, in keeping with the objective of minimum gross weight and high altitude for protection. The contractors were to supply design specifications suitable for a development contract, a recommended engine, and a list of major development problems anticipated.[2]

Seaberg and Lamar decided to bypass the big aircraft manufacturers in favor of smaller companies because, believing that production would be small, they thought the smaller firms would give the studies a higher priority. There was no bidding; Bell Aircraft, Fairchild Aircraft, and Glenn L. Martin were called in to discuss the studies, and all three were very interested. The Air Force talked to no one else. Contracts to the three were let beginning 1 July 1953 and ran to the end of the year. Bell and Fairchild were asked to design a new airplane; Martin, builder of the B-57 bomber and RB-57 reconnaissance airplane, was asked to study modifications to the RB-57 to meet the more stringent altitude requirements.[3]

Wright Field evaluated the three studies in early 1954 and had the contractors present the study results during the first part of March. Bell proposed a twin-engine airplane; Fairchild submitted a single-engine design; and Martin discussed modifications to the RB-57, including a larger wing (fig. 25). All used Pratt & Whitney J-57 engines, modified for high altitude operation and initially designated J-57-P19 (later J-57-P37).[4]

Lt. Col. Joseph J. Pelligrini, attached to a reconnaissance unit at headquarters of the Air Research and Development Command (ARDC), visited Wright Field in mid-March, saw the Martin proposal as a fast way of meeting an urgent need of the Air Force in Europe, and requested Wright Field to send ARDC headquarters a list of necessary RB-57 modifications within a week.[5] The following month, Seaberg went to ARDC headquarters in Baltimore and gave a briefing on the three studies. Attending was Lt. Gen. Thomas S. Power, who succeeded Lt. Gen. Donald Putt that month as commander of ARDC. Power was so impressed that he had Seaberg repeat the briefing at Strategic Air Command headquarters the following day. Seaberg gave a third briefing at Air Force headquarters early in May 1954.[6] Interest in high-altitude reconnaissance aircraft increased and Seaberg had every reason to believe his idea would soon become a reality. Two weeks after his third briefing, however, a new proposal for a high-altitude airplane, from Kelly Johnson of Lockheed Aircraft Company, reached Seaberg's desk with a request for an evaluation. This proposal would lead to a series of significant events in aeronautics, politics, and diplomacy.

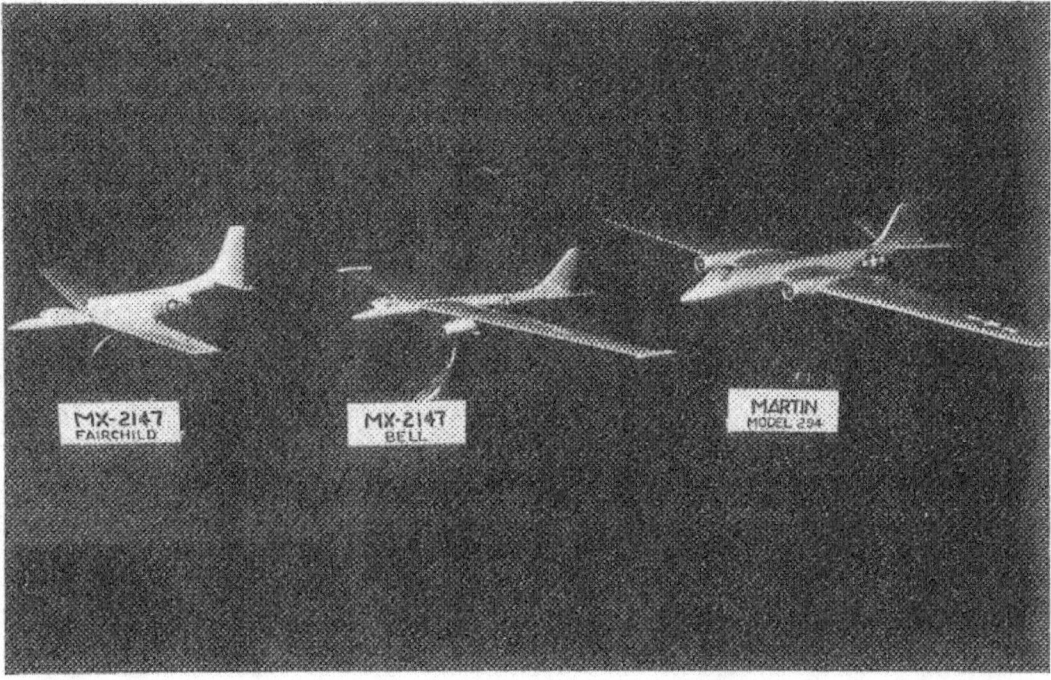

Fig. 25. Models of Fairchild, Bell, and Martin high-altitude reconnaissance airplanes from design studies conducted for the Air Force during the second half of 1953. All three used Pratt & Whitney J57-P19 engines. The Bell (X-16) and Martin (B57-D) designs were chosen for development, but only the latter was completed and is flying today as the RB-57F. (1954 photograph courtesy of W. E. Lamar and J. D. Seaberg.)

Clarence L. (Kelly) Johnson, chief engineer of Lockheed Aircraft at Burbank, California, since 1952, was already a legend among aircraft designers. He had designed and built the prototype of the first U.S. jet fighter, the F-80, in 143 days. He had gone on to design the F-90, the F-104, and many others. He had his own special brand of management and operations known as the "Skunk Works." He condensed his management philosophy to "be quick, be quiet, be on time."*

Johnson's unsolicited proposal to the Air Force came as no great surprise at Wright Field. Johnson had the confidence of and was accustomed to dealing with the highest levels in the Air Force and there was no reason for those officers to conceal their interest in very high-altitude flight from him.

As the designer of the Air Force's F-104 fighter, Johnson had proposed to use its fuselage, a larger wing to achieve an altitude of about 20000 meters, and the General Electric J-73 turbojet engine. In his review of the proposals at Wright Field, Seaberg was not impressed with Johnson's selection of the J-73 for extremely high altitude

---

*Johnson used a set of 14 operating rules for the Skunk Works including: almost complete control by project manager; a strong but small project office; an "almost vicious" restriction of the number of people connected with the project; a simple, flexible drawing system; minimum paperwork; thorough and periodic cost review; authority to subcontract; tight inspection; flight testing; prior specifications; timely funding; mutual trust; tight security; and rewards based on performance. Interview with C. L. Johnson and Ben Rich, Burbank, 2 May 1974.

flight. He felt that the more powerful Pratt & Whitney J-57, modified for high altitude operation, was required. However, it was too large to fit into the F-104 fuselage, so a modified fuselage would also be required. Since the proposals for the Martin RB-57 modification and the Bell X-16 had been approved, Seaberg saw no need to develop a third airplane and recommended against Johnson's proposal.[7] Seaberg's view was supported by the Air Force. The high-altitude B-57D was subsequently built; the Bell X-16 was initiated but cancelled in mid-1956.

When the Air Force turned Johnson down, he did not give up and a fortunate turn of events gave him a big break. In 1954, the role of the guided missile was rising rapidly, and the Department of Defense formed a number of advisory groups in mid-1954 to examine the various aspects of military planning and weapons. James R. Killian became chairman of a committee on surprise attack and was aided by several panels. One of these was on intelligence. During the course of its work, the panel learned about Johnson's proposal for a very high-altitude reconnaissance airplane and liked it. Killian was convinced of its merits and soon others, including Charles Wilson, Secretary of Defense, and Allen Dulles, director of the Central Intelligence Agency, also became convinced. It was known that the airplane had been proposed to the Air Force but that the Air Force had decided not to develop it.*

Johnson's proposal was taken to President Eisenhower during the latter part of November. As described by Eisenhower:

> Back in November 1954, Foster Dulles, Charlie Wilson, Allen Dulles, and other advisors had come to see me to get authorization to go ahead on a program to produce thirty special high-performance aircraft at a total cost of about $35 million. A good deal of design and development work had already been done. I approved this action.[8]

Eisenhower decided that the funding and direction of the project would be under the CIA and Richard M. Bissell, Jr., was selected to head it. The Air Force was to contract with Lockheed for development of the airplane, designated the U-2. Because of the sensitivity of the project, the Air Force handled its part directly from headquarters.[9] On 9 December 1954, Trevor Gardner, assistant secretary of the Air Force for research and development, visited Robert Gross and Kelly Johnson at Lockheed and told them to go ahead.[10] The Skunk Works swung into action and the first U-2 flew eight months later. It was powered by a Pratt & Whitney J-57-P37 turbojet engine, the engine Seaberg had argued was necessary.†

The U-2 (fig. 26) was capable of flying at altitudes above 21 300 meters at a speed of about Mach 0.75 (about 800 kilometers per hour at its altitude). The first operational

---

*Members of the Killian committee were briefed on Air Force plans for the B-57D and the X-16 by John D. Seaberg in the office of Lt. Gen. Donald Putt, deputy chief of staff for development, USAF, on 18 Nov. 1954. Seaberg also discussed the Fairchild and Johnson proposals and indicated that Johnson's airplane performance could be improved if the J-57-P37 engine replaced the J-73 proposed by Johnson. Letter from Seaberg to author 28 June 1976, with enclosures.

†Seaberg says that, to this day, Johnson tells him, "You had a chance to buy the U-2 and didn't do it"; and he counters with, "Kelly, you picked the wrong engine." Interview with Seaberg, 23 Nov. 1973; letter from Seaberg to author, 28 June 1976.

Fig. 26. Lockheed U-2 airplane, proposed in 1954 and in active government service continuously since 1956. Its capability for sustained flight at very high altitudes is still unmatched by any other airplane. (1963 photograph courtesy of Lockheed Aircraft Corp.)

flight occurred in the spring of 1956. The government chose research by the National Advisory Committee for Aeronautics as the cover for the covert reconnaissance operations of the airplane, but kept the NACA in the dark about its real purpose. Early U-2s carried NACA markings (and, later, NASA markings) and obtained data on high-altitude meteorological phenomena.[11] These data made significant contributions to a better understanding of turbulence, wind shears, and jet streams. In 1973, the NASA began using the versatile U-2 in its earth resources program.

In the early spring of 1954, in the midst of Seaberg's plans and before Johnson's proposal reached his desk, a British inventor brought a new and novel concept for an airplane and engine, called Rex I, to Wright Field. Unlike other airplanes, Rex I used liquid hydrogen as fuel.

Rae's Rex I Proposal

There was nothing unusual about the visitor who came to Wright Field on the chilly, overcast day of 24 March 1954. He was one of dozens who were processed through the large visitor's center adjoining the security fence to go to one of the many buildings of the huge Wright Air Development Center. Typical also was the reason for his visit. He was bringing an idea, neatly packaged in a brochure, and seeking a contract. The Center receives hundreds of unsolicited proposals annually and is geared to evaluate them. As with most such proposals, this one was destined to be rejected. What was unusual, however, was the novel solution proposed for a difficult problem, the sensitive nature of the subject, and the timing. The proposal triggered waves of interest within the government, and there followed a series of events involving hydrogen that extend to this day—events that shuttled the proposer to the sidelines and left him bewildered and embittered. His name is Randolph Samuel Rae (1914–   ).

Randy Rae is a quiet, soft-spoken man with the imagination and creativeness that mark the practical innovator and inventor. He received his engineering education at a Swiss technical school and began his career in electronics and underwater detection systems for locating submarines. He worked for the British Admiralty from 1939 to 1948, serving in four research and development groups in underwater acoustics, aerodynamics, thermodynamics, and propulsion, rising to the position of a principal scientific officer. He came to the United States in 1948 and worked in the Applied Physics Laboratory of Johns Hopkins University for four years. He started in aerodynamics and developed a supersonic diffuser for ramjet engines and later was placed in charge of the development of a complete guided missile system. More at home with technical details than overall project management, Rae soon was immersed in a difficult missile stability and control problem and devised a solution involving a gyro with a mechanical feedback. The system was put out for bid and a small company, Summers Gyroscope, won the contract. Rae met a kindred soul in dynamic, innovative Thomas Summers.[12]

The missile development that Rae was managing used a ramjet engine for propulsion. A ramjet operates at high altitudes and speeds, but as with all air-breathing engines, it is altitude-limited. The ramjet's altitude-speed limitations set Rae to thinking about other solutions to the problem in April 1953. Was there a way to operate at very high altitudes but at lower speeds, specifically in the subsonic speed range? The rocket was not the answer, for although it operates independent of the atmosphere, it is very inefficient at low speeds. Could he combine the altitude-independent feature of the rocket engine with a propulsion system efficient at low speeds? The most efficient means for aircraft propulsion at low speeds is the propeller, but it is, of course, altitude-limited. Rae conceived of using a rocket as a gas generator to drive a turbine which, through suitable gearing, would drive a large propeller. Such a propulsion system had no place in the high-speed, high-altitude operating regime of the Navy's work at the Applied Physics Laboratory. Rae became so intrigued with his concept that he left APL/JHU to work full-time on the new propulsion system. He soon learned the handicaps a lone inventor faces. He needed not only monetary support but a corporate identity as well. He turned to his friend, Thomas Summers, who very generously offered both, although propulsion was a far cry from gyroscopes and instruments.

Rae joined Summers in September 1953 and began analysis of what he called the Rex engine. The week before Christmas, Summers engaged Homer J. Wood, a mechanical engineering consultant. Wood had left the Garrett Corporation, makers of small gas turbine engines and other aircraft components, in October after ten years service during which he became assistant chief engineer in charge of turbomachinery. Wood assisted in the analysis and design of Rae's new engine.[13]

By March 1954, Rae was ready to present his idea to the government. He visited the headquarters of the Air Force Air Research and Development Command (ARDC), then located in Baltimore, and discussed his idea with Col. Donald Heaton, chief of the aeronautics and propulsion division, and Lt. Col. Langdon F. Ayers, who headed the propulsion branch. The two were engaged in planning research and development to increase the altitude capability of aircraft, and Rae's idea caught their interest. They suggested that he visit Wright Field and discuss the proposal with the specialists there.[14] This was what brought Rae to Wright Field on 24 March 1954, with brochures describing the proposal.

Rae presented his proposal to a group in the new developments office of WADC and passed out copies of his brochure. It bore the date of February 1954 and the title, "REX-1, A New Aircraft System" (fig. 27). Rae described it as "a lightly loaded low speed plane having an exceptional L/D (lift/drag) characteristic." By lightly loaded, he meant a low weight per unit area of wing; the aircraft resembled a low-powered glider. The speed of about 800 kilometers per hour would make a military airplane quite vulnerable were it not for the very high operational altitude that Rae proposed — over 24 000 meters, which was well above the capability of other aircraft and hopefully beyond the range of antiaircraft weapons. What stirred the interest of the Wright Field audience was the novel engine that Rae proposed: a three-stage turbine engine using liquid hydrogen and liquid oxygen (fig. 28). Ahead of each turbine was a small combustion chamber. All of the hydrogen and part of the oxygen were fed to the first combustion chamber. This partial combustion of the hydrogen produced a gas temperature of about 1100 K, the then practical limit for turbine materials. After

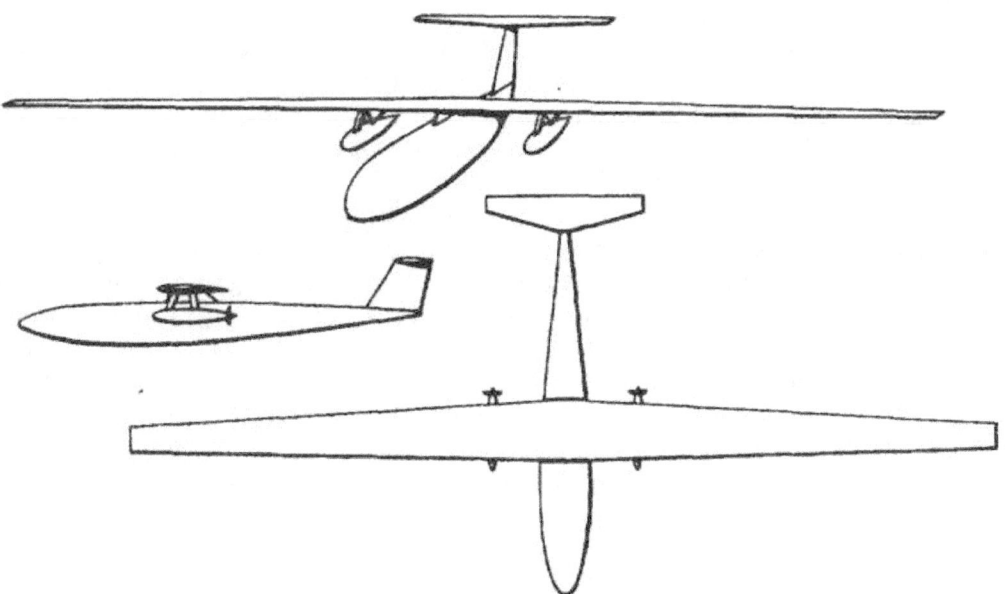

Fig. 27. Sketches of REX-1, a low-speed, high-altitude airplane using liquid hydrogen, proposed to the Air Force by R. S. Rae in Mar. 1954. Gross mass, 32 660 kg; empty mass, 16 330 kg; wing area, 434 m²; power, 1790 kW (2400 hp); take-off speed, 113 km/hr; cruising speed, 640–800 km/hr at 26 000 m altitude; range, 10 000 km. When empty, it could glide an additional 1000 km. From brochure "REX-1, A New Aircraft System," by R. S. Rae, Summers Gyroscope Co., Feb. 1954.

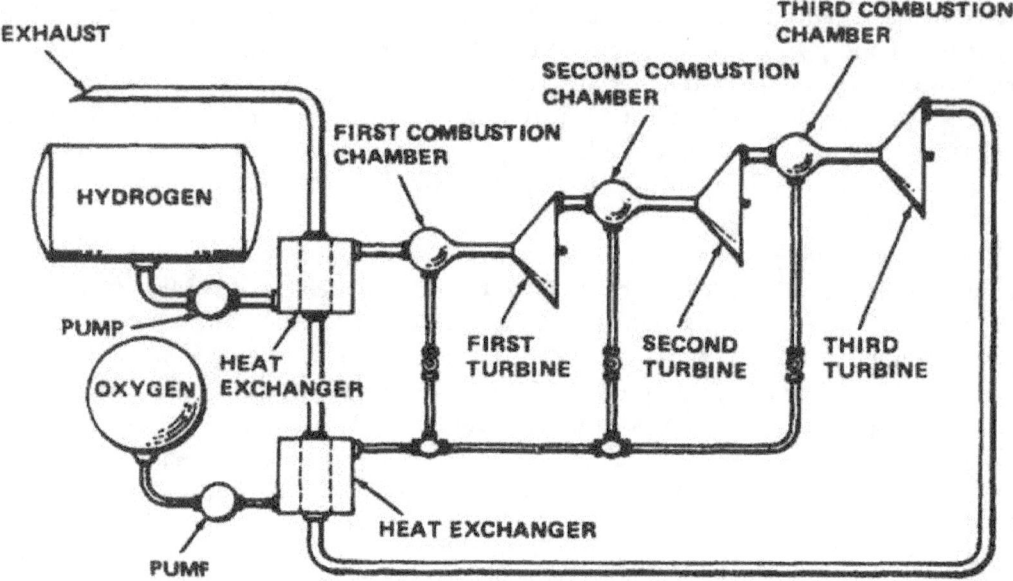

Fig. 28. Schematic of Rex 1 engine. Liquid hydrogen and oxygen, gasified by passing through heat exchangers, flow to three small combustion chambers. The hot gases drive three turbines connected to a common shaft. The gases for the second and third turbines are a mixture of the exhaust from the previous turbine and combustion gases. After the third turbine, the exhaust gases supply heat for the heat exchangers and then discharge. From the brochure "REX-1, A New Aircraft System." by R. S. Rae, Summers Gyroscope Co., Feb. 1954.

leaving the first turbine, the gases were reheated by adding additional oxygen and burning. The process was repeated for the third turbine. After leaving the third turbine, the gases passed through heat exchangers to heat the incoming liquid hydrogen and liquid oxygen.* Rae was attracted to hydrogen by its high specific heat, relatively low combustion temperature, and high energy content.[15]

The three high-speed turbines, on a single shaft, were geared down to drive a propeller. The conceptual engine was very compact (fig. 29). With both liquid hydrogen and liquid oxygen on board the aircraft, the turbine engine was independent of altitude. Rae proposed to use the turbine engine to drive a large propeller which provided the propulsive thrust by accelerating atmospheric air. The propeller, obviously, depended very much on altitude; the size of the propeller needed for thrust at high altitude later became an issue in evaluating the proposal. After pointing out the military advantages of a high-altitude aircraft, the brochure ended with a low-keyed request: "The Summers Gryoscope Company is desirous of obtaining a Government contract to develop the revolutionary REX-1 aircraft system."

As is usual in such cases, Rae left that day wondering how his proposal would be received, after the noncommittal attitude of the Wright Field listeners. In fact, his proposal caught the attention and interest of many in the Air Force and several

---

*Rae used an initial pressure of 69.7 atm, a final pressure of 0.67 atm, and a heat exchanger efficiency of 90%. He quoted an achievable specific fuel consumption of 1 lb/hp · hr (0.61 kg/kW · hr) and gave data indicating this could be attained with a four-stage turbine system with a turbine efficiency of 50%. He had analyzed both three- and four-stage turbines; by specific fuel consumption, he apparently meant both hydrogen and oxygen.

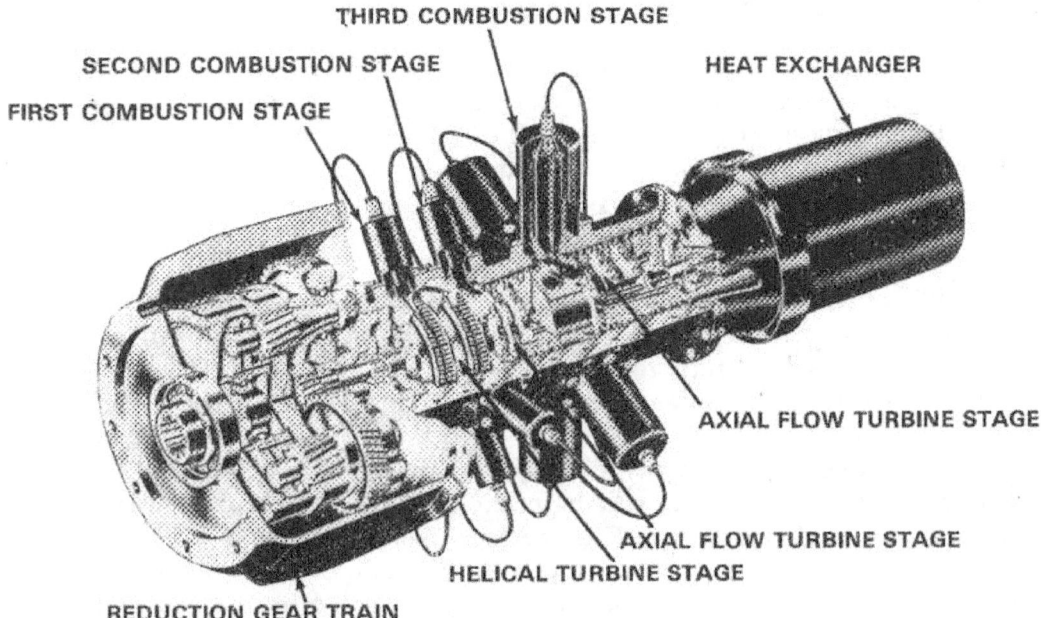

Fig. 29. Drawing of Rex I engine showing the heat exchanger, hydrogen-oxygen combustors, and three turbines. In the foreground is a reduction gear train transmitting power from the high-speed turbines to the engine application which, in the first proposal, was a propeller. From the brochure "REX-1, A New Aircraft System," by R. S. Rae, Summers Gyroscope Co., Feb. 1954.

analyses were started immediately. In response to a request for more information, Rae sent considerable detail about the proposed engine with a cost analysis. The cost was estimated to be on the order of $3 million a year for three years.[16]

In an analysis completed in May, Weldon Worth, R. E. Roy, and R. P. Carmichael examined propulsion aspects of the proposal and concluded: "There are numerous examples of optimism in the proposal but nevertheless, if the development does not bog down under adverse problems that result from impractical features, the small engine size and weight, the reasonably low fuel consumption, the high altitude combustion capability of hydrogen, and the surprising aircraft performance present a stimulating approach to a high altitude performance regime well beyond present aircraft capabilities." They added that there were other possible ways of achieving the same flight regime and discussed adverse technical factors that were based on hydrogen's characteristics and the possibility, from preliminary estimates of the propeller laboratory, that a much larger propeller than proposed might be necessary. Large, insulated lightweight tanks and a circulating gas–heat exchanger system were considered major development problems; these and a larger propeller or fan could substantially increase rize and mass of tankage, engine, and gearing between engine and propeller.[17]

In Rae's opinion, Wright Field's principal objections to his proposal centered on mass estimates and propeller size. He was kidded that his airplane would need a runway with trenches on each side of the wheels to accommodate propellers 12 meters in diameter. Rae believes he was vindicated later on both these points, but at the time he felt that the brickbats were coming at him thick and fast.[18]

Other Engines and Hydrogen Proposals

Rae's proposal to use liquid hydrogen as an aircraft fuel was, of course, not new nor was his engine the only possibility for using it. In 1937, von Ohain found that his experimental turbojet engine worked well on gaseous hydrogen (p. 73). In 1954, J. M. Wickham of Boeing studied the use of hydrogen for a strategic bomber powered by turbojets. For a subsonic cruise–supersonic dash flight, Wickham concluded that hydrogen gave a theoretical 30 percent increase in range over the use of a hydrocarbon fuel.[19]

Wright Field was also well aware of another type of engine capable of using liquid hydrogen which—like Rae's Rex 1—used liquid oxygen for combustion independent of altitude. This was the turborocket, a combination of rocket and air-breathing engine, which went back as far as a suggestion by Goddard (p. 74). During World War II, the Germans developed such an engine using a turbine, driven by decomposed hydrogen peroxide (steam and oxygen) to power an axial-flow compressor. The fuel was injected into the air stream and burned. The British had also investigated the turborocket by 1945, and Wright Field became interested after the war. Alfred M. Nelson, an analyst at Wright Field, reported his study of rocket-driven, turbine-compressor engines in December 1946 (fig. 30, top). Nelson described an engine where the rocket provides fuel-rich hot gases to power a turbine which drives a compressor. After leaving the turbine the fuel-rich gases burn in the air in the aft section of the engine. The hot gases expand through the exhaust nozzle to produce thrust. One of the best known champions of the turborocket was William C. House, who proposed a cycle in 1949 while an employee of the Aerojet Engineering Corporation.* House examined a number of bipropellants including liquid hydrogen and liquid oxygen. He apparently proposed this combination to the Air Force in September 1953 and later, but nothing came of it (fig. 30, bottom).[20]

With all this previous experience both in hydrogen as a fuel and in hybrid engines, why did the Air Force become so interested in Rae's proposal? The reasons came less from the technical interest of experts at Wright Field than from Air Force managers of research and development at Headquarters. They were under increasing pressure from other Air Force elements to develop means for increasing the operational altitude of aircraft. Rae's idea stirred interest because it was timely.

Air Force Evaluation of Rex I

In June 1954, Col. Omar E. Knox sent the WADC evaluation of Rae's proposal to ARDC Headquarters. Three laboratories, including the power plant laboratory, contributed to the evaluation. The basic engine was considered technically feasible, but considerable doubt existed regarding the technical feasibility of the propeller, hydrogen system, and airframe. If the airplane could be built as predicted, it would be

---

*Aerojet applied for and was granted a patent in Sept. 1950, but it was issued under a secrecy order because of potential military application. That order was removed and House received Patent 31 110 153 in Nov. 1963.

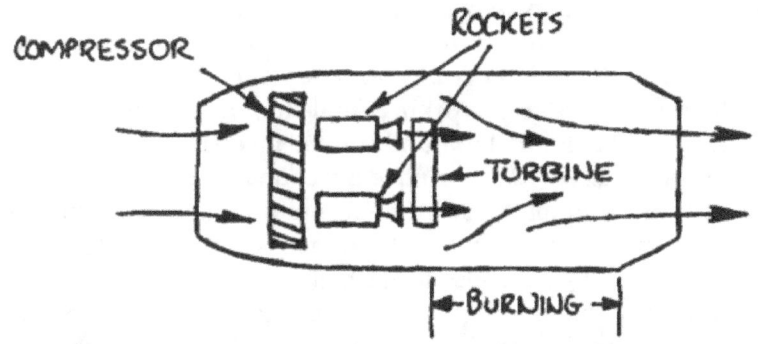

NELSON SKETCH OF TURBOROCKET, (1946). (COURTESY OF R.P. CARMICHAEL.)

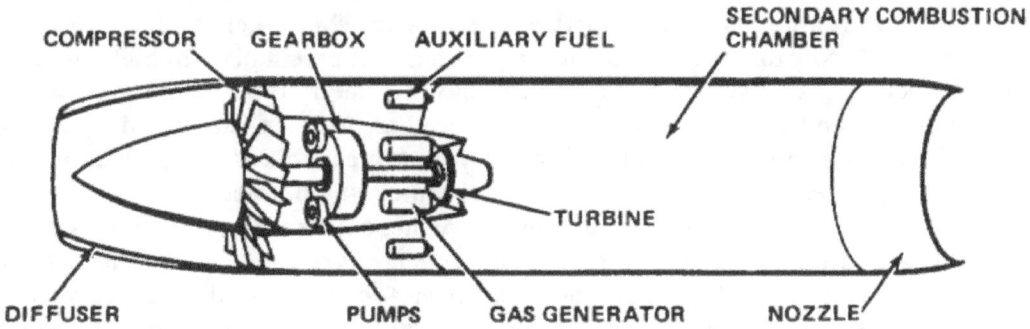

AEROJET SCHEMATIC OF AIR TURBOROCKET CIRCA 1964. (COURTESY OF W.C. HOUSE.)

Fig. 30. Schematic of air turborocket engines, also called simply turborockets, by Alfred Nelson in an Air Force memorandum in 1946 and from an Aerojet brochure ca. 1964. (William House of Aerojet championed this type of engine from 1949 into the early 1960s.) Rocket combustion gases drive a turbine which drives an air compressor. The fuel-rich turbine exhaust gases burn in the air stream and additional fuel is injected. These concepts, mentioned by Robert Goddard in 1937, were developed in Germany and England during World War II.

extraordinary in performance. Rae was praised as an imaginative and competent engineer, as evidenced by his contributions while working for the Applied Physics Laboratory and by the analysis he submitted. At the same time, however, the evaluators questioned the wisdom of placing development of an airplane with a company with so little systems capability. The cost estimate was considered unrealistic. Knox recommended against accepting the proposal, but suggested that ARDC look into overall propulsion and airframe problems of aerodynamically supported aircraft at extreme altitudes. That was exactly what Heaton and Ayers had been doing and why they were interested in Rae's proposal. They were not satisfied with WADC's negative response.[21]

In July 1954, Col. Paul Nay replaced Heaton as chief of the ARDC division of aeronautics and propulsion and was soon involved in concepts for achieving high altitude flight, including Rae's. On 9 August, Rae and Wood visited Nay and Ayers at ARDC headquarters in Baltimore and the following day Rae sent his proposal to Nay.

It contained details on program phasing and cost. Rae estimated first year costs at $1.9 million and annual costs for the next three years at $3.1 million. Included was a $50 000 sum to contract with Lockheed Aircraft Company for an airframe analysis. Rae stressed that hydrogen was more a working fluid than a fuel and that its complete oxidation was not desirable. He had done additional analysis, and the application for the long-range, high-altitude airplane included a reconnaissance radar picket as well as a bomber. Rae did not attempt to downgrade the potential problems and indeed mentioned several. He also pointed out that hydrogen could be used as a fuel in regular turbojet engines and that engine cycles using hydrogen had common elements that justified immediate component development. He requested the Air Force to provide a supply of liquid hydrogen in the Los Angeles area for component testing. To emphasize his point on the versatility of hydrogen, he sent the Air Force a report describing an engine cycle which later became known as Rex III.[22] It used air as the oxidizer and will be described later.

Rae's visit to ARDC and the revised proposal strengthened the belief of Nay and Ayers that Rae's concept should be further investigated. Three weeks after the visit, on 31 August 1954, Nay directed WADC to prepare a development plan for high-altitude engines, including the Rex I engine. He emphasized the need for long-range, high-altitude aircraft and argued that the optimum speed had not been established. This was a crucial point, for most of the emphasis was on aircraft capable of supersonic speeds, whereas the propeller-equipped Rex I was subsonic. Nay pointed out that WADC emphasis was on fans and compressors for jet engines rather than propellers, and the latter needed attention along with the hydrogen-oxygen reheat turbine cycles conceived by Rae. WADC should—as appropriate—conduct studies, experiments, and preliminary development of promising high-altitude propulsion systems, including Rex I. Summers Gyroscope was regarded as capable of analytical and experimental work with their existing facilities. WADC was encouraged to develop a working arrangement with an engine manufacturer and take maximum advantage of existing hydrogen technology, including rocket experiments. The directive was accompanied by a transfer of funds to accomplish it.[23]

The directive was clearly much broader in scope than merely contracting with Summers for the use of Rae's idea. The directive addressed the general problem of high-altitude propulsion, of which Rae's engine was one possible solution. While Summers was endorsed as capable of analytical and experimental work with their existing facilities, ARDC also suggested a working arrangement with an engine manufacturer. This constituted a limited endorsement and was not an arbitrary decision. Procurement experts had investigated Summers Gyroscope as a contractor, and the top procurement official of ARDC visited the company to satisfy himself about its capabilities for limited work on the concept.[24]

Other Reactions to Rae's Proposal

Wright Air Development Center took no action on the directive during the remainder of 1954, but there were other developments. The Fuels and Propulsion Panel of the Air Force Scientific Advisory Board considered the Rex I engine at its 29

September 1954 meeting.* The panel saw Rex I as an interesting cycle of potential importance and recommended that the development of non-air-breathing chemical engines should be actively pursued. Rex I was viewed as only one of several possibilities. The panel also recommended that a broad general study be made before development of the Rex I engine.[25]

According to Rae, the staff of the scientific Advisory Board asked him to go to the NACA Lewis Flight Propulsion Laboratory in Cleveland and give a briefing on Rex I.[26] He did so in November 1954, presenting the Rex concepts and various cycles to Abe Silverstein.†

The situation in late 1954 was tense for Rae. He had distributed fifteen copies of his brochure and backup technical data, given several briefings, and was aware that the Air Force was very interested. He had to defend his idea against a number of criticisms. He had conducted enough analysis to believe in the soundness of his approach and wanted support to develop it, but this appeared slow in coming. On the Air Force side, there was great interest in Rae's concepts—probably more than he suspected—for it touched on a critical need. The power plant laboratory, however, had reservations about the practicality of Rae's engine and how far to go with Summers Gyroscope as a contractor, and these points were clearly made in the WADC evaluation. The new development office of the weapon systems directorate, where Seaberg was pushing other high-altitude concepts, was negative about the Rex concept. Storm signals were flying for those perceptive enough to observe them.

Late in 1954, when Kelly Johnson was developing the U-2, Randy Rae was still seeking a way to get the Air Force to move on his proposal. It became clear to him that he needed to associate with a company having experience with turbines, the major component of his propulsion system. He knew Bertram N. Snow (1901–1966), dynamic vice president of the Garrett Corporation, makers of small turbines and many other components for the aviation industry. He approached Snow and later J. C. (Cliff) Garrett, founder and president of the company.[27] Garrett and Snow were very interested in Rae's ideas, but being shrewd and perceptive businessmen, they wanted to sound out Air Force interest in Rae's ideas and Garrett as a suitable contractor before they committed themselves. After assuring themselves on these questions, they began negotiations to acquire the Rex engine from Rae and Summers.[28]

WADC Response to ARDC Directive

Meanwhile, the power plant laboratory at Wright Field started actions responsive to the ARDC directive of August 1954. Four procurement requests were initiated during the first quarter of 1955. On 6 January, PR 303 was initiated with $750 000 for a

---

*Present were Prof. C. P. Soderberg (chairman), William M. Holladay, Allen F. Donovan, William D. Rannie, Addison M. Rothrock, Gale Young, and Mark Mills.

†An interesting speculation is whether Rothrock, who heard the Rex-I presentation at the 29 Sept. 1954 SAB meeting, transmitted information about it to Lewis earlier than November 1954, or asked the laboratory to investigate hydrogen, or asked the SAB staff to send Rae to Lewis. The last appears to be the most probable (p. 97). Rae's presentation intensified Silverstein's interest in hydrogen for aircraft, but was not the origin of his interest.

contract with Summers Gyroscope Company to explore the Rae concepts, including the study of an aircraft design. For unknown reasons, this procurement request was recalled and reinitiated with a new date, 10 March 1955. It became the subject of much controversy and negotiation.

On 14 January 1955, a second procurement request, PR 305, was initiated to investigate hydrogen as a fuel in conventional turbojet engines. Four engine manufacturers were listed as sources, but when the form reached Philip J. Richie, a procurement official of the power plant laboratory, he added five more to be solicited. The requests, sent out on 2 February, had a due date of 15 March. On 20 February, Richie received a puzzling directive from ARDC Headquarters: give the Garrett Corporation an opportunity to submit a proposal on PR 305. He reluctantly complied, but did not extend the due date. Garrett bid on this and later attempted to include the same kind of work in other proposals but was unsuccessful. On 15 June, PR 305 resulted in a contract with United Aircraft for $543 000.

In recognition of the unique properties of liquid hydrogen, the power plant laboratory initiated two procurement requests for studies of liquid hydrogen tanks and insulation on 25 March 1955. PR 338 resulted in a contract with Beech Aircraft in June for $172 000. PR 339 became a contract with the Garrett Corporation in October, but until then it was caught in the same web of controversy and negotiations with Garrett as PR 303.[29]

When PR 303, with Summers Gyroscope as the sole source, reached Richie in March, he decided that a talk with Thomas Summers was necessary. When Summers came to Wright Field, Richie was puzzled to find him in no hurry to submit the necessary proposal. He soon learned the reason. On 22 March, Richie was summoned to ARDC headquarters and learned that the headquarters procurement officer objected to PR 303; Garrett had acquired Summer's interest in Rex and was the company to deal with.* Richie also learned during his visit that Gen. Marvin Demler, Gen. J. W. Sessums, Col. Paul Nay, and other top officials at ARDC were very familiar with the Rex program and wanted a contract executed fast.[30]

High-Level Air Force Interest in Rex

The familiarity of top Air Force R&D officials with the Rex proposal and their desire for rapid contracting did not result solely from interest in a novel idea. The same month that Philip Richie learned of Air Force interest at ARDC Headquarters, the Fuels and Propulsion Panel of the USAF Scientific Advisory Board met at the RAND Corporation and considered superfuels.† The panel was impressed by the performance

---

*The date of Garrett's acquisition of Rex interests from Rae and Summers is not clear. An indenture and transfer agreement on the patents dated 18 March 1955 appears to be the earliest date. However, another indenture agreement was signed on 22 June 1955 from Rae to Garrett and Summers to Garrett. On 29 July 1955, an announcement was made at a meeting of Air Force and Garrett officials that Garrett had acquired the Summers interest in the Rex engine. Garrett File, AFSC, Andrews AFB.

†The attendees at the March 1955 meeting were Mark M. Mills (chairman), W. Duncan Rannie, Addison M. Rothrock (NACA), Edward S. Taylor, and Gale Young. Records of USAF Scientific Advisory Board, Pentagon.

potentials of two promising fuels—liquid hydrogen and boron hydrides. NACA experiments with hydrogen (pp. 97–98) were discussed and the panel recommended active development of hydrogen fuel systems and engine combinations, as well as preliminary design studies of aircraft to use these fuels. The panel also met with the SAB panel on intelligence to consider vehicle requirements. The fuels and propulsion panel concluded that the Rex engine might contribute to this application and recommended further study.[31]

The Air Force motivation for rapid action on high altitude aircraft stemmed from the U-2. Many in that service were unhappy having the CIA manage that aircraft. Even before it flew, there were discussions within the Air Force about a follow-on airplane. The possibility that the U-2 might get shot down was recognized early, so attention was focused on airplanes capable of higher speeds and altitudes. One of the problems foreseen for the U-2 was its vulnerability from engine flameout at high altitude.[32] If flameout occurred, the airplane had to descend to a much lower altitude—about 9000 meters—to restart the engine; at that time it was a sitting duck for antiaircraft fire.*

In addition to the flameout problem of the U-2, Kelly Johnson was faced with a problem of fuel loss from boil-off at very high altitudes. He had help on both problems from the Air Force and Pratt & Whitney, makers of the J-57-P37 engine. At the time, Col. Norman C. Appold, a combat pilot during World War II and holder of master's degrees in chemical and aeronautical engineering, was chief of the power plant laboratory at Wright Field. Earlier he had managed the Air Force contract with Pratt & Whitney for the J-57 and was very familiar with it. For this reason, and because he could draw on other propulsion and fuel experts in his laboratory, Appold was designated a "consultant" to Kelly Johnson. The father of the J-57 engine and chief engineer of Pratt & Whitney was Perry W. Pratt (no relation to the Pratt of P&W), and he too became closely involved with helping Johnson.

The J-57 turbojet engine normally operated on JP-4, a kerosene-like fuel. Johnson needed a fuel of lower volatility than JP-4 to minimize fuel loss during climb to the cruising altitude and during cruise. When the airplane took off, its fuel was at ground temperature. At high altitude, the combination of still-warm fuel and reduced pressure caused the more volatile portions of the fuel to boil away through the tank vents. Second, he needed a fuel with as high a combustion efficiency and flameout limit as he could get. Research showed that low volatility fuels had lower combustion efficiency than those of higher volatility, but this could be offset somewhat by improvements in the fuel injection system. Other research showed that fuels of low volatility had high flameout limits. In the end, Johnson, Appold, and Pratt selected a lower volatility fuel developed with the assistance of the Shell Oil Company research laboratories.[33]

During the course of studying the fuel-engine relationships for the U-2 and J-57, Appold and the fuel experts at the power plant laboratory considered a variety of fuels,

---

*On 3 May 1960, two days after Francis Gary Powers was shot down over Russia, NASA put out a press release stating in part that "the pilot reported over the emergency frequency that he was experiencing oxygen difficulties." Propulsion engineers familiar with the altitude performance of jet engines assumed Powers had a flameout and descended to a lower altitude to relight. Powers, however, insists that he was shot down at operational altitude. Gary Powers with Curt Gentry, *Operation Over-Flight: The U-2 Spy Pilot Tells His Story for the First Time* (New York: Holt, Rinehart and Winston), pp. 144, 201-202, 302, 323, 351-352.

including some of high volatility such as methane and liquid hydrogen. Methane was available in quantity, but liquid hydrogen was quickly dismissed because it was not.[34] This was the same period in which Rae was promoting the use of hydrogen in his Rex engine. Appold was well aware of Rae's Rex proposals and was involved in the actions regarding them. Sometime during the discussions between Appold, Johnson, and Pratt, the seed of the idea to use hydrogen was planted and grew.* It matured into action in 1956 as we will see later.

Long Summer of Negotiations

Stimulated by the high-level interest in Rex from his March 1955 trip to ARDC Headquarters, Philip Richie returned to his procurement duties at Wright Field's power plant laboratory expecting to let a contract on Rex within a month.[35] This was not to be. In fact, what followed was an extraordinary series of proposals by Rae and Garrett on the one hand and revisions of statements of work by personnel of the power plant laboratory on the other, with many negotiations between the two groups. These actions reverberated up and down the line, affecting virtually every level of management in Air Force research and development as far as the Assistant Secretary. At the root of the problem was a fundamental difference in approach between Rae and Garrett on the one hand and Appold and his associates at the power plant laboratory on the other. Rae insisted on a contract for the complete airplane powered by his turbine engine. This differed from the usual Air Force practice. An airframe manufacturer usually is the prime contractor for an airplane, including its tanks and fuel system, with the engine furnished either by the government or by an engine manufacturer, as a subcontractor to the airframe manufacturer. Garrett, as a manufacturer of aircraft components and small turbine machinery, had often been a subcontractor but went along with Rae's desire to obtain the complete aircraft contract. Obviously, Garrett intended, at some point, to either license or work jointly with an airframe manufacturer.

The Air Force, on the other hand, respected Rae's position as the originator of a novel solution to a difficult problem, but never viewed either him or Garrett as potential contractors for an entire airplane. The Air Force became extremely interested in hydrogen as a fuel and the Rex engine as a means for reaching very high altitudes, but was not fully convinced that either was practical. For these reasons, the power plant laboratory, not the weapon systems directorate, took the lead in initiating the purchase requests to explore the Rex concept and in dealing with Rae and Garrett. The laboratory wanted a step-by-step approach to determine the feasibility of using hydrogen and the Rex engine before initiating a large development effort. Necessary steps included a study of engine cycles, selection of the optimum cycle, and

---

*Neither Appold, Johnson, nor Pratt could recall definitely when or where the idea originated (interviews with Appold 4 Jan., with Johnson 14 Feb. and 2 May, and with Pratt 14 May 1974). The origin of the idea is less important than the interactions that occurred. Less than four months after the first U-2 flight (Aug. 1954), the NACA Lewis laboratory found that gaseous hydrogen in a turbojet combustor did not flameout as easily as jet fuel and could burn at pressures equivalent to 16000 m altitude (p. 98). No connection between the U-2 problem and the Lewis experiments has been established, but the timing is interesting.

experimental work on selected components including the fuel tank. The laboratory would review the work at each step before approving the next. This logical and conservative approach was irksome to Rae and Garrett, who were convinced they had a great idea and wanted to move fast to capitalize on it. They did decide, however, to propose a series of engines using hydrogen.

The negotiations with Garrett began on 20 April 1955 when Rae and Snow presented the Garrett proposal to Wright Field and followed it up two days later with a report. Their proposal went so far beyond what the laboratory had intended that one listener commented that it covered PR 303 "like the state of Texas covers Rhode Island."[36] Included in the proposal were three types of hydrogen-fueled engines called Rex I, II, and III.[37] All were jet propulsion engines; the propeller had disappeared. Rex I used liquid hydrogen and liquid oxygen to drive multiple turbines, with the hydrogen-rich exhaust gases dumped overboard. It was the same turbine system as Rae's original Rex I (pp. 119–121). The shaft power from the turbines was used to drive a fan which compressed incoming air (fig. 31, top). Thrust was obtained by expanding and accelerating the air through the nozzle.

Rex II was similar to Rex I except that the hydrogen-rich exhaust gases from the turbines were burned in the air in the afterburner (fig. 31, bottom). Rex II was essentially the same concept as a turborocket (p. 122).

Rex III was quite different from the other two engines. Liquid oxygen was not used, and the hydrogen served two different functions. First, heated hydrogen alone was used to drive the turbines; and second, the hydrogen was burned with air to provide the heat for the first function. This sounds like a man lifting himself by his bootstraps, but it works (fig. 32). Hydrogen from the tank is raised to a high pressure by a pump and passes through a heat exchanger where it is heated to a sufficiently high temperature to drive the first turbine. After leaving the turbine, it is reheated in a second heat exchanger and the process repeated for the third turbine. After the third turbine, the hydrogen enters a combustion chamber where it mixes with part of the engine air and burns fuel-rich. The hot combustion gases provide the source of heat for the three exchangers that heat the incoming hydrogen. After the third heat exchanger, the hydrogen-rich gases are injected and burned in the main air stream of the engine in the afterburner. The three turbines drive the compressors for the incoming engine air and the air used to burn the hydrogen.

The scope of the Garrett proposal of April 1955 became an issue between the company and the power plant laboratory, as negotiations continued. In early May, Rae complained to Brig. Gen. V. R. Haugen, director of laboratories at Wright Air Development Center, that the power plant laboratory had emasculated his program. Haugen investigated and satisfied himself that the laboratory's actions were proper and invited Rae to lunch in an effort to improve relationships.[38]

On 20 May, Garrett and Air Force officials met again. Some changes in the description of work were made by mutual consent. Garrett, willing to invest capital in developing Rex engines, sought a development contract, but Appold rejected this as untimely. Both parties, however, agreed on another matter: prompt action to ensure an adequate supply of liquid hydrogen.[39]

The government owned five acres of land within Garrett's facility at Phoenix, and this was studied as a possible site for a government-owned hydrogen liquefaction plant

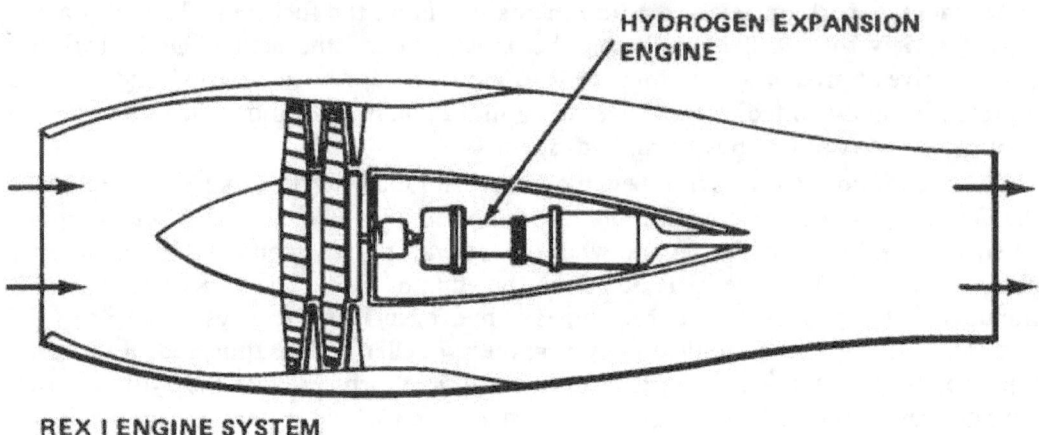

HYDROGEN EXPANSION ENGINE

REX I ENGINE SYSTEM

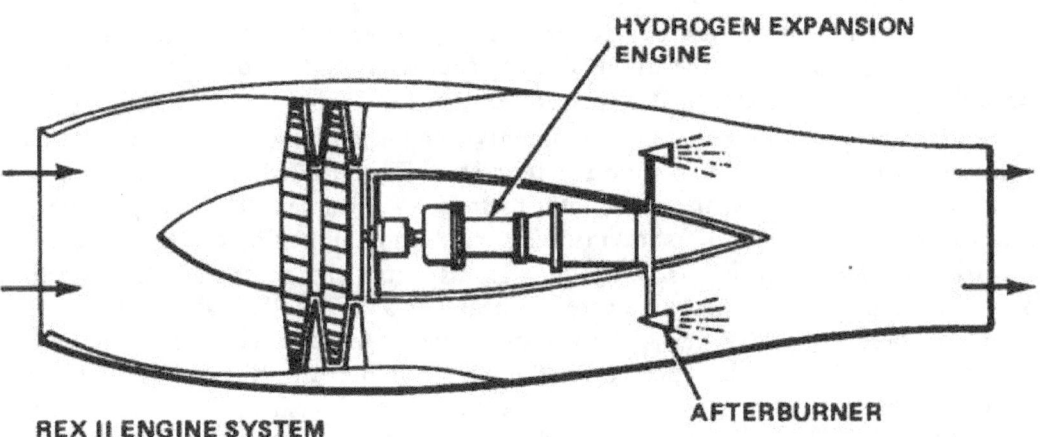

HYDROGEN EXPANSION ENGINE

REX II ENGINE SYSTEM   **AFTERBURNER**

Fig. 31. Rex I and II engine systems as proposed by Rae in 1955. The Rex I engine, first proposed in 1954, drove a two-stage air compressor and the air expanded through the exhaust nozzle for propulsive thrust. In the Rex II system, fuel was added to the airstream. Rex II was a form of turborocket that had been studied in Germany, England, and the U.S. in the 1940s. From R. S. Rae, "Various Engine Cycles Using Hydrogen as a Working Fluid and as a Fuel," Twelfth Annual Flight Propulsion Meeting, Institute of Aeronautical Sciences, Cleveland, 14 Mar. 1957.

for Garrett's experimental needs. In June, William C. Meister, a government industrial specialist, reported that the site was satisfactory. He also reported that liquid hydrogen plant details could be obtained "from standard plants built in the past."[40] He was probably thinking of the Bureau of Standards plant at Boulder or possibly the earlier Herrick L. Johnston plants, but none of these was "standard."

On 6 June the persevering Rae tried again to obtain acceptance of his original proposal for a complete airplane development but failed once more. The meeting ended with three unresolved issues: airframe work, use of hydrogen in conventional engines, and burning hydrogen in an afterburner, as proposed in Rex II.[41] On 27 June, Rae's frustrations must have reached the breaking point for in a meeting with Wright

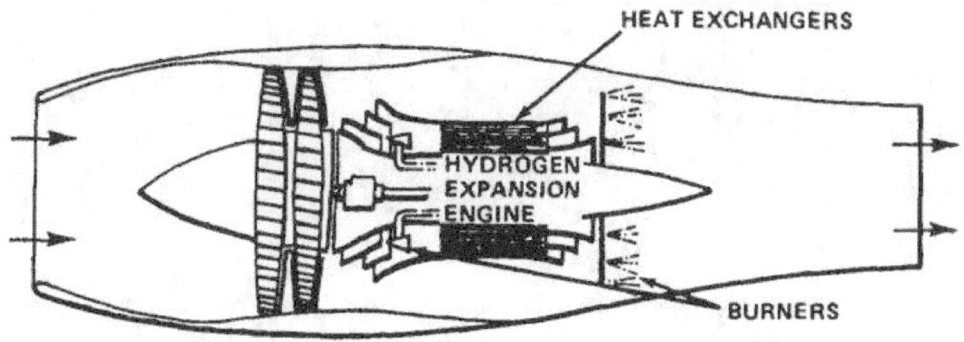

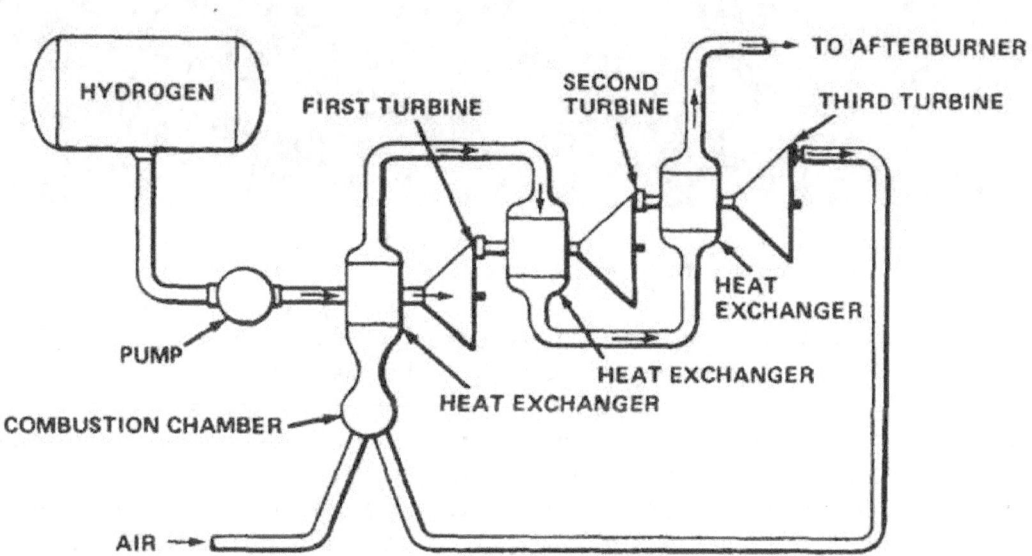

Fig. 32. Rex III engine system. Heat from burning hydrogen with air is transferred by means of heat exchangers to hydrogen on its way to the combustion chamber. The heated hydrogen drives turbines prior to combustion. There are three heat exchangers and three turbines. The turbines power a two-stage air fan or compressor. After leaving the last turbine, the hydrogen is injected and burned in the airstream in the afterburner and the expansion of the hot gas through the nozzle produces thrust. From R. S. Rae. "Various Engine Cycles Using Hydrogen as a Working Fluid and as a Fuel." Twelfth Annual Flight Propulsion Meeting, Institute of Aeronautical Sciences, Cleveland, 14 Mar. 1957.

Field officials, including a judge advocate, he refused to sign a contract with the Air Force, claiming that it neglected his patent rights.*

Meanwhile, individuals in other organizations were becoming interested in hydrogen. Silverstein of NACA had completed his analysis in April 1955 and

---

*On the same day, Rae's attorney filed a patent application for a multistage, high-altitude engine with a single combustion stage (518049). On 18 Oct. 1960, he was granted patent 2956402 for Rex III.

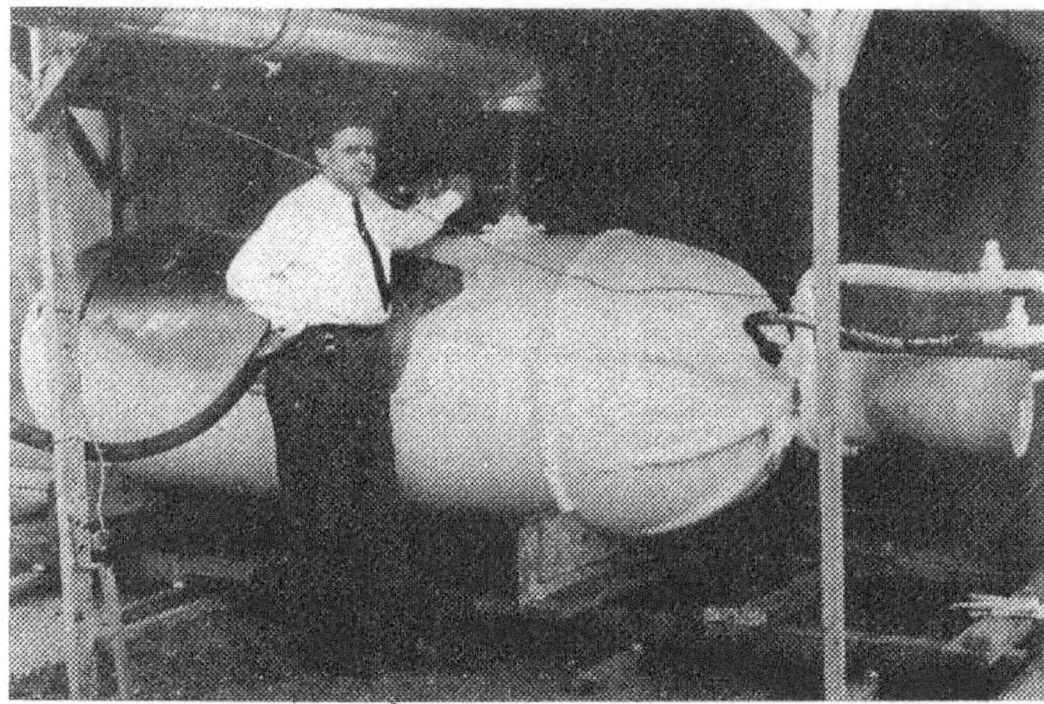

Fig. 33. Randolph Samuel Rae, creator of the Rex engines using liquid hydrogen as fuel, shown beside an experimental liquid hydrogen tank, ca. 1955. (Courtesy of R. S. Rae.)

according to one Air Force observer, "took to the road" making a circuit of high-level Air Force officials. One of these was Lt. Gen. Thomas Power, commander of the ARDC.[42] On 7 July, Power and Gen. Marvin Demler were briefed by WADC on the Rex program, with results that became evident from Appold's actions the following day. In discussions with his staff, Appold expressed concern about the Rex program, asked questions about the approach, the scope, and whether to go forward or cancel. He asked for a recommendation based on a comparison between the Rex and conventional engines available in the same time.[43] On 16 July, ARDC authorized WADC to study high-altitude engines with a two-pronged approach. One was to use conventional engines and the other, a new propulsion system for altitudes to 30 500 meters.[44] This essentially reiterated the ARDC directive of the previous year, but the sense of urgency had increased.

During July and August 1955, negotiations with Rae and Garrett continued without much success. On 25 August, Rae again refused to sign a contract, citing the inclusion of a study task as his reason. According to the notes of Frank Patella of the power plant laboratory who attended the meeting, Rae's position was: "The Garrett Corporation wants a development contract at this time and will not go along with anything less." Finally, however, after much additional negotiation and revision of work statements, two contracts with Garrett were signed in October 1955. One, coming from PR 303, was not far from what the power plant laboratory had originally specified. The other, from PR 339, was to study liquid hydrogen tanks. The two totaled $3 284 000—over four times the combined amount of the United Aircraft and Beech Aircraft contracts that had been in effect since June.[45]

When the Rex division of Garrett received its two contracts from the Air Force, after five months of hard negotiating, there was a big party and celebration. The staff was confident that they were at last firmly on the road to engine development and a great future.[46] Yet this was not to be, for the contract contained provisions that were to eventually knock Garrett out of the major competition.

Shift from Subsonic to Supersonic Aircraft

When Rae received the final work statement of the engine contract, he was disappointed to find that it specified only a supersonic airplane and a shorter range than he had been urging.[47] Rae's interest was in very long-range, high-altitude aircraft, with speed a secondary consideration. His early proposals for essentially a low-powered glider reflected this interest. His early engines used very small diameter turbines—on the order of 20 centimeters—and this was one reason he and Garrett had been attracted to each other, for small turbines were one of Garrett's specialties. During the long contract negotiations in the spring and summer of 1955, Rae still favored subsonic speeds, but the Air Force was more interested in a supersonic airplane at high altitude and would sacrifice range to get it. This was consistent with the Air Force desire for a superior airplane to supplant the subsonic U-2—superior both in altitude and in speed.

In the negotiations, Rae had gained funds for an airplane design study using the Rex engine. After receiving the work statement specifying a supersonic airplane, the Rex division prepared a "Problem Statement for Aircraft Studies" dated 7 November 1955 and negotiated a contract with Kelly Johnson at the Skunk Works. The problem statement specified Rex engines in pods for wing mounting. The size and weight of the pods with engines were given, as well as engine thrust and specific fuel consumption.[48] The pod diameter, essentially that of the engine inlet, was 122 centimeters, which meant an engine much larger than Rae's original concept; but the engineers of Garrett's Rex division did not feel that the larger sizes would be a major problem to develop.

During the course of the study, the engineers at the Skunk Works found that the thrust specified by Garrett was too low for their design needs. Agreements were reached with the Rex division on extrapolation of the engine data for engines of larger thrust and data giving specific fuel consumption as a function of Mach number. With these, the designers at the Skunk Works chose a cruise speed of Mach 2.25 and an engine with a thrust 50 percent higher than Garrett had originally specified. The larger thrust meant an engine with an inlet diameter of 150 centimeters. The engine now was considerably larger than machinery within Garrett's experience, but this did not deter the company. Some individuals within the Air Force, however, began to doubt whether Garrett was the best contractor for the engine.

The Lockheed study of a hydrogen-fueled airplane was completed and reported by the end of January 1956. Two configurations, both powered by Rex III engines of the larger thrust, were selected. The preferred design, designated CL-325-1, had a straight, thin wing and a slender fuselage. It was made of aluminum alloy with a single liquid hydrogen fuel tank in the fuselage. The second configuration, CL-325-2, used droppable auxiliary wing tanks, which reduced the airplane size and weight by about 15 percent. The CL-325 wing was lightly loaded, i.e., the wing had to support a comparatively low weight per unit area. Its thrust per unit weight was also low, which required a long runway for takeoff. The characteristics of the CL-325, taken from the Lockheed report, are given by table 4.[49]

The Garrett engine contract, which began 15 October 1955, called for engine analysis and selection of the optimum one. The Rex division of Garrett, however, had been at work for some time before the contract and had already selected Rex III and had specified it for the Lockheed study beginning in November.[50]

The first Garrett engine report, covering the first four months of the contract, was "Rex Engine Cycle and Selection," 15 February 1956, with the Lockheed report included as an appendix. The report concluded that the Rex III was the optimum engine and that Air Force mission requirements could be met. It also concluded that a Rex III with a thrust of 17800 newtons at 30500 meters altitude and Mach 2.25 was feasible; its specific fuel consumption would be less than 1360 kilograms per hour and dry mass less than 1995 kilograms. Other design data were given. The engine described had a thrust slightly greater than that selected by Johnson in the airplane design and reflected Garrett's confidence that they could develop the larger engine.

Garrett sent a team to Wright Field on 15 February to give an oral summary of the report. The members of the team were in high spirits, looking forward to a favorable reaction. They had been told that if the audience did not ask a lot of questions, it meant

TABLE 4.—*Characteristics of CL-325-1 Hydrogen-Fueled Airplane*

Dimensions	
Length, m	46.73
Wing span, m	24.35
Height, m	8.71
Wing area, sq. m	209
Wing aspect ratio	2.5
Mass, kg	
Take-off	20731
Landing	14486
Empty	13352
Payload	680
Liquid hydrogen	6553
Engines	
Number	2
Type	Rex III
Thrust, each engine, take-off and climb, N	20016
Rated thrust, each engine, at 30500 m, N	16680
Performance	
Take-off distance, m	1402
Rate of climb at sea-level, m/s	8.9
Equivalent air speed during climb, m/s	79
Cruise Mach number	2.25
Radius (to target), km	2797
Climb and descent distance, km	556
Landing distance, m	640
Stall speed at landing, m/s	37

From Lockheed Aircraft Corp. report 11195, attached to report by J. L. Bartlett, Jr., I. M. Goldsmith, and A. Shaffer, "Rex Engine Cycle Study and Selection," report RD-14-R, Rex Div., Garrett Corp., 15 Feb. 1956.

trouble. Not many questions were asked; and that night, in the hotel room, the members of the team sat around trying to figure out what had gone wrong.[51]

What had gone wrong was that members of the audience, particularly those in the power plant laboratory, were beginning to realize that Garrett was talking about a whole new ball game. Gone was the simple, ingenious, new engine with its small turbines. In its place was a large, complex engine, with compressors and turbines about the size of those in contemporary turbojet engines. In addition, there were heat exchangers using hot combustion gases, something no one had yet attempted to develop. Rex III was considerably more complex than a turbojet engine and had other problems. Frank Patella, the laboratory's contract manager for the Rex engine study, noted in his log that Garrett apparently did not realize the problems involved in the proposed three-speed gearbox or the heat exchangers.[52] The Wright Field experts may also have been troubled by Rae's assurance that liquid hydrogen production facilities would cover the development needs for the larger engine.

Garrett Loses the Fight

The 15 February 1956 presentation was the turning point. Garrett's relationship with the Air Force would be downhill from that time on. Both sides had begun the relationship with great expectations and in good faith; but step by step, the size and complexity of Garrett's proposed engines grew and ultimately destroyed the company's prospects. On the Air Force side, the need to involve a well-established engine manufacturer was seen as early as August 1954, but it took time for this position to become the dominant consideration.

Nine days after Garrett's presentation; members of the power plant laboratory staff reported to Col. Harold Robbins of ARDC headquarters. Robbins, in turn, was to brief Gen. Thomas S. Power, ARDC's commander, on 27 February. Frank Patella was among those who talked with Robbins and he noted in his diary the main points of the briefing: Garrett did not have the facilities for component tests, the tools to manufacture, or the experience needed for the large engine, Rex III. There was considerable doubt among the Air Force propulsion experts that Garrett could develop such capability in time to meet the urgent need, for all agreed that the development of a special engine for high altitude operation merited a crash program. Robbins presumably conveyed these conclusions to Power in his briefing three days later.

Back at Wright Field, power plant experts continued their analysis of the Rex engine and Garrett's capabilities to develop it. On 1 March 1956, General Haugen, commander of WADC, was briefed on the Rex situation by B. A. Wolfe of the power plant laboratory. The Rex I engine, 60 centimeters in diameter, had been considered within Garrett's capabilities; but the 150-centimeter Rex III was clearly beyond Garrett's development and production facilities. The consensus was that to proceed with the Rex III development by Garrett would be sheer folly. On 5 March 1956 another conference took place, this time between working level personnel from the development and material groups. The participants again concluded that Garrett did not then have the capability to develop the engine and that it might take 8–10 years to develop it unless a crash program were started. The power plant laboratory representatives pointed out that other engine manufacturers were now proposing essentially the same type of engine as Rex III.[53] The problem was being studied at all Air Force levels, from headquarters on down.

In the midst of these conferences, Garrett received a serious blow. The company had been urging the government to provide a liquid hydrogen plant on the government-owned land within their Phoenix facility. The city of Phoenix learned of this proposal and acted to prohibit Garrett from working with hydrogen within the city limits. This development set off a new round of conferences and staff studies within the government. On 22 March 1956, Lt. Gen. Power, commander of ARDC, recommended to the Air Materiel Command that a favorable atmosphere be sought whereby a major engine manufacturer could acquire Garrett's and Rae's interests in the Rex engine. If this could be done, the Garrett contract could be terminated and proposals solicited from major engine manufacturers.[54] Power's proposal was backed by a detailed staff study.[55] Philip Richie, at the working level in AMC, objected to the conclusions of the staff study and wrote a point-by-point rebuttal. Richie recommended that a committee be appointed to make a detailed study of the problem and use it either to convince Garrett that it was in the country's best interest to go elsewhere or use it to explain to others why the Air Force stayed with Garrett. On 10 May, Richie's recommendation was backed by his boss, Col. Merle R. Williams, WADC procurement chief, so the Air Force remained locked in an internal struggle over what to do about Garrett.[56]

Meanwhile, Bertram N. Snow, vice president of the Garrett Corporation, wrote to WADC on 12 May 1956, pointing out several problems. A remote facility was required to test with hydrogen; since none was presently available, there would be a considerable delay in carrying out the existing contract. He proposed to amend the contract to allow engine development of a prototype meeting the 50-hour test specification and to authorize the necessary facilities. If a hydrogen generating plant could not be provided by the government, Snow proposed to try for a commercial product on contract. He estimated that the prototype engine could be developed in four years, with a program and facility cost of $72.5 million. Garrett followed up this proposal with a presentation two days later: the Garrett board of directors had decided that while the company could not handle large-scale production of the engine, it could handle limited production. Two engine manufacturers had made overtures to Garrett but had been rebuffed.[57]

Sometime in the spring, the perceptive Snow sensed the changed Air Force attitudes towards Garrett. He and Rae visited General Power in Baltimore to protest. Power listened to them and promised that they would receive a reply, but that it would come from General Rawlings of the Air Materiel Command. The meeting with Rawlings was held on 18 May 1956, and Snow was told bluntly that timely and successful development of the proposed engine could be done only by a major engine manufacturer. On 18 June, Snow wrote a strong letter of protest to Assistant Secretary of the Air Force for Materiel, Dudley C. Sharpe, stating Garrett's position and included a chronology of events. He made five points: (1) Garrett owned patent rights, (2) the Air Force had encouraged Garrett to develop the Rex engine, (3) Air Force working level organizations had ignored Garrett's proprietary rights.* (4) Garrett's performance had been satisfactory, and (5) Garrett was willing to negotiate in the public interest. The thrust of Snow's letter was two proposals that eliminated the need for the government to furnish facilities. The first proposed that Garrett be given the prime contract for engine development, and Garrett would subcontract to a larger engine manufacturer any work it could not handle. The second proposed that Garrett be given the engine production contract and if production needs exceeded what Garrett could provide with its own resources, then Garrett would license a larger engine manufacturer to make the additional units.[58]

This appeal to Sharpe and visits to high-level government officials by Garrett officials did little to resolve the basic issues. Although Garrett continued working on its original contract, with several extensions, that work was essentially out of the mainstream of Air Force R&D projects. Phase I of the Garrett contract, engine analyses and selection, had been completed and presented on 15 February 1956.[59] Phase II, a thorough and comprehensive preliminary design study of the Rex III engine, was completed in May 1956.[60] Phase III, to design, fabricate, and test combustors, turbines, fuel pumps, gear boxes, and heat exchangers, continued until 1 February 1957, when the objectives were revised to be general research and development rather than specific to the Rex III engine design. On 18 October 1957, Garrett received a directive to stop all work on the contract except for the preparation of a final report. This report, in several volumes, was completed in 1958.[61]

What Went Wrong

What went wrong in the Rae–Garrett–Air Force relationship? Lt. Col. Langdon F. Ayers, who was in the midst of the Rex events from beginning to end, summarized his view in August 1956. He saw Rex as the "classical example" of the problem of exploiting innovations. He believed that established, old-time engine companies were not likely to recognize or develop innovations because of vested interests. When an individual proposed a promising engine innovation, Ayers thought that the government should move the innovation, as fast as possible, to an established engine company and reward the inventor.[62]

With the perspective of time, it is easy to see the errors made on both sides, but what can we learn from them? How can promising innovations be nurtured until they develop into a benefit for both originator and sponsor? There are no easy answers, but a few observations can be made.

An idea or concept in itself is of little or no value until it is transformed into something people need or want. In our free enterprise system, an innovator must develop his idea or else seek a suitable sponsor who then takes the risk of development. The development of an innovation as complex as an aircraft engine or an airplane requires considerable capital for facilities, equipment, and operating funds. If a sponsor already has these and is willing to use them to develop an innovation, the

---

*This was apparently with reference to an analysis by R. P. Carmichael, which will be discussed later. Garrett was also disturbed over the government's attitudes towards its patent rights.

innovator has made a fortunate alliance. Rae's case is tragic in that he twice chose sponsors who did not have the capability to meet his goals. Summers Gyroscope was clearly not suitable for more than studies and small component work. Garrett was suitable for developing small machinery and was willing to invest some of its own capital to expand, but it looked to a sponsor of its own, the government, to provide additional facilities and a development contract. The government was interested in Rae's idea but was not willing to sponsor the development of a company to exploit it, especially when there were suitable companies available. The government's position was sound, but perhaps errors of judgment were made in encouraging Garrett and later, when a contrary view became prevalent, of not promptly informing the company. Perhaps Rae, as inventor, and Garrett, as a company ready to develop the invention, erred in seeking to make too big a step in the beginning. They seemed unaware of the danger of proposing larger and larger engines until they found themselves out of the ball park. The urgency felt by the Air Force to develop an airplane superior to the U-2 settled the matter.

How can the government benefit from the ideas of lone inventors? This has been the subject of much study and a single case history can scarcely provide the answer. The Rex history does show, however, that the choice of a sponsor to exploit an innovation is all-important and that a goal may sometimes be reached better by a series of small steps rather than a gamble on one giant leap.

Other Interests in Hydrogen

During the last quarter of 1955 and concurrent with Garrett's Air Force contract, two other events occurred to broaden interest in hydrogen for aircraft.

In October, the Fuels and Propulsion Panel of the USAF Scientific Advisory Board met and considered superfuels and reconnaissance vehicles.* On superfuels, the panel noted that hydrogen was one of three main lines of attack. It was most anxious to see engine studies and preliminary aircraft design studies directed towards application of hydrogen for aircraft propulsion. Further studies by the NACA since the March meeting continued to show excellent combustion characteristics of hydrogen. The panel believed that power plant development using hydrogen would encounter minimum difficulties, but an aircraft to use low-density fuel would require substantial redesign. Also noted was a need to study the possible adaptation of hydrogen to current aircraft or missiles.[63]

Anticipating the panel's conclusions, the Air Force included $4.5 million in the FY 1957 budget request for development related to hydrogen, a substantial increase over the $1 million of the previous year.

In November, Wright Field issued a technical note, "Cycle Performance of Some Selected Engine Configurations Using Liquid Hydrogen Fuel," in which Robert P. Carmichael analyzed nine engine systems using hydrogen as a coolant and a turbine working fluid as well as a fuel.[64]

---

*The meeting, held on 21 October 1955, was attended by Mark M. Mills, chairman, W. D. Rannie, Addison M. Rothrock, E. S. Taylor, and Gale Young. SAB files, Pentagon.

He also compared the performance of these engines with a conventional turbojet. Among his conclusions: some of the hydrogen engines gave superior performance compared to conventional turbojets; precooling the incoming air with liquid hydrogen increased the mass flow through the engine; both precooling and use of hydrogen turbines increased combustion pressure and permitted operation at higher altitudes with smaller combustors than conventional jets using hydrocarbon fuels.

Carmichael's analysis was distributed to nine major aircraft engine manufacturers and seven airframe manufacturers, causing Garrett to complain that their proprietary rights had been violated (pp. 136–37).* In spite of the complexity of some of the hydrogen engines, the note must have stirred up considerable interest in hydrogen within the aeronautical community.

Summary

The Air Force began planning work to achieve very-high-altitude flight in late 1952 and this resulted in successfully modifying the Martin RB-57, a later version of which is flying today. In 1954 Kelly Johnson of Lockheed, famed airplane designer, proposed a high-altitude reconnaissance airplane that was sponsored by the government. This became the U-2, which is also still flying.

In 1954, Randy Rae proposed a novel hydrogen-fueled subsonic airplane capable of high-altitude flight. Although never built, it spawned considerable interest and activity on the potential of hydrogen as a fuel. The Garrett Corporation acquired Rae's interests and pressed the Air Force for a contract to develop the airplane and its engine. but received only a study contract and some component work. As interest grew and specifications changed from a subsonic to supersonic airplane, the required engine power increased, which meant a much larger hydrogen-fueled engine than Rae originally envisioned. The growth in engine size effectively took Garrett, a maker of small turbines and aircraft components, out of the competition. The government considered it inappropriate to set up Garrett as a manufacturer of large aircraft engines when several capable and well-established companies were willing to do the same thing. Rae and Garrett placed reliance on their patents, and their relationships with the government made a case study of the frustrations of an innovator with a single customer and needing large resources. Some benefits resulted, however; by the end of 1955, interest in using hydrogen in aircraft had grown considerably.

---

*The allegation is questionable in view of Wright Field's long background in turborockets and research on hydrogen. The official response to Garrett indicated that no proprietary data had been used in the Carmichael report and that engine cycles in general are not proprietary.

8

Suntan

The largest and most extraordinary project for using hydrogen as a fuel was carried out by the Air Force in 1956–1958 in supersecrecy. Very few people are aware of it, even now, yet over a hundred million dollars were spent—perhaps as much as a quarter of a billion dollars. Although the project was cancelled before completion, it led directly to the first rocket engine that flew using hydrogen. The project was code-named Suntan, and even this was kept secret.[1] It had all the air of cloak and dagger melodrama and indeed, its principal precursor was just that. Suntan was an effort by the Air Force to develop a hydrogen-fueled airplane with performance superior to the secret spy plane, the U-2.

Suntan had its roots in Air Force interest in very high-altitude flight during the first half of the 1950s. One approach, along conventional lines, was pushed by Maj. John D. Seaberg of the Wright Air Development Center, beginning in late 1952. This involved a modification of the Martin RB-57 and the start of the Bell X-16, although the latter was cancelled in mid-1955. A different approach, sparked by a proposal by Randolph Rae in 1954 to build a glider-like airplane powered by the Rex engine, focused on the potential advantages of using liquid hydrogen. The Air Force interest in hydrogen was supported by Abe Silverstein, associate director of the Lewis laboratory of the National Advisory Committee for Aeronautics.

By the end of 1955, the Air Force had in progress a number of research and development activities on the feasibility of using liquid hydrogen in flight. The Garrett Corporation, which bought Rae's patents and formed a Rex division with Rae as chief engineer, was three months into a contract for design studies of Rex engines and had concentrated on the largest and latest, the air-breathing Rex III. Kelly Johnson's Skunk Works at Lockheed Aircraft, past their peak effort in designing and building prototype U-2s for the CIA, was two months into a three-month design study of hydrogen-fueled aircraft for Garrett. United Aircraft (now United Technologies) was in the second quarter of a study of using hydrogen in a conventional turbojet engine, and a competitor, General Electric, was also showing interest in hydrogen. Beech Aircraft and Garrett were investigating liquid hydrogen tanks, insulation, and behavior of hydrogen in storage. The Air Force and NACA agreed that the Lewis laboratory would determine the feasibility of flying an airplane fueled with liquid hydrogen. The Air Force would provide the estimated $1 million needed, as well as lend equipment.

The driving force behind the Air Force's mounting interest in hydrogen was the determination to develop an airplane with performance superior to the U-2. Dissatisfied with its supporting role to the CIA, the Air Force sought not only to take over the operational phase of the U-2 but also to regain the initiative in equipment by developing a second-generation airplane. One prospect was the Rae-Garrett proposal, but that approach did not seem quite the right answer. In late 1955, the time was ripe for a new proposal, and soon one was made by Kelly Johnson. He was immediately seen as the right man with the right idea.

Air Force Moves Fast

The high-flying U-2 was the latest symbol of Johnson's ability to design and build a new airplane quickly in his unique and unconventional Skunk Works. Familiar with hydrogen from conducting airplane design studies for Rae and Garrett, Johnson was impressed with its potential. Early in 1956, armed with a proposal for a hydrogen-fueled supersonic airplane as a follow-on to the U-2, he visited the Pentagon where he had no difficulty seeing high Air Force officials, including Lt. Gen. Donald L. Putt, the deputy chief of staff for development.[2] Johnson offered to build two prototype hydrogen-fueled airplanes, with the first to fly within 18 months. They would fly at an altitude of 30 300 meters, a speed of Mach 2.5, and have a range of 4070 kilometers.[3] To the Air Force, which had missed the opportunity to buy Johnson's original U-2 proposal, the offer was too tempting to resist; they bought it.

New airplanes, however, are not bought without due deliberation. The Air Force went through the proper motions, but the circumstances made the outcome a foregone conclusion. After receiving the proposal, Putt called a meeting on 18 January 1956. Among those present were his counterpart for materiel, Lt. Gen. Clarence S. Irvine; Lt. Gen. Thomas S. Power, head of the Air Research and Development Command; and Col. Norman C. Appold, head of Wright Air Development Center's power plant laboratory. The purpose of the meeting was to evaluate Johnson's proposal, but in his opening remarks, Putt made it clear that the Air Force wanted a new high-altitude airplane within two or three years, whether or not it was the one that Johnson proposed.[4]

The short time that Putt specified was in keeping with Johnson's reputation, but incredibly short if liquid hydrogen, with its array of formidable problems, was to be the fuel. Engine development was considered the pacing item and the reason Appold, the Air Force's chief propulsion expert, had been summoned to the meeting.

Putt wanted six months of study and experimentation to determine the feasibility of attaining the performance goals specified by Johnson. He named Col. Ralph Nunziato, a former test pilot and a member of his staff handling intelligence-gathering equipment, to be his project officer. Power named Appold to head the ARDC team, a clear indication of the critical importance of the propulsion system to the overall effort.

Appold's first assignment was to select a qualified engine manufacturer to study a hydrogen-fueled engine and if feasible, develop it. Given a month to do this by Putt, Appold selected two candidates: the General Electric Company and the Pratt &

Whitney division of United Aircraft. He met with their representatives,* asked for and received proposals within two weeks, evaluated them, and selected Pratt & Whitney. He reported his actions at another meeting in Putt's office on 20 February, and the selection was approved.[5]

Contract negotiations with Pratt & Whitney started early in April and by the first of May, a six-month contract had been signed. Agreement was also reached with Lockheed. Officials of Pratt & Whitney, impressed with the potential of hydrogen and wishing to avoid the red-tape of a cost-plus-fixed-fee contract, agreed to a fixed cost contract. As it turned out, their costs exceeded the fixed amount and Pratt & Whitney lost money.† Lockheed held out for a provisional contract that could be renegotiated and repriced at the end of the contract. Both firms, however, were hard at work by the first of April 1956. The contracts were made retroactive, to cover the fast start.[6]

In the weeks that followed the initial meeting in February, Appold and Nunziato were very busy dealing with the two companies and consulting with specialists at the Wright Air Development Center on the feasibility of providing large quantities of liquid hydrogen. Although Appold continued as head of the power plant laboratory, it was clear that his new assignment would soon require full attention, as well as a staff. He chose Lt. Col. John D. Seaberg, the aeronautical engineer assigned to weapon systems who had started work on high-altitude aircraft in 1952 (pp. 113–14), to manage work on flight-type liquid hydrogen tanks, airframe, and complete airplane systems. Major Alfred J. Gardner, a combat pilot during World War II, holder of two master's degrees in engineering, and a propulsion specialist, was chosen to manage the engine development. Capt. Jay R. Brill, West Pointer, mechanical and nuclear engineer, would manage the logistics, including the quantity production of liquid hydrogen and its storage, transportation, and handling. The team worked initially at Wright Field and moved to ARDC headquarters in Baltimore in June, as a special projects office.[7]

Considering the highly classified U-2 and the Air Force's desire to build a superior airplane, it is not surprising that the new project was very closely held. It was given a special classification higher than "Top Secret," the highest standard category. Full access was limited to about 25 people, an extremely small number considering the size and complexity of the large research and development effort.[8]

Two compelling reasons beyond technical management and Air Force security called for a special projects office: fast contractual action and contractor security. To get an airplane developed in the two or three years that Putt demanded meant by-passing the normal, but time-consuming, management and procurement processes. Appold turned to Col. Lee Fulton, head of procurement at ARDC Headquarters and his deputy, Robert Miedel, for help; Miedel served as temporary procurement officer. They soon had a blanket "determination and findings" statement from Richard Horner, assistant secretary of the Air Force for research and development, and

---

*Jack Parker, Gen. Mgr., Aircraft Gas Turbines Div., General Electric Co., and Charles Dribble, a G.E. engineer; Wright Parkins, William Gwinn, and Perry Pratt of Pratt & Whitney.

†Pratt & Whitney received $15.3 million for the first phase of work and spent $17.1 million. Interview, Ernest Schweibert with Lt. Richard Doll, Dec. 1958.

directives from the Air Force deputy chiefs of staff for development and materiel, Putt and Irvine. These authorities allowed the Suntan team to waive normal procurement procedures and award contracts directly, with a minimum of review. This cut months from the procurement process.

Miedel bowed out in June 1956 by appointing William E. Miller as contracting officer and negotiator on all Suntan contracts, and Lt. Col. J. R. Beyers as head of contract management. Two special auditors were assigned by the Auditor General. Miller's group also handled property and contractor security.[9]

Extraordinary measures were taken to conceal Suntan from the curious and the unauthorized. The Suntan team at ARDC changed project numbers from time to time; some contracts were written through other Air Force offices, so they could not be related to Suntan. At contractor plants, Suntan workers were isolated and guarded from other units and operated as independently as possible. Special measures were taken to prevent identification of Suntan visitors by those not connected with the project.* Documentation was kept to a minimum.[10]

Lockheed CL-400

The initial contract with Lockheed called for two prototype reconnaissance aircraft, with the first to fly in 18 months. Hard on its heels, also in 1956, Lockheed received a contract for six of the aircraft. The design Lockheed selected was designated CL-400 and was capable of a speed of Mach 2.5 at an altitude of 30 300 meters.[11] The CL-400 was described openly for the first time in 1973 by Ben Rich at a symposium on hydrogen-fueled aircraft at the NASA Langley Research Center. Figure 34, taken from his paper, shows the characteristics of the CL-400. It had a fuselage diameter of three meters and a length of 49 meters to accommodate the 9740 kilograms of liquid hydrogen. The retractable ventral (bottom) fin improved directional stability at supersonic speeds.

The engines, designated 304-2, were to be supplied by Pratt & Whitney and will be described later. Each weighed 2850 kilograms, provided 42 kilonewtons at sea-level, and 27 at Mach 2.5 and 29 000 meters altitude.

The mission profile is shown by figure 35. The range was 4070 kilometers and could be extended only by a considerable increase in airplane size. Airplane sizes with lengths as long as a football field, as well as other variables, were studied at the Skunk Works. The relatively short radius of 2000 kilometers was later to become a matter of great concern.

---

*Of numerous stories of security incidents, one of the most interesting involved a good-looking female engineer of the Skunk Works who almost—and inadvertently—blew Suntan's cover. She attended a symposium on hydrogen at the NBS Cryogenics Laboratory and following established practice of the Skunk Works, registered as representing herself. Standing nearby was a male engineer who knew she worked for Lockheed but had forgotten her name. He peeked at the register and immediately grew suspicious, wondering why Lockheed was interested in hydrogen and hiding it. Interview with Col. Gardner, 19 Sept. 1973.

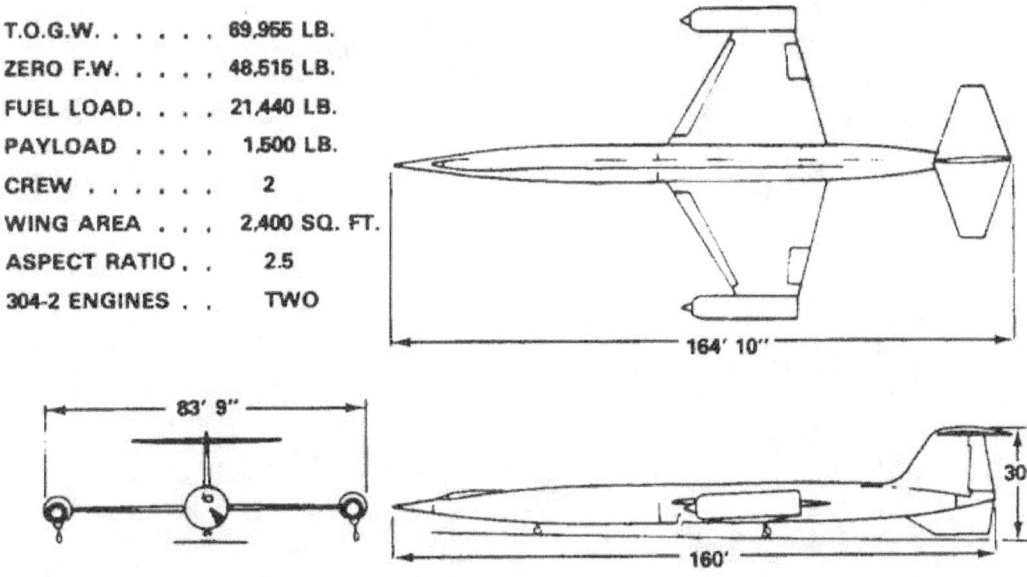

T.O.G.W. . . . . . 69,955 LB.

ZERO F.W. . . . . 48,515 LB.

FUEL LOAD. . . . 21,440 LB.

PAYLOAD . . . . 1,500 LB.

CREW 2

WING AREA . . . 2,400 SQ. FT.

ASPECT RATIO . . 2.5

304-2 ENGINES . . TWO

164' 10"

83' 9"

30'

160'

Fig. 34. Lockheed CL-400 reconnaissance aircraft using liquid hydrogen as fuel, ca. 1955. Ben R. Rich, "Lockheed CL-400 Liquid Hydrogen-Fueled Mach 2.5 Reconnaissance Vehicle," read at a symposium on hydrogen-fueled aircraft, NASA Langley Research Center, 15–16 May 1973.

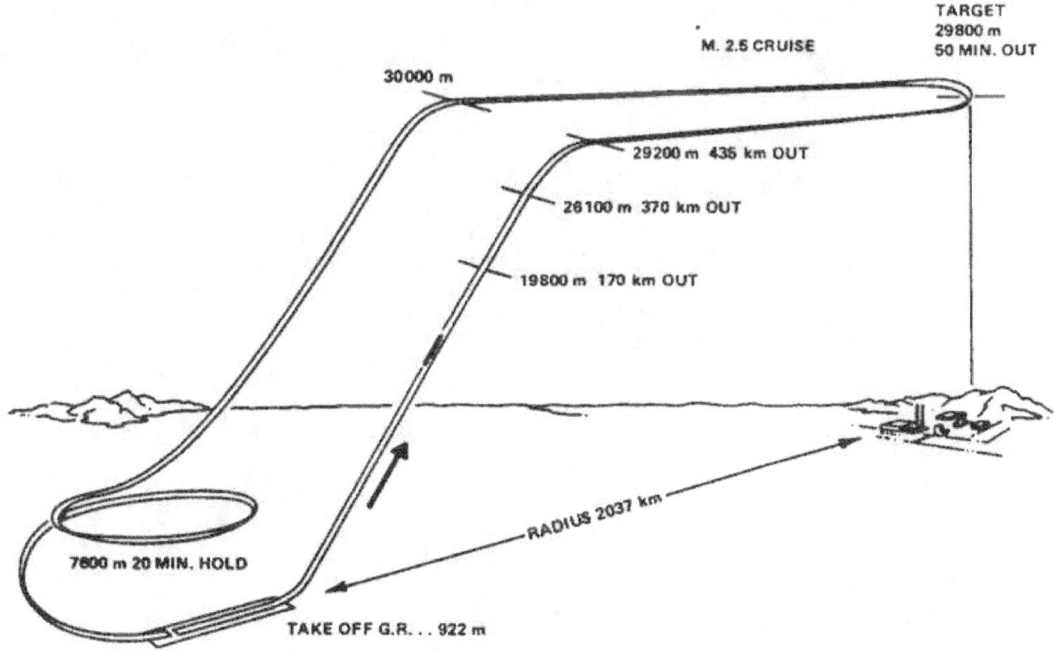

TARGET
29800 m
50 MIN. OUT

M. 2.5 CRUISE

30000 m

29200 m 435 km OUT

26100 m 370 km OUT

19800 m 170 km OUT

RADIUS 2037 km

7800 m 20 MIN. HOLD

TAKE OFF G.R. . . 922 m

Fig. 35. Mission profile for the Lockheed CL-400 using liquid hydrogen as fuel. (Source same as fig. 34.)

Fig. 36. Clarence L. (Kelly) Johnson, aircraft designer and builder extraordinary, father of the U-2 reconnaissance airplane and its first proposed successor in 1956–1958, the hydrogen-fueled CL-400. (Courtesy of Lockheed Aircraft Corp.)

Fort Robertson at the Skunk Works

Kelly Johnson saw his task as much more than designing and building a hydrogen-fueled airplane. He was also concerned about its operation, for if it was to be successful, liquid hydrogen had to be produced and shipped in quantity and be handled like gasoline. On 16 March 1956, he and his staff met with representatives of J. H. Pomeroy and Company of Los Angeles, a consulting engineering firm. Johnson wanted Pomeroy to study the engineering feasibility and cost of producing parahydrogen in quantity, and he was interested in three production rates—45000, 135000, and 225000 kilograms per day. He wanted the plant location to be in the Antelope Valley of California. Pomeroy agreed to undertake the study, and ten days later Johnson sent them a letter of intent with ground rules.[12]

At the outset of the project, Johnson assigned one of his assistants, Ben Rich, a thermodynamics and heat transfer expert, the dual responsibilities of propulsion and the handling of hydrogen. Rich, who knew little about liquid hydrogen at the time, checked Mark's *Mechanical Engineering Handbook* which stated that liquid hydrogen was an impractical fluid and only a laboratory curiosity. He was to understand why in his subsequent visits to laboratories and firms working with liquid hydrogen. Among those contacted were Professor William Giauque, University of California at Berkeley, and Russell B. Scott at the Cryogenic Laboratory of the U.S. Bureau of Standards at Boulder. Rich concluded that liquid hydrogen was mostly in the hands of highly skilled scientists, and few of them appreciated the practical problems he saw in adapting liquid hydrogen to routine use as an airplane fuel. In that application, a temperature range from the boiling point of liquid hydrogen, 20.3 K, to the frictional temperature of the airplane skin at Mach 2.5, about 670 K, had to be handled with designs and materials dictated by volume and weight restrictions. The earthbound design and construction methods used with liquid hydrogen generally were unsuitable. Moreover, Rich found that he was thinking of far greater quantities of liquid hydrogen than others; he used the unit "acre-feet" to emphasize his point. All these considerations made it obvious that the Skunk Works staff had to learn how to handle liquid hydrogen and to adapt it to the particular application. This required a liquid hydrogen test facility. During World War II, a bomb shelter revetment had been built adjacent to the Skunk Works, and it was selected as the site of the hydrogen facility. It was named "Fort Robertson" after the man who was in charge of the test operations. A Collins cryostat, capable of producing nine liters of liquid hydrogen per hour, was installed to test materials, bearings, seals, and small components. When larger quantities were needed for tank flow and spill tests, liquid hydrogen was obtained from the Bureau of Standards Cryogenic Laboratory at Boulder and stored in a 2200 liter refrigerated dewar built by the Air Force for the hydrogen bomb program. The Skunk Works also relied heavily on the experts at the NBS Cryogenic Laboratory, particularly Russell Scott, regarded as "Mr. Hydrogen," who became a consultant.

On 1 October 1956, the J. H. Pomeroy Company reported on hydrogen liquefaction plants. The report is an excellent summary of the state-of-the-art, and cites 52 references.[13] An entire plant was planned—from incoming natural gas for producing gaseous hydrogen to underground storage of liquid hydrogen. A plant of 45000 kilograms per day capacity was studied in detail, as well as multiples of it—well above the size of the Boulder installation, which had the largest capacity in existence in the U.S. Pomeroy considered the 45 000 kilogram per day capacity to be about the largest practical size. Construction cost was estimated at $45 million and operating costs at $0.386 per kilogram. A million cubic meters of natural gas per day would be required.* Pomeroy discussed an expansion engine process that would, with some additional R&D, be available. With catalysts, it would permit continuous liquefaction of parahydrogen.

Hydrogen Tanks and Systems

For a hydrogen-fueled airplane, the very low temperature and density of liquid hydrogen pose special design problems for tanks, pumps, lines, instrumentation, and other components in the fuel system. The special requirements imposed by hydrogen are recognized immediately by all who consider such designs and, of course, received major attention by the men of the Skunk Works. The CL-400 design divided the hydrogen tankage into three sections; the forward tank had a capacity of 67 000 liters; aft, 54 000; and center (sump), 15 000. The two main tanks were kept at 2.3 atmospheres pressure and the sump tank slightly lower for fuel transfer. In the sump was a booster pump, built by Pesco Products, that supplied liquid hydrogen to the engines at a pressure of 4.4 atmospheres. The engines were mounted at the wing tips, which meant that the liquid hydrogen had to pass through a hot wing with surface temperatures up to 436 K. The design provided a vacuum-jacketed, insulated line for this purpose.

There were many unknowns in the design of the hydrogen tanks and other fuel components, and numerous experiments were conducted to obtain more information. These were done at Fort Robertson and included half-scale models of the sump tank, the vacuum-jacketed lines for carrying hydrogen from the tanks through the hot wings to the engines, booster pumps, valves, controls, and other components. These were tested in thermal environments simulating flight conditions. Later a full-scale sump pump was built and shipped to Pratt & Whitney for their use in engine testing.

Is Hydrogen a Practical Fuel?

Among the first concerns of Johnson and Rich were the fire and explosion hazards of hydrogen. Could it be handled as safely as gasoline? In his early visits to laboratories using liquid hydrogen, Rich inquired about fires and explosions, but obtained little information. The laboratories went to great lengths to avoid these problems. The only previous explosions Rich learned about were some minor ones Professor William Giauque experienced when oxygen crystals formed in a heat exchanger containing hydrogen. The paucity of information led Johnson and Rich to devise a series of experiments to determine for themselves the hazards of hydrogen fires and explosions.

---

*$CH_4 + H_2O$ (steam) $\rightarrow 3H_2 + CO$

$CO + H_2O$ (steam) $\rightarrow H_2 + CO_2$

The CO_2 byproduct would be marketed to keep costs down.

For this they turned to their only testing ground, Fort Robertson—less than a kilometer from the runways of the Burbank airport.

Tests were devised in which tanks containing liquid hydrogen under pressure were ruptured. In many cases, the hydrogen quickly escaped without ignition. The experimenters then provided a rocket squib (a small powder charge) to ignite the escaping hydrogen. The resulting fireball quickly dissipated because of the rapid flame speed of hydrogen and its low density. Containers of hydrogen and gasoline were placed side by side and ruptured. When the hydrogen can was ruptured and ignited, the flame quickly dissipated; but when the same thing was done with gasoline, the gasoline and flame stayed near the container and did much more damage. The gasoline fire was an order of magnitude more severe than the hydrogen fire. The experimenters tried to induce hydrogen to explode, with limited success. In 61 attempts, only two explosions occurred and in both, they had to mix oxygen with the hydrogen. Their largest explosion was produced by mixing a half liter of liquid oxygen with a similar volume of liquid hydrogen. Johnson and Rich were convinced that, with proper care, liquid hydrogen could be handled quite safely and was a practical fuel—a conclusion that was amply verified by the space program in the 1960s. At the time, however, Johnson and Rich filmed their fire and explosion experiments to convince doubters.

The confidence of Johnson and Rich in hydrogen handling was not always shared by their hydrogen consultant, Russell Scott, who was often amazed at what he saw going on in the test areas of Fort Robertson.[14] The facility, however, was well equipped with an explosion-proof electrical system, non-sparking safety tools, hydrogen sniffers or monitors, and other safety devices. In the three years of work and the handling of thousands of liters of liquid hydrogen, there was not a single accident caused by hydrogen. There was, however, one close call. In keeping with Kelly Johnson's philosophy of austerity, the ovens used for simulating hot wing temperatures of Mach 2.5 flight were made partially of wood. There were five such ovens, and early one morning, about 2 a.m., one of them caught fire. The Skunk Works personnel, including Rich, were summoned because the fire department could not be called, for security reasons. At the time there were 2000 liters of liquid hydrogen stored in the area and Rich decided that the best course of action was to dump the liquid hydrogen on the ground. It was winter and very humid; the cold hydrogen quickly filled the revetment with fog about five feet thick. Rich and about two dozen other people were in the revetment and all they could see of each other were their heads, an eerie sight. Luckily, the hydrogen did not ignite.

Suntan at United Aircraft

United Aircraft Corporation* became involved in liquid hydrogen as a propulsion fuel in 1955 on the initiative of the power plant laboratory at Wright Field. Acting on a directive from its headquarters, the laboratory initiated a procurement request in January 1955 to investigate hydrogen as a fuel in turbojet engines. In February, invitations to bid were sent to United Aircraft and three other major engine

---

*Name changed to United Technologies Corp. in 1975.

manufacturers. Proposals were submitted in March; United Aircraft won the competition and was awarded a contract on 15 June (p. 126).

The contract was not with the corporation's Pratt & Whitney division but with the research department headed by John Lee. The work was exploratory and included cycle analyses, aircraft weight analyses, and some experiments. One of the men involved was Wesley A. Kuhrt, to whom hydrogen was no stranger. When 13 years old, he made hydrogen in his cellar laboratory by adding zinc to hydrochloric acid. Suddenly there was an explosion; glass fragments were imbedded in his chest, but he escaped serious injury. The incident neither cooled his enthusiasm for science nor created a fear of hydrogen.[15]

The Pratt & Whitney division had followed Air Force and NACA interest in hydrogen during 1955 and was also aware of Rae's Rex engines.[16] The Suntan project began for the division with a call from Appold in January 1956; by February, division officials began to believe they would win the contract for the engine. On 17 February, Perry Pratt, chief engineer, summarized what he had learned about hydrogen in jet engines. He cited six companies with experience in pumping hydrogen and described an engine that was somewhat similar to the Rex engine.* Pratt had examined the hydrogen supply problem and concluded that conversion of liquid hydrogen to its para form at time of liquefaction was feasible, and this made hydrogen storage, and shipment by truck, rail, or air practical. This optimistic report was written on Friday.[17] The following Monday, Pratt was in California visiting various people knowledgeable about hydrogen, including Kelly Johnson at Lockheed.[18] By this time, it was highly probable that Johnson and Pratt, collaborators in adapting the J-57 engine for the U-2, were aware that they would again be working together on the Suntan project.

William Sens, a Pratt & Whitney engineer, accompanied Pratt on the California trip and while there learned about Rex engines. This excited him, for six weeks earlier he had conceived an idea about hydrogen-fueled engines following a conversation with John Chamberlain, a combustion expert at United Aircraft's research laboratory. Chamberlain had pointed out that heated hydrogen was capable of a large amount of work in a thermodynamic cycle. Sens began thinking of using heated hydrogen to drive a turbine which would power an engine fan or compressor. After passing through the turbine, the hydrogen would be injected and burned in the airstream of the engine. Immediately after returning from California, Sens sent a proposal to Pratt for developing a hydrogen engine meeting the following requirements:

Altitude	30 500 m
Speed	M 2.5
Thrust	20 000 N (4500 lb)
Thrust specific fuel consumption	0.076 kg/N·hr (0.75 lb/lb thrust·hr)
Nacelle weight	2722–3175 kg
Engine diameter	155 cm

---

*Reaction Motors, Carter Pump, North American, Aerojet, Cambridge Corp., and National Bureau of Standards. Pratt mentioned an engine fan diameter of 150 cm, the same diameter that Johnson and Rae had agreed upon in the Lockheed airplane study for Garrett, and which had been officially reported to the Air Force two days earlier.

These specifications indicate that Sens was also aware of Johnson's propulsion requirements or those of the CL325-1 prepared by Johnson for Garrett (table 4, p. 134).[19]

Sens described his proposed engine as having a dual cycle, with the basic one resembling a supercharged ramjet:

> Air is . . . compressed by a low pressure ratio compressor, heated by combustion of hydrogen vapor and discharged through a . . . nozzle. In addition, heat is extracted from the air stream by means of a heat exchanger after part of the combustion of the hydrogen has taken place. This heat is used to vaporize and heat the hydrogen being used in the combustion process. In the secondary cycle the liquid hydrogen fuel is compressed to a high pressure by means of a multi-stage centrifugal type pump. The high pressure hydrogen is then vaporized and heated to a relatively high temperature in the heat exchanger located in the high temperature air stream. The hydrogen is then expanded through a multi-stage axial-flow turbine to a pressure only slightly above that of the fan discharge air. The turbine power output is used to drive the compressor used in the air cycle. Because of the large speed difference between the hydrogen turbine and the air compressor, it is necessary to use a single speed reduction gear between the two components.[20]

Sens discussed anticipated problems and the applicability of existing Pratt & Whitney experience to solve them.

Sens was not the only one in the corporation considering possible hydrogen engines. Wesley Kuhrt in the research department had been working on them for some time, and on 1 March 1956, he conceived three engine systems for which he later filed and was granted patents.[21] One was a turbofan engine (fig. 37). Air entering the inlet is compressed by the fan and flows around the centerbody to the aft section, where gaseous hydrogen is injected and burns stoichiometrically. The hot gases expand through the exhaust nozzle to produce thrust. The source of power for the air fan is a turbine driven by heated hydrogen prior to combustion. Liquid hydrogen flows to the heat exchanger around the exhaust nozzle where it gasifies and is raised to a reasonably high temperature. From the heat exchanger the hot hydrogen drives a multistage turbine which is connected to the air fan through a gear box. After leaving the turbine, the hydrogen is injected in the engine air stream and burned. Kuhrt's engine is similar to Rae's Rex III in that both employ a heat exchanger to heat the hydrogen to drive a turbine, but Kuhrt's concept is much simpler than the Rex III (p. 131).

For Kuhrt, the beginning of the Suntan work at United Aircraft was a call in early 1956 to come to the office of Wright Parkins. Present were Perry Pratt, Col. Norman Appold, and others. Appold stressed the need to get started quickly on a project to use hydrogen in aircraft engines.[22]

For Richard J. Coar, a rising, brilliant young mechanical engineer hard at work on developing the J-75 turbojet, the Suntan program also began early in 1956 when he was "yanked off his project" and assigned to the hydrogen engine work. His first task was engine analysis and learning all he could about hydrogen. He visited the Bureau of Mines, the Arthur D. Little Company, and a conference at the Bureau of Standards

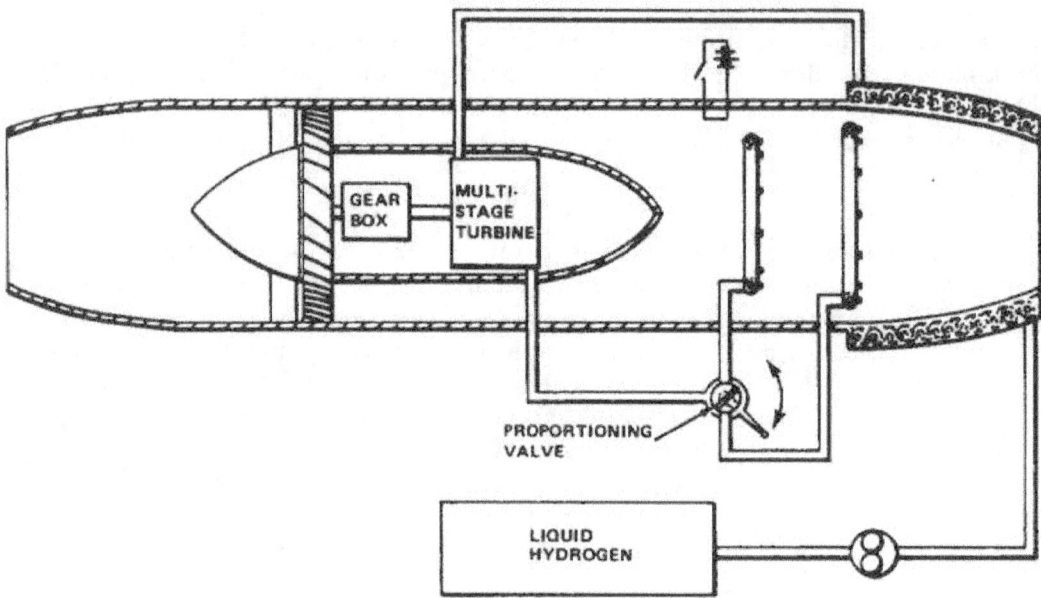

Fig. 37. Wesley A. Kuhrt's turbofan jet engine using liquid hydrogen as fuel, the precursor to Pratt & Whitney Aircraft's 304 engine. From Patent 3 241 311, 22 Mar. 1966, filed 5 Apr. 1957. (Courtesy of United Technologies Corp.)

Cryogenic Laboratory at Boulder. Inspection of the liquefaction plant convinced him that production of liquid hydrogen would be a major obstacle to military use of hydrogen. The plant was small and the laboratory techniqes required highly skilled personnel. In April, Coar went to Baltimore to negotiate a contract with the Air Force. It was on one page and technical negotiations were completed in a day—a marked contrast to the long and agonizing process that Rae, Garrett, and the Air Force had gone through earlier.[23]

Pratt & Whitney's initial approach to the problem was to analyse the various hydrogen engines that had been proposed, select one, and develop it so as to take the greatest advantage of hydrogen's unique properties. This remained their mainline approach but in a short while, they realized that modification of an existing engine would provide a quicker, though less efficient, engine for early flight experience. They proposed to modify a J-57 for this purpose, the Air Force agreed, and the contract was amended.

Shamrock

In the spring of 1956, Suntan engine activities at Pratt & Whitney were in full swing. Coar selected Richard C. Mulready, a bright young engineer, as his assistant. Liquid hydrogen handling tests began immediately with hydrogen obtained from the Cambridge Corporation in dewars. Associated with this activity were preparations for component and engine testing, including obtaining a supply of liquid hydrogen. With the help of Capt. Jay Brill, a hydrogen liquefier of 227-kilogram-per-day capacity was purchased from Herrick L. Johnston and installed in the engine test area behind the

East Hartford plant. The test area was called the "Klondike" because of the cold Connecticut winters and well-ventilated test stands that were designed to prevent the accumulation of hydrogen. Coar and Mulready also began to round up all the gaseous hydrogen tube trailers they could find to supply the liquefier.[24]

The second activity, code named "Shamrock," began in April to convert a J-57 to burn hydrogen. The design was completed in May; thereafter, component testing and engine modifications ran concurrently. The hydrogen liquefier was ready in September, engine testing began in October. The test engineers were agreeably surprised by the ease of engine operation. They ran it at full power and throttled back so far that the air fan was revolving so slowly the individual blades could be counted. Under this latter condition, the throttle could be opened and the engine would quickly and smoothly accelerate to full power. They found that the temperature distribution was good and there were no major problems. Such satisfactory results came only after careful design studies, modifications, and component testing. Among these precursory activities were the development of a heat exchanger using air bled from the compressor to gasify the hydrogen, modifications to the J-57 electronic fuel control system, and development of an oil-lubricated, liquid-hydrogen pump. Figures 38 and 39 show a schematic of the modified J-57 and comparison with the standard model.

By the fall of 1957, the J-57 experiments demonstrated beyond question that a conventional turbojet could be readily adapted to use hydrogen. Such engines could have been used to meet Kelly Johnson's tight airplane development schedule, but modifying an existing turbojet could not optimize the advantages of hydrogen. The Pratt & Whitney engineers had realized this early in their studies, as had their counterparts in the Rex division of Garrett and the Air Force. The mainline Pratt & Whitney effort from the start focused on a design of a special hydrogen engine, and its design started in April 1956 with the first contract.[25]

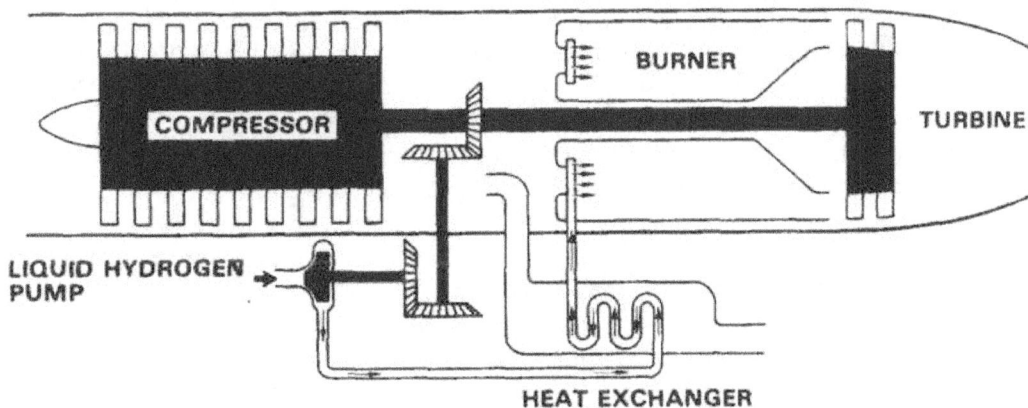

Fig. 38. Schematic of the Pratt & Whitney Aircraft J-57 jet engine modified to use liquid hydrogen as fuel, 1956. (Courtesy of Pratt & Whitney.)

Fig. 39. Comparison of Pratt & Whitney's J-57 with afterburner (about 6¼ meters long) and using hydrocarbon fuel (bottom) with the engine modified to use liquid hydrogen (top), 1956. (Courtesy of Pratt & Whitney.)

The Model 304 Engine

By mid-August 1956, Pratt & Whitney engineers had designed the new engine to use hydrogen. It was designated the "304," taken from the division's engine order number 703040, 16 April 1956.[26] It was essentially the one proposed earlier by Sens and Kuhrt and is shown schematically by figure 40. Liquid hydrogen was pumped at high pressure through a heat exchanger in the aft section of the engine. The heated hydrogen drove a multistage turbine which, through a reduction gear, powered a multistage air fan. The fan compressed incoming air, the primary working fluid of the engine. Part of the hydrogen discharged from the turbine was injected and burned in the air-stream behind the fan. The amount of hydrogen injected and burned was controlled to limit the temperature of the combustion gases which furnished the heat for the heat exchanger downstream. The remaining hydrogen was injected and burned in the after-burner section beyond the heat exchanger, and the hot gases and air expanded through the nozzle to produce propulsive thrust. The engine was similar to the Rex III but much simpler, as only one heat exchanger was used. The maximum diameter of the 304 engine was 203 centimeters, as compared to the 150 centimeters proposed by Garrett for Rex III. Nacelle length was 10.7 meters; weight 2722 kilograms; thrust at 30 500 meters altitude, 21.4 kilonewtons (4800 lb); and specific fuel consumption 0.082 kilogram/newton · hour (0.8 lb/lb · hr). These are close to the specifications in Sens's draft of 24 February 1956.

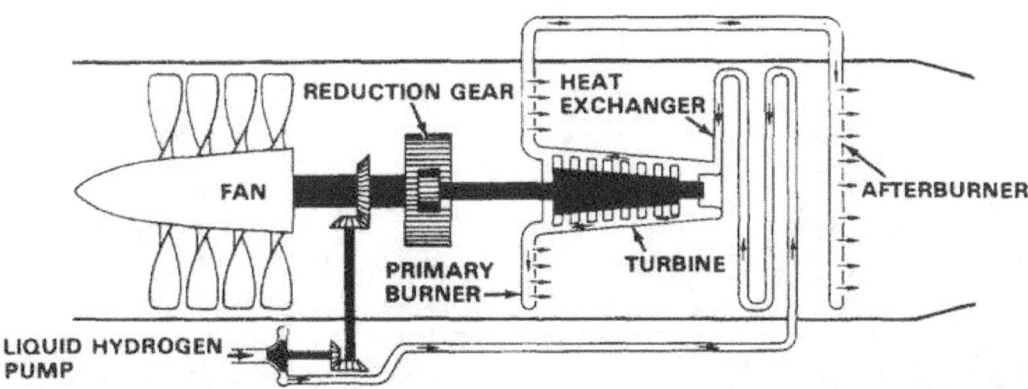

Fig. 40. Schematic of Pratt & Whitney's model 304 engine designed to use liquid hydrogen as fuel, 1956.
(Courtesy of Pratt & Whitney.)

Pratt & Whitney engineers were well experienced in all the components of the 304 engine except the liquid-hydrogen pump and the hot-gas heat-exchanger. They purchased a liquid-hydrogen pump for study, but became dissatisfied with it and proceeded to make a better one.[27] They saw two critical problems: an impeller that would handle liquid hydrogen without cavitation, and adequate sealing between the high-pressure liquid hydrogen at 20 K and the oil-lubricated bearing. Apparently they were not familiar with the work at Ohio State University on oil-free ball bearings operating in liquid hydrogen (pp. 25–26). They designed a two-stage centrifugal pump with a seal protecting conventional bearing lubrication. Figure 41 is a photograph of the pump rotor. The pump worked well and a total of 25 hours test time was accumulated in 75 tests over two years.*

The hot-gas–to–hydrogen heat exchanger (fig. 42) was the most unusual and interesting component of the 304 engine. With an outside diameter of 182 centimeters, the unit consisted of banks of 48-millimeter stainless steel tubing in an involute pattern to ensure uniform air flow. An enormous amount of tubing was used—enough to stretch over 8 kilometers; 2240 tube joints were furnace-brazed. The hydrogen passing through the heat exchanger was heated from 20 K to 1000 K, and the entering combustion gas temperature was 1500 K. The rate of heat transfer was 21 000 kilowatts (72 million Btu/hr), enough to heat 700 six-room houses.[28]

Pratt & Whitney engineers, experts in designing gas turbines, built the 304 hydrogen turbine with 18 stages, the largest of which was 45 centimeters in diameter. Operating temperature was 1000 K and power output was 8950 kilowatts (12000 hp). The turbines were tested for a total of 64 hours over a two year period. The 12-stage high-pressure group is shown by figure 43.

The first model 304 engine was assembled in East Hartford, Connecticut, by 18 August 1957—sixteen months after go-ahead (fig. 44.).

---

*One pump accumulated 4½ hours of test time with speeds as high as 25 300 rpm, pressures of 75 atm, and a flow of 1.9 kg/s.

Fig. 41. Pratt & Whitney's liquid-hydrogen pump for the model 304 jet engine. The seal between the rotor and bearings operated dry; the bearing lubrication was conventional. (Courtesy of Pratt & Whitney.)

Fig. 42. Hydrogen heat exchanger in Pratt & Whitney's model 304 jet engine using liquid hydrogen as fuel. (Courtesy of Pratt & Whitney.)

Fig. 43. High-pressure group of the 12 stages of hydrogen turbine expansion used by Pratt & Whitney's model 304 engine. The early stages operated near 1000 K. (Courtesy of Pratt & Whitney.)

Fig. 44. Pratt & Whitney's model 304 engine using liquid hydrogen as fuel. Visible inside the nozzle is the afterburner fuel injector. The engine was first tested in September 1957. The numerous wires on the engine led to sensors for measuring performance. (Courtesy of Pratt & Whitney.)

Engine Tests

The testing of the 304 engines was carried out at Pratt & Whitney's new center west of West Palm Beach, Florida. The test center, still under construction in the fall of 1957, was the result of several years of planning by United Aircraft officials to overcome the limited space for testing at their Connecticut plant. Problems of safety and noise made a more remote site desirable, and there were considerations of dispersal of facilities for defense reasons. These had led to the choice of West Palm Beach County as a desirable test site. United Aircraft acquired a large tract of land, swapped part of it for adjacent land owned by the state, and ended up with 27 square kilometers of sand, scrub pine, swamp, and alligators—well suited for remote testing of new engines. In the negotiations for the hydrogen engine contract, United Aircraft officials indicated a willingness to invest $20 million in permanent facilities at the new center if the Air Force would pay for all movable equipment, also estimated to be about $20 million.[29] The cost sharing was agreed upon in principal, if not in the exact amounts, and construction proceeded. During initial operations, the test crew often had to call for a bulldozer to clear the unpaved roads of deep ruts to allow passage; alligators were a common sight.[30]

The first 304 engine tests began on 11 September 1957 using three fluids: nitrogen, gaseous hydrogen, and liquid hydrogen. The inert nitrogen was used to check the fuel system and rotating machinery, especially bearings and seals. The first series of runs lasted through October; 4½ hours were logged, including 38 minutes with liquid hydrogen. The engine was removed for inspection and overhaul when turbine oil consumption became excessive. When reinstalled for a second series of runs on 20 December 1957, no significant failures occurred, but the engine was periodically removed, inspected, overhauled and reinstalled.[31]

Six series of runs were made through the first part of July 1958 and 5½ hours of operation with hydrogen were accumulated. Only minor problems were encountered until the last run, when there was a major failure of bearings, turbine, and heat exchanger. Meanwhile, a second engine of the same type had been installed on a twin test stand; its first run was made on 16 January 1958. Tests continued on the second engine into the first part of April, with a little over 10 hours of operation with hydrogen. The engine was removed when the low pressure section of the turbine failed.

During the testing period, Coar and Mulready designed and built a second model of the 304 engine, which had an additional (fifth) compressor stage and lower specific fuel consumption. The first 304-2 was assembled at East Hartford on 20 June 1958 and four days later was operated at the Florida test center. Tests continued for a month, with 3 1/3 hours of accumulated running time with hydrogen before the engine experienced a complete turbine failure. It was removed for repair and strengthening of the turbine disks. While this engine was in the shop, another 304 engine (presumably of the first design) was installed and operations began in mid-August. This engine operated satisfactorily through September and accumulated over 6 hours time using hydrogen. Table 5 shows a comparison of the specifications and performance of the two versions of the 304.

By the end of September 1958, the repaired 304-2 engine was back on the stand and made a short run, and another 304 engine was nearing assembly at East Hartford.

Fig. 45. Key Pratt & Whitney engineers in developing the model 304 aircraft engine and the RL-10 rocket engine using liquid hydrogen, L to R: Perry W. Pratt, Richard J. Coar, and Richard C. Mulready. Pratt was chief engineer of P&W during the development of both engines. Coar was project engineer of the 304 engine and became the chief engineer of the Florida center where the RL-10 was developed. Mulready succeeded Coar as project engineer of the 304, became project engineer of the RL-10 in 1958, and assistant chief engineer of the Florida center in 1961. Pratt has retired; Coar is vice-president for engineering; and Mulready is corporate manager for new business development.

TABLE 5.—*Characteristics of Pratt & Whitney's Model 304 Engines*

Characteristic	Model 304-1 Spec A6600	Test Performance Eng. 1	Test Performance Eng. 2	Model 304-2 Spec A-6600A	Test Performance
Sea-level static thrust					
newtons	55 600	55 422	53 429	60 048	35 028
(lbs)	(12 500)	(12 460)	(12 012)	(13 500)	(7875)
Thrust specific fuel consumption, kg/N·hr	1.10	1.252	1.220	0.900	.937
Compressor speed, rpm	3600	3630	3300	3600	2503
Pump discharge pressure, atm	—	54	42	—	34
Overall turbine efficiency	—	—	.475	-	.507

Note   Model 304-1 had 4 compressor stages.
       Model 304-2 had 5 compressor stages.

Neither was destined to run again, for time had run out on the Suntan project. In all, the engines were operated 25½ hours with hydrogen, and all indications were that the development was proceeding satisfactorily.

Baby Bear, Mama Bear, and Papa Bear

Concurrent with the engine testing was an extensive program of component testing, and the combined operations created a heavy demand for liquid hydrogen, a situation anticipated by the Air Force.

Capt. Jay Brill's primary assignment on the Suntan management team was the logistics of liquid hydrogen. In one of his first moves, he contacted the Atomic Energy Commission to scrounge the excess equipment used for the "wet" hydrogen bomb program. He was able to obtain several of the refrigerated transport dewars developed for the AEC program (p. 68). In April 1956, he began a survey of industrial firms to assess their capability and interest in building hydrogen liquefiers and producing liquid hydrogen for the program. Wright Field had prepared a specification for liquid hydrogen which was given the code name "SF-1" fuel. Brill was accompanied on his visits by Marc and Blackwell (Blacky) Dunnam. Marc was chief of the fuels and oil division of the Power Plant Laboratory and Blacky had experience with cryogenic equipment. They visited the Linde Company in New York, Hydro-Carbon Research in New Jersey, and the Air Products Company in Allentown, Pennsylvania. Brill returned to Dayton convinced that large hydrogen liquefiers could be built with existing technology. The Arthur D. Little Company was awarded a contract to serve as a consultant in hydrogen liquefaction and to study hydrogen handling and safety procedures. The Air Force also made use of the services of Russell Scott and other experts at the Bureau of Standards Cryogenic Laboratory at Boulder.[32]

In his survey of industrial firms, Brill found that there was plenty of gaseous hydrogen capacity by several processes. One firm produced excess hydrogen as a by-product in Painesville, Ohio. This was near Pesco Products division of Borg Warner Corporation, a firm Appold had involved in developing a liquid hydrogen transfer pump for the CL-400 airplane. It was also near the NACA Lewis Laboratory, which would soon need liquid hydrogen for its flight investigation. For these reasons, the Air Force contracted with Air Products, about May 1956, to build a 680-kilogram-per-day liquid hydrogen plant in Painesville. At the same time, two other contracts for similar size plants were awarded. One was to Stearns-Roger for a plant at Bakersfield, California, to support the CL-400 program at Lockheed and the other was to Hydro-Carbon Research for a plant to support Pratt & Whitney at East Hartford. The Painesville plant was named "Baby Bear" and was the first to become operational, in May 1957, at a cost of $2 million. The California plant was placed into operation in the fall of 1957, but the contract with Hydro-Carbon Research was cancelled for budgetary reasons.[33] Pratt & Whitney's initial hydrogen needs at East Hartford, over its own capacity, were supplied by truck from Baby Bear.

Another of Brill's tasks was the transportation of liquid hydrogen. Specifications for over-the-highway trailers had been drawn up by Wright Field and a contract was awarded to the Cambridge Corporation. Concurrently, permission was sought and obtained from the Interstate Commerce Commission to transport liquid hydrogen over the highway. The trailers were labeled "flammable liquid," since to reveal the true contents would blow the security cover. The U-1 semi-trailer built by the Cambridge Corporation had a capacity of 26 500 liters, with a hydrogen loss rate of approximately 2 percent per day. Figure 46 shows the U-1 and its successor, the U-2. The latter's specifications were issued on 15 March 1957 because the U-1 ran into a natural, but unanticipated, problem. The very low density of hydrogen made tandem axles on the semi-trailer unnecessary, so the U-1 had only one. During subsequent use of this equipment, there occurred an endless series of problems, all stemming from the single axle, which was unheard of for such a large trailer. It seems that each time one of these large semi-trailers went through a state weighing station, it roused suspicion, doubt about the equipment, and inquiries about the nature of the load.* The Suntan team considered painting a false second axle on the trailer but this was too obvious, and they gave in by ordering the U-2 with its second axle—one that was not needed for the load but which raised no questions on the road.[34]

To satisfy the anticipated demands for liquid hydrogen at Pratt & Whitney's Florida test center, the Air Force decided to locate a large hydrogen liquefaction plant nearby. United Aircraft obligingly deeded a tract of land to the government for the plant. The Air Force was unsuccessful in interesting private capital to put up the plant, so it funded its construction and operation by Air Products. The plant, with a 4500-kilogram-per-day capacity, was placed in operation in the fall of 1957, at a cost of $6.2

---

*In one instance, a suspicious and frustrated weighing official found one semi-trailer 45 kilograms too heavy. He ordered the driver to unload the excess but of course, the driver was powerless to do so. The Air Force had to go all the way to the governor of the state to secure a release for the load. Interview with Blackwell C. Dunnam, WADC, WPAFB, OH, 6 June 1974.

Fig. 46. The U-1 semi-trailer (top) first used to haul liquid hydrogen. The single axle, adequate for low-density hydrogen, caused so many problems with puzzled truck-weighing officials that it was replaced with the U-2 having a second axle. (Courtesy of AFSC.)

million.[35] The Suntan team called this plant "Mama Bear," but locally it was known as the APIX fertilizer plant. APIX was an acronym for Air Products Incorporated, Experimental; the fertilizer association was encouraged by the Air Force and Air Products to conceal the true identity of the product.* Mama Bear used crude oil in a chemical process to obtain gaseous hydrogen. Liquid hydrogen storage tanks at the plant were connected to the Pratt & Whitney test cells by a double-wall, vacuum-jacketed 7.5-centimeter line, 610 meters long. By April 1958, the line had carried 833 000 liters of liquid hydrogen at rates up to 1700 liters per minute for component and engine tests.

---

*Workmen were observant and soon the word spread locally that hydrogen was involved. A retired Army colonel, in his role in local Civil Defense, became alarmed that a hydrogen bomb was being manufactured in the midst of an unsuspecting community. A delegation of security officials from Washington had to visit him and convince him to keep quiet. Interview with Col. A. Gardner (USAF, ret.). 19 Sept. 1973.

Even before Mama Bear was completed, the Air Force planned a much larger hydrogen liquefaction plant to meet the anticipated testing needs of the 304 engine development. The contract was awarded to Air Products in 1957, and the plant was built a few hundred yards away from Mama Bear. It cost $27 million and when placed into operation in January 1959, had a capacity of 27 200 kilograms per day—the world's largest. Crude oil was first used to obtain gaseous hydrogen but later methane was used. This plant, called "Papa Bear," came too late for the Suntan program but served a very useful role in the space program that followed.

Suntan Fades

In addition to its technological problems, the Suntan project was the subject of conflicting technical views over its feasibility and the best way to accomplish reconnaissance. In fact, Suntan did not get very far as a wholly supported project. Within six months, a difference of technical opinion over achievable range surfaced and this contributed to the gradual demise of the project. True to its name, Suntan had no clearcut ending: it just faded away. By the middle of 1957, opposition had effectively doomed the project although it lingered through 1958 and was not cancelled until the management team, weary of waiting, so requested in February 1959. Surprisingly, one of the main opponents was the man who conceived and sold the project to the Air Force, Kelly Johnson. The main defendant was the Suntan management team, particularly Appold and Seaberg, who for some months were able to convince high officials to keep the project going in the face of mounting opposition and budgetary restraints.[36]

Johnson's change of mind apparently came during the first six months of study and experimentation on the feasibility of the hydrogen-fueled airplane. The Air Force had insisted on a minimum radius to target of 2800 kilometers and was convinced that this distance and more was feasible. Johnson, on the other hand, believed that a radius of 2000 kilometers was about the best that could be achieved. The two sides stuck to their views throughout the life of the project.[37]

Following the initial phase of study and experimentation, the project proceeded during Fiscal Year 1957 as originally planned, with an allocation of about $19 million. Lockheed ordered 4 kilometers of aluminum extrusions to build the CL-400; Pratt & Whitney went full speed in developing the 304 engine; the Massachusetts Institute of Technology contracted to provide a guidance system; and Air Products contracted to build a large hydrogen liquefaction plant adjacent to the Pratt & Whitney test center in Florida.[38]

The bottom line on how well a project is faring in government circles is the fraction of the budgeted funds that is actually allocated to it. The Air Force obtained approval for $95 million for Suntan development for the fiscal year beginning in July 1957. The first significant indication that the project was in trouble came when the Suntan management team requested release of these funds to maintain the development schedule. The request was placed on the 22 August agenda of the Air Council, the Air Force's highest management group. In preparation for this meeting, the Suntan team met with crusty, blunt Gen. Curtis E. LeMay, former boss of the Strategic Air Command who had moved up to vice chief of staff in July. It was the first time that

LeMay had received a full briefing on Suntan and his initial reaction brought dismay to the team. "What," he exploded, "put my pilots up there with a . . . bomb?"[39] LeMay not only took a dim view of using liquid hydrogen but also was apparently under pressure to find funds for other important projects. On 19 September, the team received the bad news: of the $95 million approved in the budget for Suntan, only $32.3 million would be made available for it; the remainder would be transferred to other projects. In spite of additional efforts by Gen. Samuel E. Anderson, the new head of the Air Research and Development Command, to restore the funding, the decision remained firm.[40]

Johnson's views apparently contributed to the Air Force decision to cut Suntan. Sometime during the mid-1957 period, he was visited by James H. Douglas, Jr., who had succeeded Donald A. Quarles as Secretary of the Air Force in March 1957. Douglas was accompanied by Lt. Gen. Clarence A. Irvine, deputy chief of staff for materiel and a member of the Air Council. The visitors, concerned about the short radius of the CL-400 and mindful of Johnson's ability to stretch the range of other aircraft, asked him how much margin for growth was in the CL-400. The answer: practically none.[41]

Ordinarily, range can be extended by adding more fuel or improving the fuel consumption of the propulsion system for a given thrust. Johnson could see a range growth of only a paltry 3 percent or so from adding more fuel. ". . . we have crammed the maximum amount of hydrogen in the fuselage that it can hold. You do not carry hydrogen in the flat surfaces of the wing," he explained.[42] Johnson turned to Perry Pratt for estimated improvements in the 304 engine and his answer was equally pessimistic: no more than 5 or 6 percent improvement in specific fuel consumption could be expected over a five-year period. The very low growth estimates were compounded by operational logistics problems of liquid hydrogen. As Ben Rich asked: "How do you justify hauling enough LH_2 around the world to exploit a short-range airplane?"[43]

Having exhausted their appeals by October 1957, the Suntan team drastically curtailed the project to fit the funds available. Pratt & Whitney was given $18.7 million to continue development of the 304 engine at an undiminished pace. A total of $11.6 million was allocated for hydrogen liquefaction plant construction and operation and $3 million was set aside for later use. Development of the CL-400 was cancelled, but Lockheed was asked to continue the fuel system tests; $3 million was recovered from the changes. The MIT guidance contract also was cancelled.[44]

The Suntan team, particularly Seaberg, was not convinced that Johnson's pessimism over range was justified. Contracts for additional design studies were let not only with Lockheed but also with North American Aviation, Boeing, and Convair-Fort Worth. The additional study at Lockheed did nothing to change Johnson's view. In all, 14 designs were considered, ranging from bombers to Mach 4 reconnaissance aircraft with comparisons between using petroleum fuels and liquid hydrogen. For the same range, Lockheed found that aircraft using liquid hydrogen were larger but weighed less at takeoff than those using petroleum fuels. At a given speed, hydrogen-fueled aircraft exceeded the altitude limits of petroleum-fueled aircraft by 3000 to 6000 meters.[45] By March 1958, a Boeing design appeared to be the most promising of the new studies. Powered by four engines, it would fly at Mach 2.5, 30 500 meters altitude, and have a radius of 4100 kilometers—almost twice that of the CL-400. The Boeing airplane was also considerably larger than the CL-400, with a length of 61 meters, a delta wing span of 61 meters, and a takeoff weight of 75750 kilograms.[46]

The final results of the design studies were presented to the Air Council on 12 June 1958. LeMay, who chaired the meeting, raised the same objections as previously but allowed a full discussion of the subject. The Suntan team felt that the general reaction was favorable, but this was dispelled by two significant points in the summary of the meeting. Even if a successful new reconnaissance aircraft were developed, the President might not allow its use because of international political risks. If this happened, LeMay argued, the Air Force would only be building museum pieces. The second point was even more devastating. The Air Force had given a competing project higher priority; since it was underfunded there was no justification for allocating funds to Suntan.[47]

The June meeting spelled the effective end of Suntan, but the Air Council thought that the engine work should continue for its value in advancing the technology. Since the Suntan mission was broader than the Air Force, however, the June decision was not the final word. A joint committee of the Department of Defense and the Central Intelligence Agency was formed to make recommendations regarding Suntan. The committee, headed by Edwin Land of the Polaroid Corporation,* held meetings during the summer and fall of 1958 and the Suntan management team was held together pending the results. Although not privy to the committee's findings, the team sensed the trend and terminated the Pratt & Whitney contract in November. By February 1959, with still no word from the committee or formal directive from Air Force headquarters, the team requested that the project be ended. Of the $19 million allocated for FY 1959, about half had been transferred to the Advanced Research Projects Agency for rocket projects.[48]

In retrospect, several principals of the Suntan project saw different reasons for its ending. To Kelly Johnson, designer of the aircraft, the short range and hydrogen logistics were the predominating reasons; he considered the meeting with Douglas as the effective end of the project.[49] For Norman Appold, the project manager, the end came for other reasons. Suntan was one of a variety of options for gathering intelligence.[50] The implications of flying aircraft over Russian territory, which had been on the minds of the Air Council and others since the beginning of the U-2 and its potential flameout problem, became very real with Gary Powers's experience in 1960.

For Ralph Nunziato, with access to top-level Pentagon meetings and decisions, the reasons for cancelling Suntan were purely economic. In a presentation to the Air Council, he indicated that the next phase of development would need an estimated $150 million. It was a period of stringent budgetary limitations and Suntan lost out to other projects.[51]

Even the amount spent on Suntan remains in doubt. The consensus of several involved is that about $100 million was spent and some documentation appears to support this.[52] But Appold, the project manager, firmly believes the total to be closer to

---

*Other members: Courtland Perkins of Princeton, Edward M. Purcell and H. Guyford Stever of MIT, and Allan Donovan of Space Technology Laboratories. Richard E. Horner of the Air Force and Garrison Norton of the Navy were ex officio members.

$250 million, and Richard Horner, who was assistant secretary of the Air Force for R&D, concurs.[53] Since Suntan covered many activities and since great pains were taken to camouflage the project by directing funds through various channels, the actual total cost remains unknown.

Suntan Technology and Equipment

What was learned with the Suntan project? The technology of liquid hydrogen was advanced in several ways. There is concurrently a revival of interest in hydrogen-fueled aircraft. As before, however, their potential value is controversial. NASA held a special conference on hydrogen-fueled aircraft in 1973 and has sponsored industry design studies of both subsonic and supersonic configurations. Although no specific development has started, NASA continues to sponsor research applicable to hydrogen-fueled aircraft.

On the other hand, Kelly Johnson, who turned back to petroleum fuels and designed the highly successful SR-71, remains disenchanted with liquid hydrogen. In 1974, he summed up his view: "Today, there is regenerated interest in liquid hydrogen for aircraft propulsion, but considering all phases of the problem, I do not think we will have such aircraft in the foreseeable future."[54] Seaberg, who managed design study contracts with Boeing, Convair, and North American Aviation as part of the Suntan effort in 1957, agrees with Johnson's 1974 assessment.[55] The essence of technological progress, however, is the conversion of the impossible to the possible, so the case for hydrogen-fueled aircraft remains open.

Although Suntan technology and equipment have yet to find application in aircraft, they soon found application in rocket propulsion. In 1958, the Suntan management team began searching for ways to use the technology their project had generated, as well as equipment like the boost pump and the hydrogen liquefaction plants. One result was a proposal to use liquid hydrogen in a rocket engine for the rapidly developing space program. Like a phoenix rising from the ashes, the technology and equipment of Suntan would indeed play a major role in the space program of the 1960s. To learn how this occurred, we must next consider several other developments that were running concurrently with Suntan—activities at Pratt & Whitney, General Dynamics, North American Aviation, NACA, and the Department of Defense.

Fig. 47. Suntan management team: Col. Norman C. Appold, top left; Lt. Col. John D. Seaberg, top right; Maj. Alfred J. Gardner, bottom left; and Capt. Jay R. Brill. All engineers, Appold and Gardner each held two masters degrees, Brill one. Appold and Gardner were combat pilots and Seaberg a base executive during WW II. Brill graduated from West Point 3 years after the war. Appold headed the engine laboratory at Wright Field for 5 years prior to becoming the Suntan project manager. After Suntan, Seaberg managed the Centaur development for both the Air Force and NASA, assisted by Gardner and Brill. All except Brill retired as colonels: Appold heads the C-5 project for Lockheed-Georgia; Seaberg manages remotely-piloted-vehicle R&D at Wright Field; and Gardner is an assistant to the president of Lockheed Missiles and Space Co. Brill became a brigadier general in 1975 and manages the A-10 development at Wright Field.

SUMMARY, PART II

The 1950–1957 period was one of great technological advances in the use of liquid hydrogen as a fuel in rockets and aircraft. Thermonuclear research provided the first large stimulus to hydrogen technology at the start of the period; from it came a large new cryogenics laboratory, larger hydrogen liquefiers, mobile dewars for transporting hydrogen, and other advances.

The Lewis laboratory of the National Advisory Committee for Aeronautics advocated liquid hydrogen for rockets in 1950 and for aircraft in 1954, conducted research showing hydrogen's potential, and demonstrated that liquid hydrogen could be safely used in manned flight.

The Air Force, always seeking to extend flight capabilities, took a strong interest in very-high-altitude flight in 1953, became interested in hydrogen for this purpose in 1954 as the result of an imaginative proposal of Randolph Rae, helped the Central Intelligence Agency develop the U-2 airplane using conventional fuel, and mounted a massive, crash project to exceed the U-2's performance by using hydrogen. The hydrogen airplane did not materialize, but the liquid hydrogen plants and test facilities constructed by the Air Force would find full utilization in the emerging space program.

Part III
1958–1959

PART III
1958–1959

On Friday evening, 4 October 1957, man's long dream of spaceflight became reality with the launching of Sputnik I. Most Americans were surprised, not only by the feat itself but that the Soviets had done it first.

There was no lack of public forewarning about the coming age of spaceflight, however. In July 1955, the United States had announced its intention to launch satellites as part of the scientific activities of the International Geophysical Year which was to begin in mid-1957, and this was immediately followed by press articles that the Soviets were making similar plans. In the United States, the satellite activity became Project Vanguard, authorized in September 1955.

Vanguard was the culmination of a decade of scientific research of the upper atmosphere using balloons and sounding rockets, of increasing pressure by groups who saw the feasibility of spaceflight and made realistic proposals, and finally, of the interest and backing of the scientific community through the National Academy of Sciences. The last was essential, for many earlier and sound space proposals had been treated with disdain. Some of this attitude may have come from longtime exposure to grand and impractical schemes and science fiction. Even the ideas of such scientists and engineers as Tsiolkovskiy, Goddard, Oberth, and von Braun had failed to arouse much more than transient public interest. The wartime scare caused by the German rockets had long since receded, and by the late 1940s even the the military services were hard pressed to justify space projects—in spite of the obvious advantages of reconnaissance, communications, and meteorological satellites. International interest in cooperating to study the upper atmosphere and space phenomena using high-altitude probes and satellites became a major driving force, but it evolved so gradually during the early 1950s that the public scarcely took notice. After the 1955 announcement, Project Vanguard proceeded slowly and with little publicity.

Parallel with scientific interest in space was military interest. During the heyday of ballistic missile development during the 1950s, effort was concentrated on long-range missile capability, but engineers were well aware that the same missiles, more powerful than Vanguard, could be modified to provide the additional velocity needed to launch satellites.

The Soviet Sputnik I provided the spur for action in this country. The news media reflected astonishment, dismay, and fear. American pride was hurt, competitive spirit aroused, and a determination to "catch up" with and exceed the Russians became evident.

In previous parts, we have examined the growth of liquid hydrogen technology and its potential application for rockets and aircraft. In this final part, we will examine the events leading to the decision to use liquid hydrogen in two launch vehicles for the great space accomplishments of the 1960s and 1970s. To do so, we need to understand something of the antecedents of these vehicles—the ballistic missiles of the 1950s. Also pertinent is the competition among several government organizations for a role in space and the formation of the National Aeronautics and Space Administration. Throughout this discussion, emphasis will be on launch vehicles and the considerations that led to the use of liquid hydrogen.

9

The Early U.S. Space Program

In early 1958, when the Soviet Union and the United States had each launched two satellites, it was obvious, from comparing their weights, that the Soviets were using a much more powerful launch vehicle. This led to great concern in the United States about the apparent lag in vehicle capability—a concern welcomed by space enthusiasts for it meant more support. This concern, however, was not entirely justified when all the technological gains associated with the development of intercontinental and intermediate range ballistic missiles are considered. Indeed, modifications of these vehicles provided the base for the U.S. "stable" of launch vehicles in the early years of the space program. One ICBM, the Atlas, established the feasibility of lightweight, pressure-stabilized tanks, a technology important for favorable consideration of low-density liquid hydrogen. For all these reasons, a review of the development of Vanguard, the first U.S. vehicle developed solely as a launch vehicle, and military ballistic missiles during the 1950s is helpful in understanding launch vehicle planning during 1958 and 1959.

The Navy's Vanguard

When the National Security Council endorsed the concept of a scientific satellite in May 1955, it was based on two conditions: peaceful purposes were to be stressed and no interference with the development of ballistic missiles was to be permitted.[1] Donald A. Quarles, Deputy Secretary of Defense, charged with selecting a suitable vehicle for the scientific satellite, appointed a committee headed by Homer Joe Stewart, a rocket expert at the Jet Propulsion Laboratory and professor at the California Institute of Technology. The Stewart committee recommended a vehicle proposed by the Naval Research Laboratory. It consisted of a Viking first stage,* a second stage using liquid propellants, and a third stage using solid propellants. A modification of this combination became the Vanguard launch vehicle, and the Navy managed its development.

---

*Viking, built by the Glenn L. Martin Co., was powered by a Reaction Motors rocket of 89 kN (20 000 lb thrust) and used alcohol and liquid oxygen as propellants.

Vanguard was a slender vehicle, 21 meters tall and 1.1 meters in diameter, weighing 10250 kilograms at launch. The first stage was powered by a General Electric X-405 rocket engine of 120 kilonewtons (27000 lb thrust) using kerosene and liquid oxygen. The second stage was powered by an Aerojet rocket engine of 33.4 kilonewtons (7500 lb thrust) using a hydrazine compound (unsymmetrical dimethylhydrazine—UDMH) and nitric acid as propellants. The third stage was driven by a solid propellant rocket of 13.8 kilonewtons (3100 lb of thrust).[2]

Vanguard development was treated like a second-class program from the start, particularly when it came in conflict with high-priority ballistic missile programs. Even the funding was second level, coming from an emergency fund of the Secretary of Defense for two years. By the spring of 1957, however, development was proceeding satisfactorily. Two successful test flights had been made and the third (TV-2*), with a live first stage and dummy upper stages, was on schedule. Confidentially, the Navy ordered Glenn L. Martin to make the remaining test vehicles with live upper stages and capable of launching a satellite. In the months that followed, however, the Vanguard team encountered problems and was still struggling to patch and fly TV-2 when Sputnik was launched.

The Vanguard team then found itself suddenly in the spotlight. John Haugen, the quiet, scholarly Vanguard director, briefed President Eisenhower. John Hagerty, the White House press secretary, relying on the optimistic part of Haugen's briefing, announced five days after Sputnik that the first of a series of test vehicles carrying a small satellite sphere would be launched in December 1957. Although Hagerty added that the first fully instrumented satellite would be launched in March 1958, the media emphasized the December date as the time the U.S. would match the Russian accomplishment. The Vanguard team finally launched the recalcitrant TV-2 successfully in October and on 6 December prepared to launch the three-stage TV-3. A large gathering of reporters and spectators saw TV-3 rise from the pad about a meter, fall back, and collapse into a giant fireball.[3] That was the low point in the trouble-filled Vanguard development. Success came on 17 March 1958 when Vanguard I launched its tiny but well-instrumented satellite which transmitted signals for seven years. Meanwhile, a U.S. Army team, under the technical direction of Wernher von Braun, had launched the first American satellite.

The Army's Redstone and Jupiter Vehicles

The Army's principal missile team was formed around 120 German rocket experts brought to the United States in 1945. First stationed at Fort Bliss, Texas, the Germans were transferred to the Army's Redstone Arsenal, Huntsville, Alabama, in 1950; and they were soon deeply involved in the Army's growing missile development program. The technical group was headed by Wernher von Braun, who had held the same position at Peenemünde during the development of the German A-4 (V-2) rocket, the beginning of modern rocketry.[4]

---

*TV-2 was the second of the Vanguard test vehicles; the earlier two vehicles flown were a leftover Viking and TV-1.

The first large Army ballistic missile was the Redstone, modeled after the V-2. Redstone was powered by a North American Aviation rocket engine developing 334 kilonewtons (75000 lb thrust) using an alcohol-water mixture and liquid oxygen. During eight years of research and development and 37 flights, the Redstone evolved into a 322-kilometer-range vehicle, 21 meters tall, 1.8 meters in diameter, weighing 27670 kilograms at launch.[5]

A space enthusiast since youth, von Braun proposed a satellite launch vehicle based on Redstone in 1954, a year after the first Redstone launch. In 1955, his team submitted a proposal for a satellite launch vehicle to the Department of Defense. Called Jupiter C, it consisted of a modified Redstone with two solid-propellant upper stages. This design was used in a joint Army-Navy proposal to the Stewart committee, but it was not selected. Disappointed, von Braun soon found another application—study of aerodynamic heating of a warhead reentering the atmosphere during a ballistic trajectory. Three Jupiter Cs were launched, the last less than two months before Sputnik I. After this flight, the commander of the Army Ballistic Missile Agency (ABMA) at Redstone Arsenal, Maj. Gen. John Bruce Medaris, ordered the remaining Jupiter equipment into storage. As enthusiastic a space proponent as von Braun, Medaris was waiting for the right opportunity to show what ABMA could do in spaceflight.[6]

The perfect opportunity soon came. Medaris and von Braun were dinner hosts to visiting Neil McElroy, who was succeeding Charles Wilson as Secretary of Defense, when word came that Sputnik I was launched. The rest of the evening and the following morning were devoted to what ABMA could do. On 31 January 1958, the Medaris-von Braun team launched Explorer I, first American satellite, using a modified Jupiter C vehicle.*

The Air Force and the Ballistic Vehicle Build-Up

Although the Army had shown great initiative in ballistic missile development, the Air Force became the dominant military service in long-range, ballistic missiles. The Air Force had the responsibility for developing the Atlas and Titan intercontinental ballistic missiles (ICBM), the Thor intermediate range missile, and later, the Minuteman, an all-solid-propellant missile.

The Atlas and Titan had a range of about 10000 kilometers and a payload capability of 700 kilograms. The Atlas was powered by two 667-kilonewton (150000 lb thrust) first-stage engines plus a 267-kilonewton (60000 lb) sustainer engine. At launch, all three engines operated and at the end of first-stage operation, the two large engines were jettisoned leaving the sustainer engine to continue to operate during the second phase. Propellants for all three engines came from common tanks which constituted the bulk of the structure. These tanks were made of thin-gage stainless steel and depended upon internal pressure for structural stability. Since Atlas jettisoned only its first-stage engines, it was called a 1½ stage vehicle. Titan I, on the other hand, was a

---

*It consisted of a modified Redstone first stage and three upper stages of solid rockets. The three upper stages used 11, 6, and 1 Sergeant solid rockets, respectively. The Sergeant, 11 cm in diameter, was developed by the Jet Propulsion Laboratory, which teamed up with ABMA in building and launching the first satellite.

conventional two-stage vehicle which jettisoned both first-stage engines and associated tankage. Its tanks were of the more conventional design with internal ribs for structural stability. Titan I's engines were similar to those for Atlas, and both vehicles used a jet-grade fuel similar to kerosene, with liquid oxygen as oxidizer.*

The Atlas, developed by the Convair division of General Dynamics Corporation, is of special interest to our story.† As one of two contractors studying 8000 kilometer vehicles for the Air Force in 1947, Convair chose a ballistic missile over a winged, subsonic vehicle—a bold decision at the time. A key to long-range ballistic missiles was achieving very light structures and an imaginative Convair engineer, Karl J. Bossart proposed several bold innovations for light structures. By the end of 1948 and three test flights, Bossart was able to incorporate his innovations into the Atlas design. One was the use of integral, thin-wall, pressure-stabilized tanks previously mentioned. Although Oberth had proposed such balloon-like tanks in 1923 and both the Glenn L. Martin Company and North American Aviation had used the concept in satellite designs for the Navy (pp. 41, 44), such tanks had never been built and flown. Bossart had independently conceived the idea during design calculations when he found that the tank pressure needed for the inlets of the engine's pumps was greater than the internal pressure necessary for the tanks to remain stable under aerodynamic forces and vehicle loading. Bossart also dispensed with insulation for the liquid oxygen tank and used swiveling rocket nozzles to control the pitch, yaw, and roll of the vehicle during flight.[7]

Bossart's innovations and the initial Atlas project aroused little interest until the early 1950s when the Air Force swung away from air-breathing propulsion and winged missiles in favor of ballistic missiles. The break for Atlas came in 1954 when the Air Force Strategic Missiles Committee recommended that it be developed with some changes. The committee also recommended that a new management group be established to accelerate ballistic missile development. This resulted in the formation of the Ballistic Missile Division of the Air Research and Development Command under Brig. Gen. Bernard Schriever, and the ballistic missile program began to accelerate. In fiscal year 1953, funding was $3 million; in FY 1954, $14 million; in FY 1955, $161 million. In February 1955, another advisory committee recommended additional development of intermediate range ballistic missiles (IRBM) with a range of 2800 kilometers. By the summer of 1955, the Air Force had two ICBMs (Atlas and Titan) and one IRBM (Thor) under development. The Army won approval to develop the Jupiter IRBM.‡ The Navy turned to solid propellants and the Polaris missile was initiated. The Air Force also became interested in solid-propellant missiles and in 1957 began development of the Minuteman. During this period, funding continued to climb: in FY 1956, $515 million; in FY 1957, $1.3 billion.[8] Thus, by the time of Sputnik, six U.S. missiles were being developed with the highest national priority, and

---

*The later Titan II used storable propellants, UDMH and N_2O_4.

†Consolidated Aircraft became Consolidated-Vultee (Convair) in 1943 and was absorbed by General Dynamics in 1954. The division building rockets has been named Convair, Astronautics, and now Convair-Aerospace.

‡Jupiter, developed by the Chrysler Corp. for the Army, is not to be confused with Jupiter C developed in-house by ABMA.

all were larger and had greater payload capability than Vanguard. By 1958, development of liquid-propellant missiles not only provided the basic technology applicable to future launch vehicles but also the vehicles themselves were to become the greater part of the first generation of launch vehicles.

A key technology responsible for achieving low structural weight of the Atlas missile was Bossart's thin-gage, pressure-stabilized tanks. This concept met with considerable skepticism during the development of the Atlas. Opponents pointed out that if pressurization should fail, the tanks—and the missile—would collapse of their own weight. An equal concern was how well the tanks could withstand high aerodynamic loads during the early part of a flight. Doubt was sufficiently widespread that Titan, the second ICBM, was built with tanks of conventional design. An unanticipated severe test of the pressure-stabilized tanks came with the first test flight of Atlas in 1957. Hot exhaust from the turbopump burned through control wiring and the vehicle began to tumble while still in the atmosphere, placing excessive loads on the tanks. They held, and anyone viewing the film of the flight could easily become a convert to Bossart's concept. In spite of this, however, some engineers remained unconvinced, and prominent among them were those in the von Braun team.* This attitude was important to their later consideration of liquid hydrogen, as we will see.

During the build-up of missile capability in the 1950s, President Eisenhower and the Department of Defense kept booster programs closely related to surface-to-surface military requirements, much to the disappointment of space enthusiasts. A prevailing attitude was that spaceflight was not yet practical, and work to make it so was far too costly to be taken very seriously. To be sure, there was tolerance for research on high-energy propellants and other means for achieving high rocket velocities, but it was peripheral to the main task of developing long-range ballistic missiles. Up to the time of Sputnik, talk of spaceflight was very unpopular in the halls of government and proponents had to tread very lightly. "Space Cadets" were frowned upon and the use of the word "space" in a proposal in pre-Sputnik days invited budget cuts within the executive branch and in Congress.

Competition for the Space Role

Sputnik unleashed all the pent-up desires of U.S. space proponents in both the military and civilian sectors. The military advantages of satellites for reconnaissance and communications were obvious, but plans ranged far beyond these applications. For years the Air Force had quietly been preparing for manned flight into space. The Army was more aggressive, speaking of moon bases as the ultimate "high ground." The Army also had the superb missile development team of ABMA with over three

---

*Bossart and his colleagues at General Dynamics staged a demonstration in an attempt to show engineers from ABMA the toughness of the thin-wall tanks. They pressurized a discarded Atlas tank and invited one of the engineers to knock a hole in it with a sledge hammer. The blow left the tank unharmed, but the fast rebounding hammer nearly clobbered the wielder. In another instance, von Braun expressed his attitude towards the tanks during a good-natured exchange on using the Atlas for Project Mercury: ". . . John Glenn is going to ride on that contraption? He should be getting a medal just for sitting on top of it before he takes off!" Interview with K. J. Bossart, 27 Apr. 1974; group interview with Grant Hansen, K. E. Newton, Deane Davis, Donald Heald, and Bossart, Convair Aerospace Div., San Diego, 29 Apr. 1974.

thousand engineers and technicians to provide sound, detailed proposals. Maj. Gen. J. B. Medaris, ABMA commander, was a strong space advocate and had the backing of those above him, especially the blunt and aggressive Secretary of the Army, Wilbur Brucker. Navy space enthusiasts lacked high-level support, hence the Navy was not a strong competitor. The Air Force, with responsibility for intercontinental ballistic missiles, viewed space as a logical extension of its airspace. It was already skirmishing with the Army over the Thor and Jupiter IRBMs and this extended into their bid for a role in space.

With the big money in the military and their traditional role of spearheading costly flight developments, a strong civilian role in space appeared remote. Even the first scientific satellites were managed and controlled by the military, although the scientific community had access to the resulting data. Almost everyone assumed that the same arrangements would characterize future U.S. space efforts.

The only civilian government group seriously in the space role competition was the normally quiet and timid National Advisory Committee for Aeronautics (NACA), which had almost missed the boat on jet propulsion twenty years earlier. NACA had smart, eager young men as well as wise old officials, and some of both groups were determined not to miss the opportunities offered by space exploration. The NACA lacked the money and clout of the military services and traditionally cooperated with the military on expensive development projects, such as the X series of experimental aircraft. The military provided the funds, managed the development and initial operations, while the NACA provided the instrumentation and analyzed the experimental results. Eventually, the aircraft were turned over to NACA. In its first proposals for space exploration, NACA's director Hugh Dryden envisioned the same sort of working relationship, but both he and the military reckoned without the will of the ex-military man in the White House.

President Eisenhower was well aware of the interservice rivalries as well as the international implications of a peaceful effort for space exploration. In response to Sputnik, he had allowed the Army to proceed with a back-up to Vanguard, but he had not accepted the concept of a significant military effort in space. In November 1957, he appointed an old and trusted friend and advisor, James R. Killian, president of the Massachusetts Institute of Technology, to be his special assistant for science and technology. Killian and his science advisory committee played a key role in influencing the policy for space research during the months that followed. That policy turned in favor of a civilian space program.

Consolidation of Military Space Projects

In his 9 January 1958 State-of-the-Union message, President Eisenhower spoke of the need for a single focal point for advanced military projects, including anti-missile and satellite technology. Four days later, Secretary of Defense Neil McElroy told the House Armed Services Committee that he was establishing an Advanced Research Projects Agency (ARPA) responsible to him for anti-missiles and outer space projects. ARPA was formally established on 7 February with Roy W. Johnson, a former executive vice president of the General Electric Company, as the director and Rear Admiral John Clark as his deputy. A month later, Herbert F. York, director of the

Livermore Laboratory of the Atomic Energy Commission and associate director of the department of physics of the University of California, was appointed chief scientist. ARPA had authority over all military space activities.[9]

On 27 March, President Eisenhower approved ARPA's plans for space exploration as announced by Secretary of Defense McElroy. When a new civilian space agency was organized, it would take over the non-military space programs. ARPA's plans included earth satellites and space probes for scientific investigations, the latter as part of the International Geophysical Year program. Losing no time, ARPA authorized the Air Force Ballistic Missile Division to launch three lunar probes with Thor-Vanguard vehicles and the Army Ballistic Missile Agency to launch deep space probes with the new Jupiter IRBM equipped with the same cluster of solid rocket stages that had placed Explorer I in orbit. The original FY 1959 budget request of $340 million for ARPA was raised to $520 million.[10]

Not long after he went to ARPA, York met David A. Young of Aerojet and invited him to work with him. Young, who had worked with liquid hydrogen and oxygen in the late 1940s (p. 33), accepted.[11] He was among the first of a number of highly competent rocket and missile experts recruited for ARPA by Johnson, Clark, and York. These experts, hired and paid by the Institute of Defense Analysis, a private firm that provided technical and administrative services for ARPA, received the same salary as they did from their former employer. Young recruited Richard B. Canright from Douglas Aircraft, where he had been assistant chief engineer of missile systems. Prior to that, Canright had conducted research at the Jet Propulsion Laboratory of the California Institute of Technology on liquid and solid propellant rockets. Canright, a knowledgeable and experienced propulsion expert, was well familiar with hydrogen. He had operated a rocket on gaseous hydrogen and oxygen in the early 1940s (fn, p. 34) and wrote a paper in 1947 on the relative importance of specific impulse and density for long-range rockets (pp. 47–48). He and David Young were members of the NACA subcommittee on rocket engines and staunch supporters of the NACA high-energy rocket program. From the Air Force came dynamic and aggressive Richard S. Cesaro. He was recommended to York by Richard Horner, assistant secretary of the Air Force for R&D, and came to ARPA in June.[12] Cesaro, a long-time employee of NACA in propulsion research and committee management, had moved to the Air Research and Development Command headquarters in January and was the technical director for aeronautics and astronautics. Cesaro was a master at maneuvers in government decision-making processes, a technical gadfly, and an aggressive proponent for using advanced technology.[13]

By early June, a number of experts were working for ARPA and York assigned them to a number of ad hoc panels to plan and initiate military space programs.*

Canright organized an informal panel on vehicles and he persuaded some of his colleagues—Cesaro, Youngquist, Irvine, and Young—to serve on it. In mid-June 1958,

---

*John F. Kincaid to solid propellant chemistry. Samuel B. Batdorf to man-in-space. Charles R. Irvine and Capt. R. C. Truax to Project 117-L, Arthur J. Stosick to large engines. David A. Young to communication relays. Roger B. Warner to meteorology. Richard B. Canright to scientific satellites. Col. Dent L. Lay to Project ARGUS. Robertson Youngquist to exploratory research, and Richard S. Cesaro to satellite tracking.

Canright and Young were asked by Johnson to present ARPA's plans to a panel of the National Security Council. Canright made two recommendations on vehicles and engines: use a cluster of proven rocket engines for large vehicles, and use hydrogen and oxygen as propellants in upper stages. To Canright, using multiple rocket engines for space vehicles was an extension of aircraft practice, where the redundancy of multiple engines was a tried and proven method of achieving reliability. According to Canright, the panel accepted his recommendations, and he later took advantage of this apparent endorsement to push for his ideas in ARPA planning. Richard Cesaro, long a proponent of high energy fuels for air-breathing engines and rockets, also favored the use of hydrogen-oxygen as propellants for upper stages. He, like Canright, supported the use of multiple engines for large vehicles.[14]

NACA Takes the Initiative

During its 43-year life, the NACA had been content to be a government aeronautical research organization. At the end of World War II, there were three major research laboratories with a staff of about 8000. It contracted a modest amount of research with universities and non-profit institutions. As a non-competing service organization, the NACA was close to and strongly supported by both the military and the aeronautical industry. In the 1950s, the laboratories had started research on missiles; by 1957, such work constituted from a quarter to a third of total research.[15] Among the staff were rocket and space enthusiasts who saw Sputnik as merely an endorsement of what they had been advocating. The majority of the staff, however, were deeply committed to aircraft powered by air-breathing engines, and Sputnik produced an ambivalence. As Bruce T. Lundin, then a division chief at the Lewis Flight Propulsion Laboratory and now its director, saw the situation, "we were divided into two strong camps . . . half were afraid we were going to get sucked into space [research] and the other half were afraid that we were going to get left out." Lundin's view was that the future of the NACA lay in responding to the national need and to "use our unique capability to bring our nation into space. It was either us or the military and I really felt that the United States should go into space as a peaceful civilian activity rather than carrying a sword."[16]

Early in December 1957, Hugh Dryden, NACA's director, summoned the directors and associate directors of the research laboratories to Washington to discuss the future posture of the NACA with respect to space. Abe Silverstein, associate director of the Lewis Flight Propulsion Laboratory, went with Ray Sharp, the director. Silverstein asked Lundin, who had been expressing his views freely, to put them in writing as the Lewis position. Lundin spent a Sunday afternoon writing a paper which Silverstein used, with minor revisions and additions, at the Washington meeting. Dryden opened the meeting with some introductory remarks about the possible courses for the NACA and then asked the research directors for their opinion. The Lewis position was the most enthusiastic response that Dryden got from the Center people. Henry J. E. Reid and Floyd Thompson from the Langley laboratory were not very enthusiastic about building up the NACA to move into space activities. Smith DeFrance from the Ames laboratory was opposed to it, fearing that it would destroy the whole concept on which the NACA was based.[17]

The Lundin-Silverstein paper presented compelling arguments for an active role in space activities. After discussing and discarding two possible roles, the suggested one was:

> . . . a bold and visionary approach based on (a) the importance of space exploration, (b) its urgency and importance to our national survival, (c) the importance of research at this time, (d) our traditional role of leadership in independent aeronautical research, (e) the obvious need for consolidating present Governmental research on space problems into a single agency, (f) our future needs in the way of staff and facilities, and (g) examples of what new knowledge is needed and how the foregoing would provide it.[18]

Following the meeting of the laboratory directors, the chairman of the NACA, James Doolittle, and Dryden took a very unusual step. They invited about 30 research laboratory middle managers—division chiefs in the main—to Washington for cocktails and dinner and an unfettered discussion of what they thought the course of the NACA should be. The 18 December 1957 event, known as the Doolittle dinner by some and the Young Turks dinner by others, was an affair with no holds barred. One or two took the opportunity to berate NACA management for their ultra-conservative position in the past and to air old grievances. Most, however, were with Walter T. Olson, chief of the fuels and combustion division at the Lewis laboratory, when he stood up and made a strong argument for moving boldly into space. Doolittle and Dryden got the message: the younger NACA staff members were enthusiastic for space.[19]

The long established procedure of the NACA, when faced with the prospect of entering a new field, was to form a special advisory committee to look into the matter. This not only obtained the services of prominent and knowledgeable people but formed a basis of support. In November 1957, NACA authorized a special committee on space technology which was organized in January 1958. H. Guyford Stever, associate dean of engineering of the Massachusetts Institute of Technology, was the chairman.* This committee formed seven working groups involving many persons who later became prominent in space activities.

The NACA staff also began its own studies of space technology and desirable research objectives. On 14 January 1958, Dryden released a NACA staff study entitled "A National Research Program for Space Technology." It called for a space effort based on cooperation between government agencies. NACA would step up its space activities, build new facilities, and add to the staff, but would limit its work to basic research. Large vehicles would be flown by the Department of Defense with technical assistance by NACA. This was similar to past arrangements, particularly for research

---

*Among its sixteen members were Norman C. Appold, the manager of the Suntan project; Wernher von Braun, technical director of the Army Ballistic Missile Agency; and Abe Silverstein, associate director of the NACA Lewis Flight Propulsion Laboratory. Also on the committee were J. R. Dempsey, manager of General Dynamics–Astronautics and responsible for developing the Atlas, and S. K. Hoffman, general manager of the Rocketdyne Division, North American Aviation.

aircraft such as the X-15, and would offend no one. Two days later, the main committee of the NACA met and passed a resolution on space flight calling for a national program involving research in space technology, development of scientific and military space vehicles, and research on higher atmosphere and space phenomena. A cooperative program between NACA, DoD, the National Academy of Sciences, and the National Science Foundation was seen as the best way to implement the program. The resolution emphasized that the NACA role in space was one of coordination and research on space technology requiring expansion of its current activities and called on the special committee on space technology to review needed research and help formulate a program for the NACA.

On 10 February 1958, the NACA staff reinforced the resolution with a report entitled "A Program for Expansion of NACA Research in Space Flight Technology: With Estimates of the Staff and Facilities Required." The report described a program bold in concept and broad in scope.*

The implications of Sputnik for civil space activities were considered by the Congress, and by early 1958, several proposals were pending—including one that would put the space program under the Atomic Energy Commission. None of these suited the administration and in February, President Eisenhower asked his scientific advisor, James Killian, to devise a plan. The result was a recommendation in March, approved by the President, to give the civil portion of the space program to the NACA and strengthen and rename it. On 2 April, less than a week after giving the ARPA the green light on its space plans, Eisenhower sent Congress a bill to establish the National Aeronautics and Space Administration. At the same time, he directed NACA and DoD to discuss current space programs and decide which should be transferred to NASA when it came into being. The Bureau of the Budget thereafter took a very active role in pressing for decisions on the transfer of programs and funding from DoD to NACA-NASA. Thus, in the period from 2 April until NASA was formally in business on 1 October, the new organization had plenty of clout to contend with the ARPA, Army, and Air Force in jockeying for a role in space.

Following the President's 2 April directive, ARPA's Roy Johnson and Herbert York met with NACA's J. W. Crowley, Ira Abbott, and Robert Gilruth to discuss the transfer of programs. It was obvious that purely scientific space programs would be transferred and that reconnaissance satellites would remain with the military, while the disposition of manned spaceflight and launch vehicles was uncertain. Both the military and civilian sides saw a need for large launch vehicles and both included such vehicles in their planning, leaving the precise responsibility for later resolution. It was inevitable, however, that ARPA and NACA were on a convergent course with respect to launch vehicles and propulsion.

---

*Among the facilities proposed was one for chemical rockets up to a thrust of 4.5 MN (1 million lb). A smaller test stand equipped with an ejector to reduce ambient pressure at the nozzle for altitude simulation was also proposed. Other facilities included nuclear rockets, pumps, gas generators, and smaller-scale rocket stands. Liquid hydrogen was named as one propellant to be used. These facilities, estimated to cost $380 million, would be built over a 5-year period. The plan also called for an operating budget increase of $100 million annually and more than doubling the staff (to 17 000).

During the spring of 1958, NACA's director of research Hugh Dryden sought a strong leader within the NACA staff to come to Washington and help him formulate a civilian space program. He found his man in Abe Silverstein.*

Silverstein (1908-   ) was a sharp, aggressive, imaginative, and decisive leader. He could be charming or abrasive. He was a hard bargainer at the conference table in technical and management matters but very warm-hearted in personal relationships. He could cast work aside like a cloak and radiate such warmth and empathy for people that those who had felt his lash in a technical discussion earlier could forget their chagrin and respond to him with equal warmth. Many damned his ways but liked the person. He had an uncanny technical intuition, or feel, for the right approach: and those who were dismayed at his methods could scarcely question his judgment. To a casual observer he might appear to be a one-man show, one who would not delegate, or one who liked to participate in all technical decisions, large and small. Yet this is not a complete picture, for Silverstein and his methods were far more complex. In conferences with his superiors or peers, Silverstein was a restrained yet highly skilled proponent for his cause. He was a good moderator or chairman. On the other hand, in conferences on matters where he was directly responsible for the outcome, he was far more direct and aggressive. At the latter, he liked to gather together a group of his subordinates—and, later, contractors—about the conference table and engage in a free-for-all argument over various technical merits or weaknesses of a program or proposed action.

Sometimes he displayed a near-mania for winning the argument, especially on rare occasions when it became rather obvious that he was on the wrong side. Wise associates never pressed him too hard when he painted himself into a corner, for he would never admit it and more time would be lost. Even then his amazing sixth sense in engineering was functioning and absorbing all inputs, and he never followed a bad argument with a bad decision. He respected those who stood up to him and stoutly made a good technical point, but woe to him who made a weak argument, for Silverstein could be relentless. Strangely enough, this dominant personality seldom produced lasting antagonism and did not diminish the growth of strong and competent subordinates; many went on to distinguished careers.†

Silverstein strove always for excellence and he inspired the same in those who worked with him. He attributed this trait to his mother, whom he described as "ideally trained to do space work because she knew the importance of perfection." He eschewed politics, or its equivalent, in management and disdained image building, which is probably why he did not regard himself as a "headquarters-type" person.

---

*Silverstein almost lost the opportunity. He recalls that very early in 1958 Dryden asked him to work with him on space planning, but Silverstein refused because he was "not a headquarters-type person." Silverstein, however, was intrigued with the opportunities offered by space and about a month later, approached Dryden with an organizational plan for NACA. He remembers Dryden looking at him coldly and saying, in effect: "Silverstein, I invited you up here to work with me on this thing. If you are willing to come up, fine, otherwise, forget it." Silverstein took the job. Silverstein interview, 29 May 1974.

†Three of the best known are: George M. Low, past deputy administrator of NASA and now president of Rensselaer Polytechnic Institute; Edgar M. Cortright, past director of the Langley Research Center and now an executive in industry; and Bruce T. Lundin, director of the Lewis Research Center.

Fig. 48. Abe Silverstein, director of space flight programs, National Aeronautics and Space Administration, 1958–1960, who had a decisive role in the use of liquid hydrogen as a fuel in the upper stages of the Saturn launch vehicle. (Photo ca. 1955.)

In the spring of 1958, Silverstein transferred to NACA headquarters and began to assemble a staff to help him plan a space program. Others from the three laboratories joined in to help as called upon. By mid-year, personnel at the laboratories and at headquarters were assigned to 11 program elements.* By mid-July, a FY 1960 budget proposal had been prepared by Silverstein, Robert Gilruth, Morton Stoller, Edgar Cortright, and Newell Sanders. The space vehicle portion was $349 million and included vertical probes, 12 small satellites and 3 larger ones for scientific experiments, a satellite for an astronomical telescope, 3 communications satellites, 4 lunar probes, inter-planetary probes, 4 manned space capsules weighing 1140 kilograms each, a 4450 kilogram manned satellite for biological and life science studies, and a winged vehicle for a recoverable space vehicle. The budget called for $80 million for propulsion systems, including $30 million for a single engine of 4.5 meganewtons (1 million lb. thrust), $15 million for nuclear rockets, $12 million for high energy propellants, $15 million for a clustered rocket of 4450 kilonewtons (one million lb thrust), $5 million for solid rockets, and $3 million for solid propulsion components. The budget also called for $26.7 million for a spaceflight staff of 1700 and $50 million for facilities, including a space projects center.[20] On 29 July 1958, President Eisenhower signed H.R. 12575 making it the National Aeronautics and Space Act of 1958. The NACA was absorbed, along with its laboratories and personnel, when NASA officially began operations on 1 October 1958. The new space agency was humming with activity and Silverstein was its chief planner and director under administrators Glennan and Dryden.

---

*Unmanned satellites: P. Purser, A. J. Eggers, A. Zimmerman, F. O'Sullivan; manned spacecraft: M. Faget, H. Henneberry, Eggers; astronomical telescope: Zimmerman, O'Sullivan, R. T. Jones; meteorology: E. Cortright, M. Stoller, Zimmerman, O'Sullivan; communications: N. Sanders, Stoller; lunar probes: Brown, Stoller; internal power: Cortright, O'Sullivan, von Doenhoff; advanced rockets: A. O. Tischler, G. Thibodaux; range: E. Buckley, Stoller, F. Smith; guidance: Sanders, Stoller; other projects: Zavasky, Cortright.

10

Early High-Energy Upper Stages

During the 1950s, interest in high-energy propellants for upper stages of rocket vehicles had steadily mounted. Such propellants were initially seen as a means for increasing the range of ballistic missiles, but this shifted early towards increasing the capability of rockets to launch satellites and space probes. The principal candidates were hydrazine-fluorine, hydrogen-oxygen, and hydrogen-fluorine, but none had reached the development stage. The coming of Sputnik and U.S. plans for a strong space program quickly produced action; the first high-energy upper stage authorized for development used liquid hydrogen–oxygen. The decision, and subsequent development, owed much to an earlier program, which had faltered—Suntan, the program to develop a high-altitude reconnaissance airplane fueled with liquid hydrogen (chap. 8).

Legacy of Suntan

The Pratt & Whitney Aircraft division of United Aircraft Corporation was operating the first 304 hydrogen-fueled engine in the initial series of Suntan tests at the Florida test center when Sputnik was launched. At that time there was no indication that Suntan was soon to end, but neither this possibility nor Sputnik caught the astute Perry Pratt napping. About two years earlier, he had recruited C. Branson Smith from the Hamilton Standard division of United Aircraft, and one of his first assignments was to study the possibilities for Pratt & Whitney's entry into the rocket field.[1] With the Rocketdyne division of North American Aviation and Aerojet-General Corporation the giants in the large liquid-propellant rocket engine business, and Bell Aircraft and the Reaction Motors division of Thiokol smaller but very aggressive, it was obvious that a newcomer would need to do something new and different. As the division began to move fast on the hydrogen-fueled engine for Suntan in early 1956, Smith found his answer. In April 1956, he jotted down in his work log two subjects of potential interest: hydrogen as a high-energy fuel and pentaborane as a storable fuel. Smith's approach to rocket work was methodical: educate the staff, make an evaluation of rockets to establish the best type on which to concentrate, and propose an experimental contract to gain experience and advance the basic technology. In May 1956, Smith briefed Pratt on early results. From this meeting came a decision to summarize the status of hydrogen rocket engine work, to consider rocket engines for aircraft auxiliary propulsion, and to consult with Wesley Kuhrt on the air turborocket. Smith visited the Los Alamos Scientific Laboratory and learned about their nuclear rocket. Although ammonia was to be the working fluid in early tests, Los Alamos was also interested in using hydrogen; the Livermore Laboratory, also working on nuclear rockets, had definitely planned to use hydrogen. Smith learned a great deal about hydrogen at Los Alamos and at the Bureau of Standards cryogenic laboratory in Boulder. He submitted his report on rockets to Pratt in June and revised it the next month. He concluded that: (1) in the next decade the rocket would become capable of performing most military missions at greater speeds and altitudes than gas turbine powered aircraft, (2) the most immediate application for Pratt & Whitney was an auxiliary thrust rocket to increase aircraft performance, and (3) in the missile field the ultimate fuel was hydrogen. Smith saw the first step for the division as a general educational program; the second step, the development of a small rocket for boosting aircraft in combination with gas turbine engines; and the third step, to propose hydrogen for long-range ballistic missiles. Smith added that the division's present and anticipated experience with hydrogen offered an opportunity to overtake existing rocket competitors.[2] Although this is what happened later, Smith's recommendations did not produce immediate action. Pratt & Whitney was fully engrossed in the Suntan program to use hydrogen in a modified J-57 and in developing the 304 engine. By the fall of 1956, Smith was making calculations of hydrogen and oxygen as coolants, and in November he visited NACA's Lewis laboratory to learn about work on hydrogen as a rocket fuel with oxygen and fluorine as oxidizers.

On 4 April 1957, Smith and Pratt made a presentation on hydrogen-fueled rockets at a management meeting of United Aircraft officials.* As a result, Pratt asked Smith to prepare a proposal for the first phase of a hydrogen rocket program. He was to make a cost estimate, note the limited availability of hydrogen, and consider a possible substitute fuel. Apparently, the United Aircraft management was willing to get involved in rocket work, but not wholly convinced that liquid hydrogen was the best fuel. In mid-April Smith was among a group of engineers who met with the Navy on boron fuels. After the meeting, Pratt took Smith to visit Col. Norman C. Appold, the Suntan manager, at the Air Force's Air Research and Development Command headquarters in Baltimore. They proposed that Pratt & Whitney develop a liquid hydrogen–liquid oxygen rocket engine, but Appold was not receptive; he wanted Pratt & Whitney to concentrate on Suntan objectives.

In July 1957, Smith summarized a year's thinking about rockets. In propellant evaluation, he indicated that hydrogen led for vehicles requiring maximum performance, such as ICBMs, satellites, and space ships. Hydrogen was well-suited for long-range missiles and could halve the gross mass of those being developed using kerosene and oxygen. He also saw advantages for hydrogen for shorter-range missiles, particularly stages that must accelerate to very high velocities. The substitution of hydrogen for kerosene in the second stage of an ICBM would increase the payload 50 percent without increasing the gross mass. Alternately, ICBMs then being developed could place payloads into a satellite orbit by using hydrogen in the second stage. He recommended to Pratt that a vigorous effort be mounted to develop a hydrogen-

*H. M. Horner, L. S. Hobbs, W. P. Gwinn, W. A. Parkins, and B. McNamara.

oxygen rocket engine. Smith also considered boron fuels and solid propellants in his review and proposals to the Navy and Air Force.[3]

On 25 July 1957, Smith visited the power plant laboratory at Wright Field and made a presentation featuring hydrogen as a rocket fuel for a 267 kilonewton (60000 lb thrust) second stage for an ICBM. Because of the classification of Suntan, he omitted telling his audience about the considerable experience Pratt & Whitney had amassed with liquid hydrogen.* The reaction of the Wright Field group to Smith's presentation was that his results appeared reasonable, but hydrogen had not been pursued previously because of its high cost and low availability. Smith was encouraged enough by the reactions to complete a proposal in August, but nothing came of it. It was the same month that the first 304 hydrogen expander engine was assembled at the East Hartford plant for shipment to the Florida test center and two months before Sputnik I.

The Air Force had ample precedent, in previous work at Ohio State University in the 1940s, to be interested in hydrogen for rockets. The fuels and propulsion panel of the Air Force Scientific Advisory Board foresaw the need for a hydrogen-fueled rocket a year before Sputnik I. At the panel's meeting on 14 November 1956, high-energy propellants were considered for upper stages of high-performance rockets. The panel, of which Abe Silverstein was a member, recommended that two rocket engines be developed in the 111 to 222 kilonewton (25000–50000 lb thrust) size using high-energy chemical propellants.† Liquid hydrogen–oxygen was singled out as being a particularly attractive high-energy combination. The panel was aware of the Air Force's plants which had been built to produce liquid hydrogen in quantity for the "air-breathing super fuel program" (i.e., Suntan) and that ample quantities would be available for testing. The ballistic missile program used vast amounts of liquid oxygen, so the panel thought the time for using the hydrogen-oxygen combination had come.[4]

Although the minutes do not single out the contributions of individual members, the influence of Abe Silverstein is unmistakable. He had been intensely interested in liquid hydrogen as an aircraft fuel since 1955 and was directing a strong research program on it. As early as 1950, he had organized a meeting of government and industry rocket experts on the subject of high-energy rocket propellants for long-range missiles (p. 76). At the 1950 meeting, the rocket group at the NACA Lewis laboratory had recommended liquid hydrogen as their first choice for fuel, with hydrazine and ammonia as alternatives. Fluorine was the favored oxidizer, with oxygen as the alternate. At the time of the panel meeting, hydrazine or ammonia with fluorine, and hydrogen with fluorine or oxygen, were the high-energy combinations of greatest interest in the country. Prior to its meeting, the panel had visited the NACA Lewis laboratory, the Air Force power plant laboratory, and Pratt & Whitney.

The Air Force waited over a year (until December 1957) to reply to the board's recommendations. With respect to rocket engines using ammonia-fluorine, the Air Force cited a contract with Bell Aircraft on experiments with a 156 kilonewton (35000 lb thrust) chamber using ammonia-fluorine. With regard to hydrogen-oxygen, the Air

---

*Col. Appold, the Air Force's Suntan manager, had visited the laboratory the previous day and presumably had informed the staff about Pratt & Whitney's hydrogen experience.

†Dr. Mark M. Mills was chairman; other members attending, besides Silverstein, were Dr. W. Duncan Rannie of JPL and C.I.T. and Dr. Edward S. Taylor of M.I.T.

Force commented that six months previously (about June 1957), Wright Field had prepared a procurement request for a liquid-hydrogen engine, but did not send it to industry because of NACA work with this combination. The November 1957 NACA firing of a liquid hydrogen–fluorine rocket (p. 92) was cited. The Air Force intended to follow through when NACA completed its exploratory work—an indication that the Air Force felt no great rush for action, two months after Sputnik I.[5] This was consistent with the disappointments Pratt & Whitney was experiencing in trying to sell the Air Force a hydrogen-oxygen engine. The stimulus needed was to come later in a negative way—the demise of Suntan.

During the remainder of 1957 and the early part of 1958, Smith and his group at Pratt & Whitney continued to work on the analysis of hydrogen cycles and the layout of engines with thrusts ranging from 31 to 133 kilonewtons (7000 to 30 000 lb). Exploratory meetings were held with the Air Force, and on 4 March 1958, Perry Pratt sent to the ARDC a preliminary design and proposal for a 68 kilonewton (15 000 lb thrust) advanced rocket engine using liquid hydrogen and liquid oxygen as propellants.[6] It was intended for the Air Force's growing astronautics program and specifically for applications being developed by the Missile Systems Division of Lockheed Aircraft Corporation. Lockheed was studying advanced versions of its WS117L reconnaissance satellite, and their work indicated a thrust level of about 31 kilonewtons (7000 lb). Lockheed, however, wanted to use fluorine instead of oxygen with hydrogen.

The hydrogen-fluorine combination produces peak performance using a smaller proportion of hydrogen than the hydrogen-oxygen combination; when this is coupled with the greater density of fluorine over oxygen, the result is a much more compact stage for the same thrust and duration. (Lockheed was also very interested in an even denser combination, hydrazine-fluorine.)

By the spring of 1958, the Suntan management team at ARDC decided that the time had come for a rocket engine using liquid hydrogen to power an upper stage.[7] They were aware of the Lockheed studies and the efforts of Krafft Ehricke of Convair-Astronautics to sell ARDC a Mars probe using a hydrogen-oxygen stage on top of the Atlas intercontinental missile.

The Suntan team may have hedged on the selection of oxidizer. Liquid oxygen was the safest choice, but fluorine was also of interest.* The team coordinated the proposal with Brig. Gen. Marvin C. Demler, deputy commander of ARDC, and it was signed by Lt. Gen. Sam Anderson, the commander. The proposal was addressed to Gen. Thomas C. White, Air Force chief of staff, but the air staff decided to pass it to Richard E. Horner, the assistant secretary of the Air Force for research and development.[8] Horner had followed the Suntan work closely and was aware of its coming termination and the desirability of finding an application for the new technology. The Suntan management team arranged a briefing for him and brought in Pratt & Whitney representatives to strengthen their presentation. Horner favored the proposal, but decided it should be

---

*In addition to Lockheed's interest in fluorine, three of the Suntan team—Col. Norman C. Appold, Lt. Col. John D. Seaberg, and Capt. J. R. Brill—had attended the NACA conference in November 1957 that was devoted largely to liquid hydrogen as a fuel, including data from the firing of a liquid hydrogen–fluorine rocket.

sent to Roy Johnson, director of the Advanced Research Projects Agency which was heading military space coordination. Johnson and his staff were briefed on 13 June 1958 (the day following a briefing of the Air Council on Suntan). Horner also arranged for ARPA staff members to visit the Suntan liquid hydrogen facilities in July 1958. Richard Canright and David Young, both well acquainted with hydrogen, were greatly impressed at seeing liquid hydrogen being pumped through nearly a half kilometer of piping at Pratt & Whitney's Florida test center. The Air Force made a persuasive case for ARPA to choose Pratt & Whitney for the development of a liquid hydrogen rocket, based on the division's experience in building a hydrogen expander engine for Suntan and the ready availability of a large supply of liquid hydrogen.[9]

While the Suntan team was making a bid within the government to develop a hydrogen-fueled rocket, Smith and others at Pratt & Whitney revised their March 1958 proposal to the Air Force to conform with Lockheed Aircraft desires and resubmitted it on 5 May 1958. In the following weeks, the two companies cooperated in analyses and layouts of five different propulsion systems, all for a proposed advanced reconnaissance satellite. During this period there were many reviews of the work by Air Force and ARPA representatives. At one such meeting on 9 July 1958, Air Force, ARPA, Lockheed, and Pratt & Whitney representatives unanimously agreed to select liquid hydrogen and liquid oxygen as the propellants and a thrust level of 53.5 kilonewtons (12000 lb). Nine days later, Pratt & Whitney engineers drafted engine specifications and a proposal for company approval before sending them to ARDC. The engine was to be developed in 18 months at a fixed price of $19.8 million.[10] Essentially the same engine was in fact developed later as the RL-10.

In August 1958, the ARDC was authorized to proceed with the development of a hydrogen-oxygen engine. Its application was not for a Lockheed-built stage, however, but for a stage proposed by General Dynamics–Astronautics.

Origins of Centaur

The first rocket stage to fly using liquid hydrogen and liquid oxygen as propellants was the Centaur stage on top of an Atlas intercontinental ballistic missile. Centaur was the brainchild of Krafft Ehricke. For nearly three decades, Ehricke had prepared himself for the space age; when it dawned with Sputnik, he was ready. Within a month, he proposed a hydrogen-oxygen stage for use with the Atlas missile. Ehricke was able to move rapidly because previous work on the Atlas missile and the ideas of others about hydrogen-oxygen upper stages had laid the groundwork.

Ehricke became a space enthusiast at the age of eleven when he was captivated by Fritz Lang's "Girl in the Moon," shown in Berlin in 1928. Advanced in mathematics and physics for his age, he appreciated the great technical detail that Hermann Oberth had provided to make the film realistic. Young Krafft became acquainted with Tsiolkovskiy's space rocket using hydrogen-oxygen, which he read about in Scherschevsky's *Die Rakete fuer Fahrt und Flug*. He also tackled Oberth's *Wege zur Raumschiffahrt* in his early teens, but was slowed by the mathematics. Ehricke graduated from the Technical University in Berlin (aeronautical engineering) and took postgraduate courses at the Humboldt University in celestial mechanics and nuclear physics. He was conscripted into the army, served in a Panzer division on the

Russian front during World War II, but was recalled and reassigned to rocket development work at Peenemünde in June 1942. There he came under the strong influence of Walter Thiel, in charge of rocket engine development, who was killed in the first British air raid on Peenemünde in October 1943. Peenemünde, under Maj. Gen. Walter Dornberger and Wernher von Braun, his technical director, had a single purpose—the rapid development of specific weapons—and there was no official tolerance of work not directly related to the main goal. In spite of this, Thiel shared Ehricke's desire to look beyond the immediate future to greater possibilities. Thiel himself drew plans for testing rockets larger than any yet dreamed of—on the order of 5–14 meganewtons (1–3 million lb thrust). He wanted to use natural gorges in Bavaria as testing sites. He talked to Ehricke about resuming his own earlier experiments with liquid hydrogen in small rocket thrust chambers. The experiments of Heisenberg and Pohl with a nuclear reactor using heavy water excited Thiel. When he heard that Heisenberg was planning to operate a turbine with steam heated by the heavy water reactor, Thiel urged Ehricke to study the possibilities of using nuclear energy for propulsion. Ehricke considered several working fluids, but both he and Thiel favored hydrogen and believed it was a fuel with a future.[11]

As the war was ending, Ehricke helped move Peenemünde records into Bavaria, to keep them out of Russian hands. He made his way on foot to Berlin where he found his wife and went into hiding until the Western Allies moved in. He was located by the U.S. Army, given a six-month contract, and came to the United States to rejoin the von Braun team as part of the Paperclip operation.*

Ehricke and von Braun recalled another time they had considered hydrogen. In 1947, von Braun asked Ehricke to check a report by Richard B. Canright of the Jet Propulsion Laboratory on the relative importance of exhaust velocity and propellant density for rockets of the V-2 size and larger (pp. 47–48). It had caught von Braun's attention because he and two associates had written a paper the previous year which Canright had cited.[12] Von Braun had found, under the assumptions of fixed tank volume and a relatively heavy structural mass, that propellants with the highest densities and reasonably high exhaust velocities had the greater ranges. Canright, on the other hand, found that for large rockets and his assumptions (which included a variable tank volume and relatively light structural mass), exhaust velocity was decidedly more important than density. Canright's analysis showed hydrogen to be superior to other fuels when using the same oxidizer. Both Ehricke and von Braun, familiar with Oberth's case for using hydrogen-oxygen in upper stages of rockets (appendix A-2), agreed that hydrogen had a good potential for certain applications. Practical experience with liquid hydrogen in rockets at that time, however, was still very small and its handling problems large. The Army, for whom von Braun and Ehricke worked, wanted practical propellants that could be stored and handled safely in the field. This convinced von Braun to stick to well tested and denser propellants, but Ehricke felt less restrained and hydrogen's potential remained prominent in his thinking.

---

*Ehricke wanted to work for the Americans, and he hid each time someone knocked on his door, waiting for the right caller. One day his wife answered the door and routinely said, "I don't know where he is." As she did so, she recognized the insignia of a U.S. Army officer and immediately began screaming, "He's here! He's here!" Interview, 26 Apr. 1974. Paperclip was the project for bringing German rocket experts to the United States.

Fig. 49. Krafft A. Ehricke, father of the first hydrogen-oxygen stage, Centaur, shown receiving the Loesser award at the 1956 International Astronautical Congress, Rome. (Courtesy of F. C. Durant, III.)

In 1950, Ehricke moved with von Braun to Huntsville, Alabama, but grew restless with both the climate and von Braun's conservative engineering. He joined Walter Dornberger at Bell Aircraft in 1952. Bell, then busy developing the Agena upper stage, also proved to be unable to offer Ehricke the opportunity for a breakthrough he was looking for; and in 1954, when K. J. Bossart of Convair contacted him to work on the Atlas ICBM, he was ready to move. Soon after, the Air Force established the Ballistic Missile Division under Brig. Gen. Bernard Schriever to accelerate development of the Atlas. Schriever insisted on total dedication to the job at hand and Ehricke found himself again in the same atmosphere as at Huntsville and Peenemünde. In Charlie Bossard, however, Ehricke found a man of kindred spirit who, like Walter Thiel at Peenemünde, encouraged him to think beyond the immediate task. To imaginative Ehricke, this attitude and the climate and atmosphere of southern California were heaven.

By 1956, Ehricke was conducting in-house studies of vehicles for orbiting satellites. He approached the Rocketdyne Division of North American Aviation to obtain preliminary design data on various rocket propulsion systems employing turbine-driven pumps. He did not have much luck in getting government interest for his proposals, although he was a passionate believer in space exploration and a very persuasive person. The first week in October 1957, he visited Maj. George Colchagoff at ARDC headquarters. Ehricke had gotten wind of the Suntan project and was hoping to gain support for launching a satellite. It was the austere period under Secretary of Defense Charles Wilson, when "space" was out of favor. Although Colchagoff was receptive and personally convinced of the value of spaceflight, the official position made it difficult for Ehricke to round up support. On the Monday following Sputnik I's flight, however, Ehricke found many who indicated they had always favored spaceflight and now felt free to talk to him.

Excited by the new atmosphere, Ehricke returned to San Diego and began to streamline his plans. A. G. Negro, a Rocketdyne applications engineer, visited Convair on 11 October 1957 and returned with a request for information on a small pressure-fed rocket engine using liquid hydrogen and liquid oxygen. It was to produce 31 kilonewtons (7000 lb of thrust) and be capable of restart at altitude. By the end of October, Negro established the design characteristics of the engine, including a combustion pressure of 4 atmospheres and exhaust velocity of 4030 meters per second.[13]

In December 1957, General Dynamics–Astronautics submitted a proposal to the Air Force entitled, "A Satellite and Space Development Plan." It was for a four-engine, pressure-fed hydrogen-oxygen stage with each engine developing 31–33 kilonewtons (7000–7500 lb of thrust). According to Ehricke, "The reason why we selected this engine system in teamwork with the Rocketdyne Division of N.A.A. was simply that we wanted to avoid the delay by what we thought would have to be a brand new pump development. We were, for security reasons, not aware of the Pratt & Whitney's pioneer work in this field."[14]

The Air Force did not buy the specific General Dynamics–Astronautics proposal, but in the following months activities within the government clearly foreshadowed an emerging space program. Among these, General Dynamics–Astronautics and Pratt & Whitney were brought together by the Air Force and the Advanced Research Projects

Agency and in August 1958 were authorized to proceed with the development of the Centaur stage, the first to use liquid hydrogen and liquid oxygen.

NASA Plans, ARPA Acts

The ink was hardly dry on the President's signature establishing NASA on 29 July 1958 when Abe Silverstein established a committee to coordinate government plans for propulsion and launch vehicles. He called it an informal technical advisory committee for propulsion with himself as chairman. It had a flexible membership to meet the needs that arose.*

At the first meeting on 7 August 1958, the agenda included high-energy upper stages. Recognizing that a choice of the best high-energy propellant combination depended upon the application and stage design, a working group was appointed to review available information and present it to the committee for evaluation. Headed by A. O. Tischler, a propulsion researcher Silverstein recruited from the NACA Lewis laboratory, the working group was instructed to give particular attention to hydrogen-oxygen and hydrogen-fluorine.† In so doing, the Silverstein committee also agreed to defer a contract then under consideration by the Air Force to develop a hydrazine-fluorine engine of 356 kilonewtons (80000 lb thrust).[15]

At the second meeting of the Silverstein committee on 14 August, discussion of the high-energy propellant stages centered around a hydrazine-fluorine engine of 53 kilonewtons (12000 lb thrust), being developed by Bell Aircraft for the Air Force, and a contract just awarded by Wright Field to Aerojet-General to study the feasibility of a hydrogen-oxygen engine of 445 kilonewtons (100000 lb thrust). However, no actions were taken on these.[16]

By the time of the third meeting of the Silverstein committee on 28 August 1958, the Tischler working group on high-energy upper stages reported progress.‡ Hydrazine-fluorine engines in thrust ranges from 27 to 90 kilonewtons (6000 to 20000 lb) were marginally superior to hydrogen-oxygen engines for a well-designed stage using tank pressurization to force propellants to the engine. A hydrogen-oxygen engine using a turbopump, however, was superior to a hydrazine-fluorine engine of the same thrust using tank pressurization instead of pumping. Hydrogen-fluorine engines were lighter than engines using the two other propellant combinations.[17]

According to Silverstein and Tischler, this meeting provided the impetus for final actions by ARPA on a hydrogen-oxygen engine. Cesaro reportedly slipped out of the meeting and telephoned his associates to move fast on the Centaur proposal.[18]

---

*Attendees at the initial meeting were: for ARPA, Dr. Arthur Stosick and Richard Cesaro; for the Air Force, Col. Donald Heaton, ARDC, and C. W. Schnare and William Rogers, WADC; for NASA, William Woodward and A. O. Tischler.

†Other members: R. B. Canright and R. S. Cesaro of ARPA; Joseph Rogers and Alfred Nelson, WADC; Alfred Gardner, ARDC; and M. L. Moseson, NACA-Lewis. Nelson was a propulsion analyst at Wright Field; Schnare was the chief rocket engine expert at Wright Field and Joseph Rogers worked for him. Moseson was a design specialist at NACA-Lewis.

‡Attendees: Dr. Jack Irvine, Richard Cesaro for ARPA; Col. Donald Heaton, C. W. Schnare, Joseph Rogers, Richard Shaw, and B. Chasman for ARDC; Lt. Col. Nils Nengtson for AOMC; Dr. Abe Silverstein, William Woodward, and A. O. Tischler for NASA.

Whether for this reason or by coincidence, ARPA issued order 19-59 the following day (29 August), directing the commander of ARDC to initiate a high-energy fuel stage for use with a modified Atlas missile. The propellants were to be liquid hydrogen and oxygen. The propulsion system was to be either pressure-fed or pump-fed, with a total thrust of 133 kilonewtons (30 000 lb) in single or multiple units. The final design would be determined after detailed studies were made by the propulsion and vehicle contractors and review by ARPA. Preliminary flight rating testing was to be 18 months from go-ahead. The sole source for the engine contract was Pratt & Whitney Aircraft Division of United Aircraft Corporation, and the sole source of the vehicle was Convair-Astronautics Division of General Dynamics Corporation. ARPA would review and approve the design, development, and financial plan; provide policy and technical guidance; arrange technical direction; prescribe management and technical reports; and receive credit for technical and scientific information released on the project.[19]

In essence, ARPA bought Krafft Ehricke's modified Centaur proposal, and the wording of its order suggests that a fast decision was made before final proposals and designs were determined. The commander, ARDC, designated the special projects office, then headed by Lt. Col. John D. Seaberg, as responsible for implementing the order. This was the office formerly headed by Col. Norman Appold, who managed the Suntan project. Seaberg and others in the office, including Majors Alfred J. Gardner, Jay R. Brill, and Alfred J. Diehl, had all been a part of Suntan. Two days after the ARPA order, Pratt & Whitney conducted the tenth and. final series of tests with the hydrogen-fueled 304 turbojet engine. Suntan became a thing of the past and Centaur, a hydrogen-oxygen rocket stage on top of Atlas, rose as its replacement. All the plant, equipment, and technology of Suntan could now be brought to bear in assuring that Centaur would succeed.

The impact of ARPA's order for Centaur was not immediately apparent to NASA, and Tischler's working group continued its study of high-energy upper stages, as a coordinated government effort. At the fourth meeting of the Silverstein committee on 11 September 1958, the working group on high-energy propellants had not heard from all pertinent contractors or assimilated all the data, but Tischler reported to the parent committee on the tentative results. He compared three propellant combinations— hydrogen-oxygen, hydrazine-fluorine, and hydrogen-fluorine—and systems using pressurized tanks versus systems using turbopumps. With payload capability as the criteria, conclusions were: (1) for pressurized systems, hydrogen-oxygen and hydrazine-fluorine were about equivalent; hydrogen-fluorine was 10–15 percent better in payload capacity; (2) systems with turbopumps were 5–15 percent better than pressurized systems; (3) hydrogen and oxygen had both been pumped successfully but pumping of fluorine needed further research; this reduced the comparison to pumped hydrogen-oxygen versus pressurized hydrazine-fluorine where the former has a 10–20 percent greater payload capability; and (4) a 53–62 kilonewton (12000–14000 lb thrust) rocket engine appeared best for a Thor first stage and two such engines would be suitable for an upper stage of the Atlas.[20]

At the fifth meeting of Silverstein's committee (25 September 1958) the agenda concerned liquid hydrogen pumping and storable propellants. Richard Coar and Walter Doll of Pratt & Whitney Aircraft discussed their experience with liquid hydrogen pumps. One pump developed for "another purpose" (i.e., Suntan) had a flow rate of 2.2 kilograms per second with 17 hours of operation; a second pump had a flow rate of 45 kilograms per second with a delivery pressure of 54 atmospheres. They offered to deliver the latter pump in 18 months at a cost of $4.5 million. Stanley Gunn and Merle Huppert discussed Rocketdyne work on liquid-hydrogen pumps. They envisioned a six-stage, axial-flow pump. Tests of single stages of such a pump were scheduled for November, as part of the firm's work on the nuclear rocket. These presentations gave further evidence to the working group that pumping liquid hydrogen was not a major obstacle to the development of a pump-fed, hydrogen-oxygen rocket engine.[21]

Other than information exchange, the Silverstein committee took no action to initiate development of a high-energy upper stage. Tischler drafted plans for two sizes of hydrogen-oxygen engines but they were tabled. Alfred Nelson summarized high-energy propellants, engines, and stage designs. Thrust levels varied from 31 to 600 kilonewtons (7000 to 135000 lb). In NASA's first ten-year plan (November 1958), mention was made that hydrogen-oxygen upper stages in the 45 to 445 kilonewtons (10000 to 100000 lb thrust) range would be available.[22]

One of the eight organizations whose high-energy propellant data Tischler's working group had studied for the Silverstein committee* was the NASA Lewis laboratory, where both men still had close ties. Since the Lewis research continued to influence the former Lewis men in NASA headquarters, a summary of it during 1958–1959 is pertinent.

Lewis Hydrogen Rocket Experiments, 1958–1959

After their initial success in operating a hydrogen-cooled, hydrogen-fluorine rocket engine in November 1957 (p. 92), Howard Douglass, Glen Hennings, and Howard Price, Jr. continued the experiments until February 1959. Fourteen runs were made using the showerhead and triplet type of injectors with comparable results. A maximum exhaust velocity of 3455 meters per second was obtained at a flow rate that was 14 percent liquid hydrogen with a combustion pressure of 20 atmospheres. This was 97 percent of the maximum theoretical performance. The experimenters reported no problems relative to engine operation, starting, or stability of combustion. They did, however, have a number of minor problems with the injectors and with operating the thrust chamber beyond its design limits. Following this series of experiments, another team of researchers made 26 more runs with the same type of engine over a range of combustion pressures and exhaust nozzle expansion ratios. Earlier, Vearl Huff had suggested the technique of exhausting the rocket into a properly proportioned duct closed at the rocket end. The high-velocity rocket exhaust pumped the air from the duct, reducing the pressure in the immediate vicinity of the rocket nozzle and thereby simulated high altitude. The exhaust duct needed for silencing and for removing hydrogen fluoride from rockets using fluorine was ideal for the new purpose, so the one duct served three purposes. The nozzle altitude simulation

---

*Bell Aircraft, North American Aviation, Aerojet-General, General Dynamics–Astronautics, Martin, Space Technology Laboratories, Wright Air Development Center, and the NASA Lewis laboratory.

technique was used to test a rocket with a nozzle area ratio of 100 and the measured exhaust velocity was 4730 meters per second (at a combustion pressure of 49 atm), one of the highest performance values obtained by a chemical rocket engine.[23]

Cell 22, with its two parallel test stands capable of handling engines up to 22 kilonewtons, was the workhorse cell for high-energy propellants through 1957. A new, larger facility for high-energy propellants and engine thrusts up to 89 kilonewtons was ready for its first hydrogen tests on 14 November 1957. The initial run used gaseous hydrogen and a water-cooled chamber which leaked, causing ignition problems and a minor explosion, or a "hard start" in the rocket engineer's vocabulary. The chamber was repaired and five days later satisfactory starting was achieved, but other troubles arose. A malfunctioning indicator led the operator to increase propellant flows, and after a second of operation at a pressure of 30 atmospheres in a chamber designed for 20, the chamber burst. It was not a very auspicious start for the new facility, which continued to be plagued with propellant system, control, and instrument problems for several months. By mid-March 1958, fluorine was being used in the new facility and in the first week of May, liquid hydrogen. The climax to the series of facility problems also came in May when an experiment with gaseous hydrogen and liquid fluorine was conducted. Three successful runs were made and during a pause, with the propellant tanks still at high pressure, fluorine demonstrated its reactivity. A slight leak in a flanged joint at the top of the tank allowed fluorine to escape, and it immediately found substances with which to react. These reactions quickly heated the heavy stainless steel flange and pipe until they also reacted with the fluorine and with a swoosh, a column of fluorine shot upward, reacting with everything in its path, including water vapor in the air.[24] Fortunately, a wind quickly dispersed the fluorine compounds. The joint that leaked contained a soft aluminum seal that had been thermally cycled many times over a period of months with no leakage. It had been tested just prior to the ill-fated experiment and found satisfactory. These kinds of problems are normal in research where new ground is being plowed. The flange problem was solved by using welded joints, but the accident and the subsequent delay caused a shift in research plans. Work with hydrogen-fluorine at 22 kilonewtons in Cell 22 was proceeding well, and a decision was made to concentrate on hydrogen-oxygen at the new facility.

As regeneratively-cooled thrust chambers at 89 kilonewtons were not available, the first series of tests with gaseous hydrogen—liquid oxygen was made with uncooled chambers. The gaseous hydrogen was no handicap in this situation for it simulated the same physical state at the injector as liquid hydrogen after absorbing heat in a regeneratively-cooled jacket. Nineteen runs were made during 1958, with performance ranging between 94 and 99 percent of theoretical.[25]

Meanwhile, the 89 kilonewton, regeneratively-cooled engine became available and by June 1959, 32 runs were made with liquid hydrogen—oxygen. Run times varied up to 102 seconds and all showed satisfactory cooling and high performance. Later, an additional 14 runs, equally successful, completed the investigation and the results were reported in April 1960. Exhaust velocities up to nearly 3300 meters per second were obtained with a nozzle designed for sea-level operation, so even higher velocities were possible with a larger nozzle and operating at simulated altitudes. The investigators used a small quantity of gaseous fluorine flowing ahead of the liquid oxygen to spontaneously ignite with the hydrogen and provide a smooth start.[26] Figure 50 shows a comparison of the 22 and 89 kilonewton, regeneratively-cooled engines used for the liquid hydrogen–fluorine and liquid hydrogen–oxygen experiments during the 1957–1959 period and later.

The injectors designed for both propellant combinations at both thrust levels followed the general concepts agreed upon at the August 1957 design conference (p. 89). They generally employed a large number of elements to promote vaporization and intimately mix the propellants. Some were impinging jets in triplet or doublet arrays, some were showerheads with either parallel or converging jets. All gave high, satisfactory performance. Figure 51 shows one of the injectors used with the 89 kilonewton engine.

The greatest value of research of this type is in advancing technology and getting someone else interested in using it for further advances. In the latter, the Lewis rocket research on hydrogen during the 1950s made two contributions. First, it influenced the views of Abe Silverstein, who began planning the NASA spaceflight program in the spring of 1958. He was greatly interested in hydrogen as a fuel, not only for rockets but for other applications. Silverstein followed the rocket work even after he went to Washington, and the May 1958 fluorine accident, which occurred while he was still commuting to the laboratory on weekends, reinforced his conviction that the performance gain by using the denser hydrogen-fluorine combination over hydrogen-oxygen was not worth the additional problems. The regeneratively-cooled hydrogen-oxygen operations at 89 kilonewton thrust in 1959 further convinced him that hydrogen-oxygen was by far the most attractive of the several propellant candidates

Fig. 50. The author with NACA's 22- and 89-kN regeneratively cooled rocket engines used for performance experiments of liquid hydrogen–fluorine and liquid hydrogen–oxygen, 1957–1959 and later.

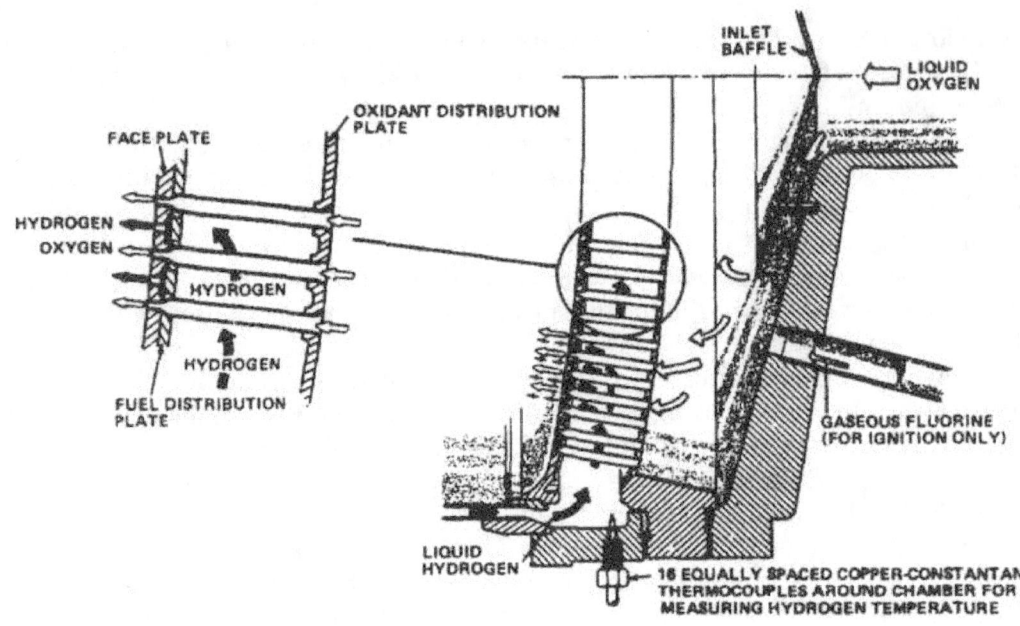

Fig. 51. Cross-section of an injector used in experiments with an 89-kN rocket engine using liquid oxygen and regeneratively cooled by the liquid hydrogen. A lightweight design, the injector was a converging showerhead type. Measured performance was 93 percent of theoretical. From Tomazic, Bartoo, and Rollbuhler, NASA TMX-253, Apr. 1960.

for high-performance rocket stages. His convictions were to have an important bearing on decisions made at the end of 1959, decisions that have determined the course of space vehicles to this day.

The second value of the Lewis hydrogen research was the influence it had on other rocket engineers. During 1959, 92 people from 42 organizations made 60 visits to the Lewis rocket laboratory. While not all were interested in hydrogen, the two major rocket engine manufacturers, Rocketdyne and Aerojet, each made three visits; Pratt & Whitney, with a go-ahead in August 1958 from the Air Force to develop a hydrogen-oxygen rocket engine for flight, made three visits during 1959. In fact, Pratt & Whitney representatives began visiting the Lewis rocket laboratory in 1957, much to the surprise of the laboratory officials who had previously found the company aloof when it came to exchanging information about aircraft engines.[27]

Transfer of Centaur

In October and November 1958, the Air Force let contracts with Pratt & Whitney and Convair to develop Centaur, the hydrogen-oxygen upper stage for Atlas. Also in October, NASA Administrator Keith Glennan requested ARPA Director Roy Johnson to transfer Centaur to NASA. This was agreed upon in principle by Deputy Secretary of Defense Donald Quarles by November. The Air Force, however, had missions requiring the Centaur vehicle and wanted to retain management control. As a consequence, both the Advanced Research Projects Agency and the Air Force mounted strong efforts during the first part of 1959 to reverse the initial transfer agreement. NASA was well aware of the problem and on 6 May 1959, Glennan proposed a compromise to Quarles. Glennan's plan kept the contracts in the Air Force's name with NASA supplying the funds. The Air Force management team was to remain in the Air Force's pay but be physically located at NASA headquarters and report to NASA. Since the Vega and Centaur space vehicles both used an Atlas as the first stage, Glennan further proposed a coordinating team consisting of a NASA chairman and a representative from each of the following: ARPA, Air Research and Development Command, and the Jet Propulsion Laboratory, plus the Vega and Centaur project managers. NASA accepted responsibility for developing six Centaur upper stages and launching the Atlas-Centaur vehicles; payloads remained the responsibility of the mission agency.[28]

Glennan's plan required very close cooperation between NASA and the Air Force. What remained utterly inconceivable to some Air Force officers, however, was a military project essentially in the hands of civilians. They urged that control remain with ARPA and the Air Force. The stalemate remained until 10 June 1959 when Richard E. Horner, then NASA associate administrator, wrote to Herbert York, director of research and engineering, with a new plan to resolve the differences. Horner proposed that the Air Force establish the position of Centaur Project Director, filled by an Air Force officer, to be located at the Ballistic Missile Division of the Air Force at Los Angeles. The director would report to a Centaur program manager in NASA headquarters. The Ballistic Missile Division would provide only office space and administrative services. NASA would provide the Air Force project director with technical assistance and assign technical experts to him. Horner emphasized the need for a single line of authority from NASA headquarters to the project director. He also proposed a joint program management coordinating committee made up of a NASA member, an Air Force member, an ARPA member, and the project directors of Vega and Centaur, with Abe Hyatt of NASA as the chairman. This committee would periodically review the progress and resolve interface problems in addition to serving as a communication channel to the organizations involved. On 19 June, J. B. Macauley, York's deputy, responded to Horner and essentially agreed to his proposal. He assumed that the FY 1960 budget provided for six Atlas vehicles and recommended that the director of the military communications satellite program at the Ballistic Missile Division be an ex officio member of the coordinating committee. Horner quickly agreed and on 1 July 1959 Centaur was transferred from ARPA to NASA with scarcely a ripple. Lt. Col. John D. Seaberg remained the project director and Milton Rosen, the NASA program manager.[29] About the same time, Pratt & Whitney ran the RL-10 engine using liquid hydrogen and oxygen for the first time.

Summary

During the 1950s, development of high-energy rockets centered around the choices of hydrazine or ammonia with fluorine, and liquid hydrogen with fluorine or oxygen. Pratt & Whitney, a latecomer to rocket engine development, began to study hydrogen-fueled rocket engines in early 1956—the same time that the company began development of a hydrogen-fueled jet engine in the Suntan project. Despite a growing

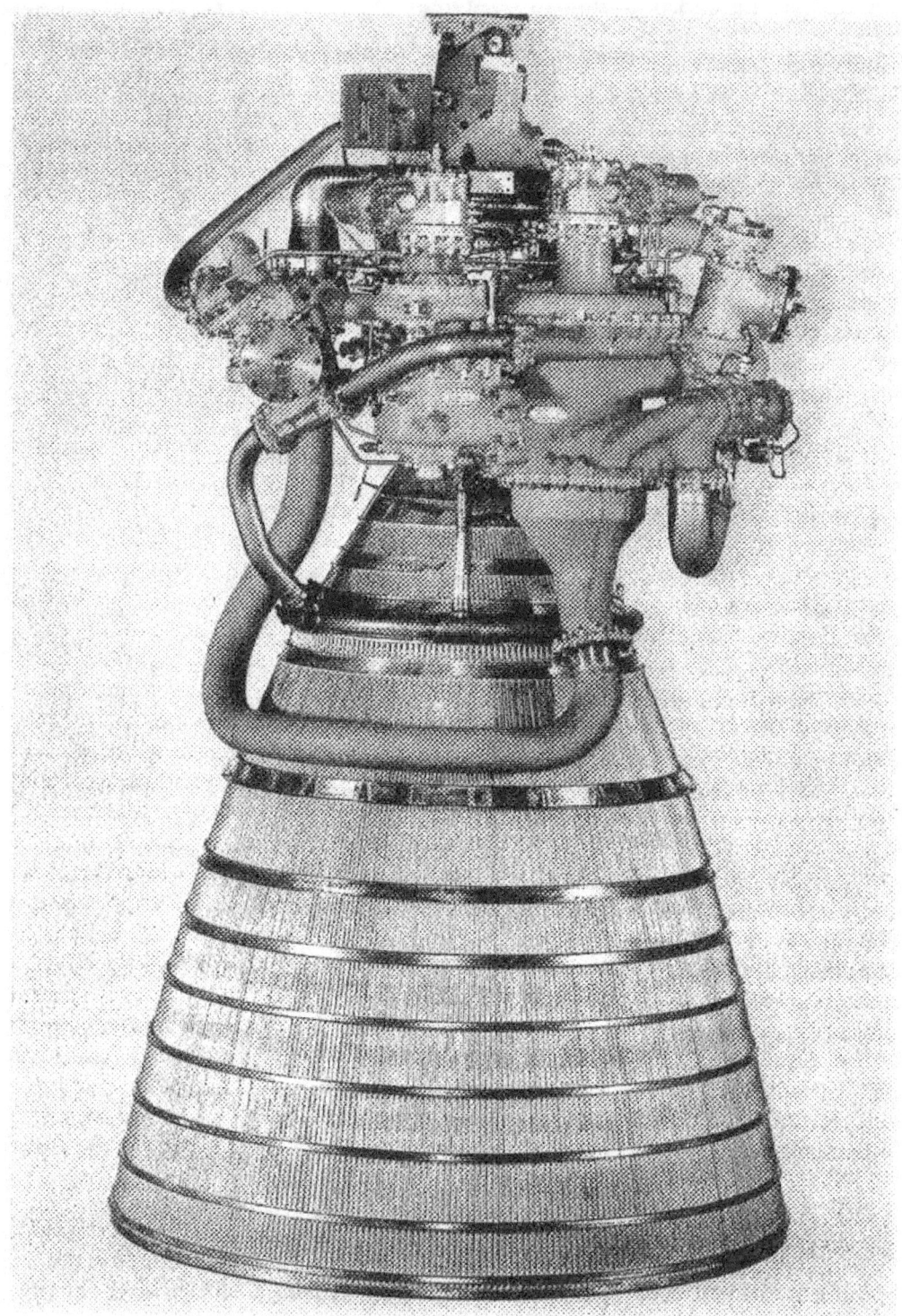

Fig. 52. Pratt & Whitney's RL-10 rocket engine, the first to use liquid hydrogen. Thrust, 67 kN at altitude; exhaust velocity, 4245 m/s; exit diameter, about 1 m. First engine run, July 1959; two of these engines powered the Centaur stage.

interest in high-energy rockets and a specific recommendation in late 1956 by the Air Force's Scientific Advisory Board to develop high-energy rocket engines, little was done until a series of stimuli about a year later: the faltering of the Suntan project, Russian satellite accomplishments, growing awareness of the military advantages of satellites, and the emerging role of the civilian space agency.

The first high-energy stage resulted from the efforts of Krafft Ehricke of General Dynamics–Astronautics. In December 1957, he proposed that the Air Force develop a hydrogen-oxygen upper stage for the Atlas using an engine designed by the Rocketdyne Division of North American Aviation. In the spring of 1958, the Air Force Suntan team proposed a hydrogen-fueled rocket engine using the Suntan experience of Pratt & Whitney and available facilities. Meanwhile, Abe Silverstein, a proponent for using liquid hydrogen in aircraft and rockets who was directly involved in such research since the early 1950s, was brought to Washington to head civilian space planning and projects. In the summer of 1958, spurred by the Air Force hydrogen engine proposal, military space needs, and Silverstein's planning activities, the Advanced Research and Projects Agency ordered Pratt & Whitney to develop a hydrogen-oxygen engine and General Dynamics–Astronautics to develop an upper stage for Atlas using the engine. This was to become the Centaur stage, intended to serve both military and civilian space needs. Despite some military objections, management control of Centaur was transferred to NASA in mid-1959. Silverstein's interest in hydrogen, aided by continued experiments at NASA Lewis laboratory, was a key factor in later decisions involving upper stages for the Saturn launch vehicle.

11

Large Engines and Vehicles, 1958

During the mid-1950s, the Air Force sponsored work on the feasibility of building large, single-chamber engines, presumably for boost-glide aircraft or spaceflight. This work provided the basis for fast response when the nation felt the need to catch up with the Russians in launch vehicle capability.

In 1956, the Army's missile development group, under the technical direction of Wernher von Braun, began studies of large launch vehicles. The possibilities opened up by Sputnik accelerated this work and gave the Army an opportunity to bid for the leading role in launch vehicles. The Air Force, however, had the responsibility for the largest ballistic missiles and hence, a ready-made base for extending their capability for spaceflight. One example of this was Centaur, the hydrogen-oxygen upper stage for the Atlas ICBM.

During 1958, actions taken to establish a civilian space agency, and the launch vehicle needs seen by its planners, added a third contender to the space vehicle competition. In this chapter, we will examine these activities during 1958 and how they resulted in the initiation of a large rocket engine and the first large launch vehicle.

Early Air Force Interest in Large Engines and Vehicles

The development of the Atlas intercontinental ballistic missile had hardly begun to accelerate when the Air Force research and development arm began considering larger rocket engines for larger vehicles. In 1955, the Air Force contracted with the Rocketdyne division of North American Aviation to study the feasibility of a single-chamber engine with a thrust of 1.3 to 1.8 meganewtons (300 000–400 000 lb). Rocketdyne designated this engine the E-1 and the same year announced that a single-chamber engine of 4.5 meganewtons (1 million lb of thrust) was also feasible.[1] There were no specific requirements for these large engines, but presumably the Air Force was looking ahead to the need to carry larger ballistic payloads and perhaps to manned spaceflight or boost-glide hypersonic aircraft concepts such as Dynasoar.

At the November 1956 meeting of the fuels and propulsion panel of the USAF Scientific Advisory Board (p. 189), large rocket engines were considered. The panel recommended that the Air Force study the feasibility of very large rocket engines on the order of 22.3 meganewtons (5 million lb of thrust). This was far larger than any that had been considered; the minutes do not reveal the panel's reasons for such interest.

The Air Force waited over a year before replying to this recommendation. The reply mentioned the work begun at Rocketdyne in 1955 and indicated that future Air Force requirements for thrusts greater than 4.5 meganewtons could probably be met more efficiently by clustering "appropriately-sized" smaller engines. A vehicle requirement for 22.3 meganewtons could be met in the same manner. The Air Force reply left unclear what size engines it was interested in, but the same month Wright Field initiated a design competition for a single-chamber engine of 4.5 meganewtons. The proposals were evaluated and a contract awarded to Rocketdyne in June 1958. The large engine was designated the F-1.[2]

Transfer of Large Engine to NASA

When Abe Silverstein came to NACA headquarters in early 1958 to organize a space program, one of his immediate concerns was increased launch vehicle capability. Consequently, his proposed FY 1960 budget, completed on 19 July 1958, contained $30 million to initiate development of a 4.5 meganewton single-chamber engine and $15 million for clustering existing ICBM engines to achieve the same total thrust (p. 185).

By late July it became obvious that the large engine work sponsored by the Air Force would be transferred to the new space agency. To deal with this and other launch vehicle matters, Silverstein organized an informal propulsion committee in early August (p. 195). At the 14 August meeting of this committee, the Air Force disclosed that its contract with Rocketdyne on the 4.5 meganewton engine would run out of funds in the fall and that $2 million more, to be supplied by NASA, would be needed by 1 October to continue the work for an additional five months. Since contract negotiations took 5 to 8 weeks, a decision by NASA was urgently needed. Silverstein, however, resisted this pressure for NASA to make an immediate commitment.

The problem of developing a large engine was further complicated by the need for facilities to test it. This matter was considered at the 28 August meeting of Silverstein's committee. Air Force representatives revealed that contracts would be let by the end of the month for a test stand at Rocketdyne's test facility capable of handling 4.5 meganewton engines. The Air Force already had a test stand capable of handling this size engine at Edwards Air Force Base, but it was tied up with Atlas missile development. Silverstein and his propulsion assistant, A. O. Tischler, were concerned that the Air Force plans essentially committed the large engine development to Rocketdyne. Silverstein decided at the meeting that any development of a large engine by NASA would be through competitive bidding. Richard Cesaro of ARPA argued that bidding should start immediately, but again NASA officials resisted the pressure to act at that time.

When the Silverstein committee met for the sixth time on 9 October, NASA was formally in business and moving. Tischler, placed in charge of the large engine, announced that requests for competitive bids would be out within two weeks. Five days later, NASA sent invitations to bid to seven contractors and a briefing on what was wanted was held a week later.

The invitations called for a single-chamber engine of either 4.7 or 6.7 meganewtons (1 or 1.5 million lb thrust), but at the contractors' briefing Tischler made it clear that the higher thrust was wanted.* By 24 November, NASA had received proposals and appointed a technical and a management team to evaluate them. On 9 December the two evaluation teams reported to the Source Selection Board; and three days later, the Board recommended to Administrator T. Keith Glennan that Rocketdyne be awarded the development contract.† Glennan approved and the selection was made public the same day. In less than a month (9 January 1959), NASA signed a definitive contract with Rocketdyne for the development of the F-1 engine with a sea-level thrust of 6.7 meganewtons.[3]

The Army's Bid to Develop Large Launch Vehicles

Although the Air Force took the initiative in sponsoring studies of large rocket engines, the Army Ballistic Missile Agency took the lead in proposing specific large vehicles. These began with studies by Wernher von Braun's missile development team in 1956 and led eventually to the Saturn vehicles developed during the 1960s. By the time the first Saturn was authorized by the Advanced Research Projects Agency in 1958 and a decision made about which propellants to use in its upper stages late in 1959, large launch vehicle concepts had undergone a number of changes. Von Braun's team initially opposed the use of hydrogen and oxygen in the second stage of the Saturn. To understand why and to follow the evolution of Saturn in its early phases, a few observations about von Braun and his team are helpful.

In 1930, when 18, Wernher von Braun was working with Germany's rocket pioneer Hermann Oberth, and von Braun's entire subsequent career was devoted to rockets and spaceflight. As technical director at Peenemünde, he was responsible for developing the V-2, the beginning of modern liquid-propellant rocketry. He headed the 120 Germans brought to the United States by the government at the end of World War II. In 1950, the Germans became the core for an expanding organization assigned to the development of Army guided missiles at Redstone Arsenal, Alabama. By 1956, the guided missile development division at Redstone, with von Braun as technical director, numbered over 2000, of whom 350 were Army officers. Over 200 of these officers were graduate engineers who strengthened the civilian staff of engineers and technicians. By 1958 the division (then called development operations) had a complement of over 2800, about 80 percent of the ballistic missile agency.[4]

As head of large engineering organizations both in Germany and the United States for almost a quarter of a century, von Braun managed by committee or group decision. At Redstone, his division consisted of ten laboratories representing various technical aspects of missile development, each headed by a highly competent member of his old German team. He used these men as a council for decision making; at meetings, von Braun assumed the role of chairman or moderator. He knew how to listen, maneuver,

---

*Tischler prepared the invitation with only the higher thrust value but included the lower value when Hugh Dryden, NASA's deputy administrator, pointed to prior agreements between NASA and the Air Force. At the bidder's briefing, Tischler made it clear the higher value was preferred and in later negotiations, Silverstein confirmed it. Interview with Tischler, 25 Jan. 1974.

†Silverstein chaired the Board with J. W. Crowley, Abe Hyatt, R. E. Cushman, and R. G. Nunn as members; the author was a member of the technical evaluation team.

and persuade; proposed actions were thoroughly thrashed out until mutual agreement was reached. Thereafter, all united behind the decision to make planned actions a success.

The loyalty and competence of the von Braun team were outstanding. The core of hand-picked German engineers had worked for von Braun in developing the V-2. They had suffered through the Allied air raids together, escaped the advancing Russians in the closing days of the war, and migrated to a new land and new life in 1945. At Fort Bliss, Texas, they were enemy aliens who, though well treated, could not go into El Paso without a military policeman as escort.[5] These experiences tied the group together—loyal to each other and to von Braun as their leader. As excellent engineers, they were determined to prove their worth.

A third observation is about von Braun's ability to sell himself and his ideas. A man with charisma, he knew how to deal with bureaucracy,* how to compromise, and how to maneuver to achieve his objectives. He used his talents to fire the imagination and stimulate interest in spaceflight unabashedly, to gain support for his team and his ideas. The publicity given von Braun seems not to have bothered his German colleagues, who worked as much in obscurity as he did in the limelight. The team understood and appreciated von Braun's ability in public relations and willingly assisted him in building up his reputation and image, because the group shared in the rewards of increased support.

Von Braun was as conservative an engineer in actual design and construction as he was a bold innovator in concepts. The design of the V-2, Redstone, Jupiter, and Saturn all reflect the conservatism of von Braun and his team. They looked askance at such lightweight structural innovations as Bossart's thin-wall, pressurized tanks for the Atlas ICBM, which they jokingly referred to as "blimp" or "inflated competition." They preferred husky, sturdy structures which Krafft Ehricke characterized as "Brooklyn bridge" construction. Their structural designs were sound, if somewhat on the heavy side. This conservative design philosophy mitigated against the use of liquid hydrogen which, more than conventional fuels, depended upon very light structures to help offset the handicap of low density.[6]

The final observation about von Braun and his team stems from their alliances. By fate and by choice, these engineers were aligned with the military in Germany and in the United States; those alliances were both an advantage and a handicap. The advantage lay in pressing military requirements in both countries, which assured the team virtually a blank check in developing rocket missiles. Emphasis was on achieving success rapidly and seldom, if ever, on minimum cost. But the same reasons that gave the team liberal support also restrained them from deviating from the immediate task at hand. This meant little tolerance for indulging in schemes for spaceflight, von Braun's greatest interest. He was arrested and jailed in 1944 for alleged sabotage of the

---

*At a dinner honoring von Braun at his departure from NASA in 1972, Eberhard Rees, his longtime deputy and associate, spun a yarn about German bureaucracy. Peenemünde purchase requests had to be approved by Army headquarters, and a request for a gold-plated instrument mirror was rejected as insufficiently justified. Rees, attempting to write a technical justification, was stopped by von Braun. Just tell them we want it because a solid gold one would be too expensive, he advised. Rees did and the request was promptly approved. Interview with D. D. Wyatt, Bethesda, MD, 31 Aug. 1975.

A-4 missile he was developing because he was overheard speculating on spaceflight.* At the U.S. Army's Redstone Arsenal, von Braun was under similar restraints, although he soon found a kindred spirit in Maj. Gen. J. B. Medaris, commander of the Army Ballistic Missile Agency.

Von Braun wanted to adapt existing missile equipment to launch a satellite as early as 1954. He lost out to Vanguard in a 1955 bid to launch satellites for the International Geophysical Year, but by 1956 he had assembled equipment capable of launching a satellite. Sputnik I gave him the long-awaited opportunity and he succeeded with Explorer I on 1 February 1958.

Explorer I was the opening gun in the Army's campaign for a strong role in space. Following the initial Russian and American satellites, it became clear that Russian launch capability far exceeded that of the U.S. and the von Braun team was quick to respond to the U.S. outcry for larger launch vehicles. Among those envisioned was one of multiple stages; the first stage, a cluster of 4 engines, would develop a total of 6.7 meganewtons (1.5 million lb of thrust). The report on this study was submitted to the Department of Defense on 10 December 1957: "A National Integrated Missile and Space Development Program." It was the first of several bids for a space role by von Braun and Medaris.

The December 1957 report was updated in March 1958; it described 11 launch vehicles starting with the Navy's Vanguard and Army's Juno I, and continuing to the very large vehicle of 6.7 meganewtons (table 6). Two of the proposed vehicles used high-energy upper stages with hydrogen-oxygen as one of the candidate propellant combinations.[7] One of these was the stage that Krafft Ehricke had proposed in December 1957 (p. 194).

The March 1958 report also recommended the development of 14 propulsion systems including two large engines (table 8, p. 216). One was a cluster of 4 Rocketdyne E-1 engines of 1.8 meganewtons (400 000 lb of thrust) each, using kerosene-oxygen; the other, Rocketdyne's F-1 engine of 4.5 to 6.7 meganewtons (1–1.5 million lb of thrust), also using kerosene and oxygen.†

The Army Ballistic Missile Agency proposed that hydrazine be considered as an alternative to kerosene for first-stage engines. Also recommended was an array of upper stages and engines: large-thrust engines using space-storable (non-cryogenic) propellants, hydrazine-fluorine, and nuclear fission; and small-thrust engines using electric or solar power. These advanced engine concepts indicated that the von Braun team was not at all conservative when it came to planning and proposing.

---

*Walter Dornberger, former commanding officer of Peenemünde, described the incident in his book, *V-2* (New York: Viking, 1958), pp. 200–207, quoting Field Marshall Keitel: "The sabotage is seen in the fact that these men have been giving all their innermost thoughts to space travel and consequently have not applied their whole energy and ability to production of the A-4 as a weapon of war."

†According to H. C. Wieseneck, Rockwell International, Rocketdyne conducted a series of rocket engine studies during 1957 and 1958 in support of the Juno vehicle studies at ABMA. Among options considered was the use of 8 existing ICBM engines that led to Rocketdyne's H-1 engine, which was used in Saturn I. Wieseneck to M. D. Wright, NASA, 6 Feb. 1976.

Fig. 53. The Centaur stage (left), 3 m in diameter, was the first to use liquid hydrogen. The Atlas-Centaur (right), 37 m tall, was first flight-tested in 1962 and within a decade 25 flights had been made, 19 of which were successful. The vehicle is still in use, but may be replaced at the end of the 1970s or early 1980s by the shuttle. (1965 and 1967 photographs.)

Fig. 54. Wernher von Braun, father of modern rocketry and developer of Saturn launch vehicles. Shown
    with a Saturn IB, 43 meters tall, used to launch the first flight test of the Apollo lunar module on 22 Jan.
    1968.

NACA Working Group on Launch Vehicles

In the first part of 1958, when von Braun and his team were proposing an integrated national missile and space vehicle program to the Department of Defense, von Braun was also participating in a study of space technology for the National Advisory Committee for Aeronautics (NACA) and making similar proposals to it. He was a member of the NACA special committee on space technology chaired by Dr. H. Guyford Stever (p. 181). Von Braun was also chairman of a working group on launch vehicles for the Stever committee. Abe Silverstein and Col. Norman C. Appold were members of the Stever committee and of von Braun's working group.[*]

During the course of its study, the Stever committee met periodically and heard progress reports from the chairmen of its several working groups, including von Braun. One such meeting was called for Monday, 17 March 1958, at NACA's Ames aeronautical laboratory in California. "I have put a substantial amount of work into the preparation of such a [vehicle] program," von Braun cabled S. K. Hoffman, Abraham Hyatt, Silverstein, and Appold, "but do not wish to present it to the committee without your prior approval." He suggested a meeting at a motel near Ames for Sunday the 16th.[8]

Assisting von Braun on his NACA assignment, but remaining behind the scenes, was Francis L. Williams. He had left Wright Field to join von Braun at the Army Ballistic Missile Agency in February 1958 and was familiar with the December and March proposals that the agency had made to the Department of Defense for an integrated vehicle program. Young and handsome, ambitious and smart, Frank Williams was not content to remain faceless behind the scenes like von Braun's German colleagues. He wanted part of the action, specifically to accompany von Braun to the NASA meetings. Aware of von Braun's work habits, he devised a strategy for the 17 March meeting that worked. He prepared a vehicle program, wrote himself travel orders, stowed his bag nearby, and made an appointment with von Braun before time to depart for California. As expected, time ran out before von Braun had reviewed the program. Williams, of course, was ready to accompany him on the flight to continue the discussion. In California, Williams persuaded von Braun to let him present the program so that von Braun would be free to comment on it like the other members. Von Braun agreed.[9]

The bold plans of the Ballistic Missile Agency delegation evoked plenty of comments at NACA meetings, but this did not deter the proposers. On 1 April 1958, von Braun's group issued a document that astounded the quiet, conservative people in NACA headquarters. Soon all hell broke loose. On the report cover was printed "Interim Report to the National Advisory Committee for Aeronautics, Special Committee on Space Technology: A National Integrated Missile and Space Vehicle Development Program: by the Working Group on Vehicular Program." Inside was the same proposal the Ballistic Missile Agency had made to the Department of Defense. A 23-year spaceflight program was laid out with rows of launch vehicles ranging from small

---

[*]Other members of the vehicle working group: Abraham Hyatt, Navy Bureau of Aeronautics; Louis Ridenour, Lockheed Aircraft; M. W. Hunter, Douglas Aircraft; C. C. Ross, Aerojet-General; Homer J. Stewart, JPL; George S. Trimble, Jr., Martin; Krafft Ehricke, Convair-Astronautics; S. K. Hoffman, Rocketdyne; and W. H. Woodward, NACA, secretary.

to huge. The flight missions included satellites ranging from small unmanned scientific ones to a 50-man permanent satellite with a mass of about 450 metric tons. There were also flights to the moon, interplanetary probes, and expeditions to Mars and Venus. Total cost was estimated at $30 billion.[10]

The bold and imaginative plan was too much for the NACA to swallow, and NACA's director, Hugh Dryden, moved to dissociate his organization from it. The headquarters copy bore a red tag with the notice: "IMPORTANT—that this Interim Report . . . not be allowed outside the NACA headquarters building under any circumstances—unless by specific approval of Dr. Dryden." A staffer attached a comment to the report that the Ballistic Missile Agency was "apparently advertising it rather broadly to get implication of NACA approval for von Braun's pitch."[11]

At Huntsville, Williams received calls for copies of the report and asked NACA headquarters for permission to distribute it. Dryden replied that he had no objection, provided that "A statement should be attached to each copy indicating that the report has not been approved by the NACA Working Group on a Vehicular Program and, therefore, cannot be considered to be an official recommendation of the Working Group or of the NACA Space Technology Committee."[12]

The report contained a number of sound, timely recommendations; among them was "that a development program be initiated immediately for a large engine, in excess of one million pounds thrust [4.5 MN], and the required test facilities with emphasis on early availability of the engine for flight test and operational use." The report was prophetic when it recommended a spaceflight program "with particular emphasis on a manned lunar landing within the next 10 years." Another recommendation was "that long-range vehicle responsibility be assigned to individual development teams without delay under the direction and coordination of a central group." There was little doubt that von Braun had his own team in mind. He was recommending the same vehicle program to the military and civilian sides of the government and courting both to get the vehicle responsibility.

On 18 July 1958, a revised and toned-down version of the earlier interim report was issued by the NACA working group on vehicles. Gone was the recommendation to initiate development of a large engine and in its place was "A development program be initiated immediately for a booster in the 1.5 million pound thrust [6.7 MN] class, with emphasis on early availability."[13] In the months that followed, development of both the large engine and the large booster was initiated—steps which the Stever committee merely endorsed in its final report, without including details that had been submitted by the von Braun working group.[14]

In the time between the April interim and 18 July 1958 final report of the vehicle working group, von Braun had correctly sensed the direction political winds were blowing. The recommendation on vehicle responsibility now read "under the direction and coordination of the NATIONAL AERONAUTICS AND SPACE AGENCY in conjunction with the ADVANCED RESEARCH PROJECTS AGENCY."[15] He was still taking no chances.

The report of the NACA working group recommended 15 vehicles in five generations of development; with some additions and revisions, these were along lines similar to previous recommendations of ABMA as can be seen by comparing tables 6 and 7. The first three generations that NACA recommended comprised 11 vehicles and

TABLE 6. — *ABMA's Proposed National Integrated Missile and Space Development Program, March 1958*

No.	Vehicle	Operational Date	Payload kg
I	Vanguard	1958	2–10
Ia	Juno I	1958	8–16
II	Juno II	58–59	27–45
IIa	Thor + 117L stage	58–59	90–140
III	Juno III	59–62	140–320
IV	Atlas + 117L	61–63	700–900
V	Atlas + H_2-O_2 pressurized stage	61–64	1100–4000
VI	Juno IV	62–64	230–450
VII	Titan	60–80	450–1400
VIII	Titan + Polaris	62–80	1400–2300
IX	Mod Titan (1st stage recoverable; 2d & 3d stages N_2H_4-F_2 or H_2-O_2)	65–80	2300–4500
X	Mod Jupiter (1st: 4×1.7 MN, RP-O_2 recov.; 2d: 1×1.7 MN, RP-O_2 or H.E. prop.; 3d: 356–445 kN, N_2H_4-F_2)	63–70	11000– 16000
XI	Large orbital carrier of 2 recoverable stages (1st: 2×6 MN* N_2H_4-O_2, delta wing; 2d: nuclear with NH_3 or H_2)	69–80	23000

Source: "A National Integrated Missile and Space Vehicle Development Program," 2d ed., report D-R-16, Dev. Oper. Div., ABMA, Redstone Arsenal, AL, 14 Mar. 1958.

*Correction by author of obvious misprint.

were based on current missile developments with high-energy stages added. In the fourth generation, an alternate vehicle was added that used 9 ICBM engines in its first stage, a configuration—favored by the Advanced Research Projects Agency—which was a forerunner of Saturn I. In the fifth generation, vehicles requiring thrusts as high as 27 meganewtons (6 million lb) were recommended for a recoverable first stage. The hand of Silverstein and the 1956 recommendations of the Air Force's Scientific Advisory Board appear to have been at work for this large thrust vehicle, a forerunner of the 5-engine first stage of the expendable Saturn V developed during the 1960s.

The NACA working group also recommended 17 propulsion systems which were essentially a revised and expanded version of the ABMA recommendations, as can be seen by comparing tables 8 and 9. Among the NACA additions was an engine with a thrust of 2.2 meganewtons (500000 lb) using hydrazine-fluorine or a "similar high-energy propellant." This would be a follow-on to a 53-kilonewton (12000-lb-thrust) engine using hydrazine-fluorine, being developed for the Air Force by Bell Aircraft, and the recommended 356–445-kilonewton (80000–100000-lb-thrust) engine using the same propellants. Both ABMA and the NACA working group appeared initially to favor hydrazine-fluorine over hydrogen-oxygen, but this was to be reversed within 18 months.

The day following the issuance of this report, Silverstein, in his spaceflight role at NACA headquarters, completed his FY 1960 budget request, which included funds for a large engine, the clustering of ICBM engines, and high-energy propulsion systems (p. 185). Ten days later, on 29 July 1958, President Eisenhower signed the bill

TABLE 7.—*NACA Working Group's Recommended Space Vehicles, July 1958*

Group	Type	Vehicle	Operational Date	Payload kg
I	IA	Vanguard	1958	2–10
	IB	Juno I	1958	8–16
II	IIA	Juno II	58–59	45–90
	IIB	Thor + 117L stage	58–59	90–180
	IIC	Juno IV	59–80	230–1130
III	IIIA	Atlas + 117L and/or	59–63	900–1400
	IIIB	Titan	60–62	450–1400
	IIIC	Mod. Atlas + 89 kN H_2-O_2 and/or		1400–4100
	IIID	Mod. Titan + 53 kN N_2H_4-F_2	62–64	1400–2700
	IIIE	Uprated Atlas —3 × 668 kN eng. + high-energy upper stage and/or		
	IIIF	Uprated Titan + high-energy upper stage 1st stage recoverable	63–80	2300–4500
IV	IVA	Basic large carrier—(1st: 6.7 MN, recov.; 2d: 2.2 MN; 3d: 356 kN high energy) and/or	63–70	11000–16000
	IVB	1st: 9 × 668 kN Atlas eng.; 2d: 3 × 668 kN; 3d: 178 kN high energy	63–70	11000–
V	VA	Recov. booster (1st: 2 to 4 × 6.7 MN; 2d: 1 × 6.7 MN)	68–80	23000–68000
	VB	Recov. booster (1st: 2 to 4 × 6.7 MN; 2d: nuclear)	68–80	45000–113000

Source: Working Group on Vehicular Program. "Report to the National Advisory Committee for Aeronautics Special Committee on Space Technology." 18 July 1958

creating the National Aeronautics and Space Administration; and on the next day, he asked Congress for $125 million for NASA operations. Silverstein's spaceflight budget reflected confidence that NASA would develop large engines and launch vehicles for manned flight and high-energy upper stages for unmanned vehicles.

ARPA Initiates First Large Launch Vehicle

The Advanced Research Projects Agency (ARPA), established since February 1958 and having a budget, could have acted immediately on the large launch vehicles proposed by the Ballistic Missile Agency in the December 1957 and March 1958 proposals to the Department of Defense, but did not. Instead, on 17 April 1958, ARPA requested that the Army Ordnance Missile Command study an advanced satellite carrier vehicle patterned after Juno III.* The new vehicle, designated Juno IV, was

*Juno I was a modified Redstone with three upper stages of solid propellant rockets. Juno II was a modified Jupiter IRBM with the same upper stages as Juno I. In Juno III, the solid propellant rockets in the upper stages were slightly larger. Juno I launched the first U.S. satellite (Explorer I) and two others (Explorers III and IV). Juno II launched two space probes (Pioneers III and IV) and two satellites (Explorers VII and VIII). Juno III was not built.

TABLE 8.—*ABMA's Recommended Engine Developments, 1958*

No.	Thrust N(lb)	Propellants	R&D
1	1.8 MN (400 000 lb) sea level	RP-O_2	1956–61
2	Cluster, 4 × 1.8 MN, SL	RP-O_2	1958–63
3	356–445 kN (80 000–100 000 lb) in vacuum of space	N_2H_4-F_2	1957–61
4	2.2 MN (500 000 lb), vac.	N_2H_4-F_2 or similar	
5	45–90 kN (10 000–20 000 lb), vac.	space storable (non-cryogenic)	1957–61
6	134 kN (30 000 lb), vac., pressurized tanks	H_2-O_2	1958–60
7	4.5–6.7 MN (1–1.5 million lb), SL	RP or N_2H_4-O_2	1960–66
8	445 kN, vac.	space storable	1960–65
9	2.2 MN, vac.	space storable	1960–65
10	1.3 MN (300 000 lb), vac.	nuclear fission	1957–65
11	4 N (1 lb), vac.	ion*	1957–66
12	45 N (10 lb), vac.	solar power	1957–64
13	0.9–2.2 MN (200 000–500 000 lb), vac.	arc-thermodynamic*	1958–?
14	0.9–2.2 MN, vac.	magnetohydro-dynamic*	1958–?

Source: "A National Integrated Missile and Space Vehicle Development Program," 2d ed., report D-R-16, Dev. Oper. Div., ABMA, Redstone Arsenal, AL, 14 Mar. 1958.

*Requires electric power source.

based on a modified Jupiter IRBM as the first stage with the addition of upper stages.[16] ARPA earmarked $46 million for the project.

In the months following the Juno IV order, interest at ARPA shifted to alternative vehicles. During this period David Young, Richard Canright, and Richard Cesaro began discussing larger launch vehicles based on using a cluster of existing engines for the first stage. Canright, on loan from Douglas Aircraft, had examined the desirability of using multiple rocket engines in launch vehicles for redundancy and reliability, following much the same philosophy used for large aircraft. He was, therefore, an instant and strong advocate for a large launch vehicle using a cluster of engines. He differed from the Ballistic Missile Agency, however, in that he wanted to use existing engines—the tried and proven rocket engines powering the Atlas ICBM and Thor IRBM. Each of these produced a thrust of 670 kilonewtons (150 000 lb), but both were capable of a 25 percent increase in thrust. This meant that a cluster of 8 or 9 could produce a total thrust of 6.7 meganewtons (1.5 million lb). Cesaro, a former NACA propulsion researcher at the Lewis laboratory, also favored large launch vehicles using multiple engines.[17]

In addition to large vehicles, Canright also began to consider smaller launch vehicles that could use existing missiles as first stages. In these studies, it is not surprising that he favored the Douglas-built Thor over the Chrysler-built Jupiter. He argued that Thor not only had the capability of the Jupiter, but cost much less. Word of his considerations of Thor reached ABMA, home of Jupiter, where naturally there was some unhappiness over the turn of events. ABMA was also well aware of Air Force interest in large vehicles, evidenced by a June 1958 contract with Rocketdyne for a

TABLE 9. — *NACA Working Group's Recommended Engine Developments, 1958*

No.	Thrust N(lb)	Propellants	R&D
1	1.7 MN (380000 lb), sea level	RP-O₂	1956–61
2	Cluster, 4 × 1.7 MN, SL	RP-O₂	1956–64
3	6.67 MN (1.5 million lb), SL	RP or N₂H₄-O₂	1960–64
4	Cluster, 2 or 4 × 6.67 MN, SL	RP or N₂H₄-O₂	1960–65
5	27 kN (6000 lb) in vacuum of space; vernier	space storable (non-cryogenic)	1958–59
6	200 kN (45000 lb), vac., pressurized tanks	N₂H₄-N₂O₄	1958–61
7	445 kN (100000 lb), vac.	space storable	1960–63
8	2.2 MN (500000 lb), vac.	space storable	1960–66
9	53 kN (12000 lb), vac.	N₂H₄-F₂	1958–63†
10	89 kN (20000 lb), vac.	H₂-O₂	1959–60
11	356–445 kN (80000–100000 lb), vac.	N₂H₄-F₂	1958–63
12	2.2 MN, vac.	N₂H₄-F₂ or similar	1960–65
13	2.2–4.5 MN (0.5–1 million lb)	nuclear with hydrogen	1957–66
14	4–4450 N (1–1000 lb), vac.	ion*	1957–?
15	4–4450 N, vac.	arc-thermo-dynamic*	1958–?
16	4–4450 N, vac.	magnetohydro-dynamic*	1958–?
17	4–4450 N, vac.	thermonuclear	1958–?

Source: Working Group on Vehicular Program, "Report to the NACA Special Committee on Space Technology," 18 July 1958

*Requires electric power source
†Under development at Bell Aircraft for the Air Force.

study of large engines. There was plenty of competition building up over who would be responsible for developing launch vehicles.

One day in mid-1958, Roy Johnson, ARPA's director, sent Canright to represent him at a meeting in the office of Wilbur Brucker, Secretary of the Army. Involved were Brucker, Maj. Gen. J. B. Medaris of ABMA, ARPA chief scientist Herbert York, David Young, and others. Brucker, a blunt, outspoken Michigan attorney and vigorous proponent for the Army, lost no time in coming to the point: ARPA had sold out completely to the Air Force, ignoring the Army's superb missile team at Huntsville, as well as the equally superb missile, Jupiter. Canright attempted to state the reasons for selecting Thor over Jupiter, but Brucker interrupted and in colorful language made it amply clear that the Army's capability should not be ignored. After the meeting, Medaris told York and Canright that von Braun's operations required about $90 million a year and if ARPA would pay half that amount, the Army would be satisfied. Canright was incensed over the Army's pressure tactics, but York apparently saw little else that could be done. Years later Canright believed that this meeting was a major factor in the assignment of ABMA to develop a large launch vehicle.[18]

The meeting with Brucker did not resolve the issue of the configuration for the large launch vehicle. Canright went to Huntsville and told von Braun and his associates what ARPA wanted: 7 or 8 Rocketdyne H-1 engines in a cluster for the first-stage propulsion system. At the time, von Braun still favored the Juno V configuration using a cluster of 4 larger engines, the E-1, still on the drawing board. Canright recalls Medaris taking him into his office along with von Braun and saying, in effect, that trying to make 8 engines of such complexity work together was totally impractical. Canright, however, remained firm; he cited the favorable reaction of the National Security Council's panel and indicated that if ABMA was not willing to cluster the engines, a contractor could be found who would. The meeting left von Braun still unsatisfied with the 8-engine cluster, and he continued to argue for the use of fewer and larger engines.[19]

The planning of Silverstein at NACA and the Air Force's June 1958 contract with Rocketdyne for feasibility studies of a 4.5-meganewton engine increased the pressure for ARPA and ABMA to resolve the stalemate over using the cluster of existing ICBM engines for a large vehicle. According to Richard Cesaro, a crucial meeting occurred at the Pentagon in mid-1958. Medaris and von Braun represented ABMA, and Roy Johnson, David Young, and Cesaro represented ARPA. With control of the purse strings, the ARPA men laid their views on the line in forceful language and had their way. They also made it clear that ARPA was not going to serve merely as a money conduit, but intended to manage the work, a far cry from the blank check approach that ABMA had enjoyed in the past.[20]

Competition from another direction faced ARPA: civilian space planning led by NACA's Silverstein. When Silverstein organized his propulsion and vehicle coordinating committee (p. 195) with its first meeting on 7 August 1958, the ARPA men sprang into action. The day of the committee meeting, Young and Canright went to Huntsville to discuss the possibility of von Braun's starting immediately on the cluster engine. They proposed using some Juno IV funds for this as an expediency. Eight days after Young and Canright returned to Washington, Johnson signed ARPA order 14-59. It directed the Army Ordnance Missile Command and ABMA to provide a development and funding plan for a large launch vehicle and to demonstrate its feasibility in a full-scale, captive test by the end of 1959. Initial funding was $5 million; the same day, Johnson signed ARPA orders 15 and 16 for Juno IV development under reduced funding.[21]

ARPA order 14-59, 15 August 1958, was the start of the first U.S. large launch vehicle, which would later be named Saturn. With ABMA assigned to build a large launch vehicle, Medaris and von Braun began to escalate the funding needed. By the end of August, ARPA agreed to triple the funding, although this was not formalized until December. The name of the new vehicle was changed from Juno IV to Juno V, because the former had been widely identified with the cluster of four E-1 engines.

In September, a member of von Braun's staff made a tactical error. The team was accustomed to thinking big, and in a briefing to visiting NASA administrator T. Keith Glennan, a cost analysis was shown which used the firing of a hundred Juno Vs as a mission model. It was only an arbitrary assumption for a cost analysis, but on learning about it, Johnson of ARPA grew very concerned that the ambitious von Braun was getting out of hand and that the whole program might be cancelled as too costly before it was well started. The President's National Aeronautics and Space Council was meeting on 24 September, and Johnson summoned Medaris to Washington the day before in order to reach an understanding about the project. After a two-hour meeting, the two agreed upon $13.4 million for FY 1959 and $20.3 million for FY 1960 for research and development. An additional $1.6 million to modify a Huntsville test stand and $7 million for Atlantic Missile Range facilities brought the FY 1959 funding to $22 million—quadrupling the initial $5 million in five weeks. This was still prior to ABMA's submission of a development and funding plan.

In October 1958, the September agreement hit a snag. On 10 October, ABMA submitted a formal request for the $1.6 million to alter its test stand. It moved through government channels smoothly until it reached the Bureau of the Budget. On 1 October, NASA was formally in operation and on 14 October, Glennan requested the Department of Defense to transfer the Jet Propulsion Laboratory and the space activities of ABMA to NASA. The Bureau of the Budget was a party to this request, so when it received the ABMA request for $1.6 million for the test stand, it withheld approval until the Juno V project was clarified as to its scope and the responsible agency.

The enterprising staff at ARPA took the Bureau of Budget disapproval as only a momentary setback. An analysis was prepared showing that Juno IV was really not needed and its funds could be diverted to support Juno V. Johnson cancelled Juno IV and ordered a maximum recovery of those funds from ABMA. The ARPA staff was confident that the recovered funds, some $8 million, could be switched to support the clustered engine project, Juno V. Young and Canright hurried to Huntsville to see if the amount was sufficient to cover the proposed work, which included upper-stage design studies, additional component testing, and purchase of long-lead-time equipment. Von Braun's engineers convinced them that more money was needed and submitted two plans: one at $17 million and the other at $11 million. ARPA considered these and decided to allocate the $8 million for design studies, component testing, and testing another "battleship" (non-flightweight) first stage. An additional $3.4 million was allocated for purchasing equipment with long delivery times. It was now the end of October and the promised funding for Juno V in FY 1959 had climbed to $33 million. In planning for the next fiscal year, ARPA requested $40 million for Juno V work at ABMA and $14 million for guidance equipment.

Both ABMA and ARPA must have been pleased with the upward trend of funding, but on 13 November they got a shock. During that week, the Bureau of Budget had found that both ARPA and NASA had requested funding for a large launch vehicle in FY 1960. Clearly the problem of who does what needed resolution. On the 13th, James Killian, the President's science advisor, met with DoD, ARPA, and NASA officials to discuss, among other things, deleting Juno V funds from the ARPA budget. The question of transferring the large launch vehicle from ARPA to NASA was raised, but Glennan was noncommittal, so the issue remained unresolved. On 19 November, Secretary of Defense McElroy and his deputy, Donald Quarles, agreed to include $50 million in the DoD budget for the clustered engine stage, subject to further discussions with Killian and the Bureau of the Budget. This remained intact through the budget review and was in the FY 1960 budget submitted to Congress in January 1959.

The ARPA men were elated over the McElroy-Quarles action and two days later amended order 14-59 to increase the funding to $13 million, as promised in September. The same day Johnson urged Quarles to help in securing Bureau of Budget approval for the $1.6 million for the Huntsville test stand. Also the same day, ABMA submitted a proposal to ARPA for increasing FY 1959 funding for the clustered engine project to $32.9 million, in accordance with the development plan, which included one vehicle for static firing and four more for test flights. The funding for FY 1960 was estimated at $60 million—$10 million more than McElroy and Quarles had agreed to include only two days earlier.

Quarles tabled the $1.6 million request for the Huntsville test stand until the FY 1960 budget was clarified. This occurred on 3 December and Quarles told Johnson that the DoD budget would contain $50 million for the clustered engine stage. Soon after, the Bureau of the Budget released the held-up funding for the test stand. Both ARPA and ABMA had reason to rejoice on another matter resolved on 3 December. An agreement of that date left ABMA with the Army but "immediately, directly, and continuously responsive to NASA requirements."[22]

Summary

During the mid-1950s, the Air Force contracted with the Rocketdyne Division of North American Aviation to study rocket engines larger than those in intercontinental ballistic missiles. This began with the E-1, about three times larger than an ICBM engine, but Rocketdyne believed that an engine with a thrust of 4.5 meganewtons (1 million lb)—over six times larger than an ICBM engine—was feasible. In late 1956, the Air Force's Scientific Advisory Board was even bolder and recommended studies of engines up to 22 meganewtons (5 million lb of thrust). The Air Force, however, believed that such a large thrust was best attained by clustering smaller engines. In mid-1958, the Air Force contracted with Rocketdyne for design studies of the F-1 engine, with a thrust of 4.5 meganewtons. Shortly thereafter, responsibility for developing a large engine was transferred to NASA; in October, NASA opened the competition to other contractors and indicated a preference for 6.7 meganewtons (1.5 million lb of thrust). Rocketdyne won the competition and a development contract was signed early in 1959.

It was the Army, however, which took the initiative in proposing large launch vehicles using E-1 and F-1 engines, beginning with studies in the mid-1950s. In late 1957, the Army missile development team, under the technical direction of Wernher von Braun, submitted a national integrated missile and space development program to the Department of Defense. Included was a vehicle with a thrust of 6.7 meganewtons. In early 1958, the National Advisory Committee for Aeronautics formed a vehicle working group as part of a space technology committee. The working group was headed by von Braun and included Abe Silverstein, soon to become the chief planner at the new civilian space agency. The NACA group modified and extended the Army's recommended vehicles and propulsion systems. The favored high-energy propellant combination in both the Army and NACA plans appeared to be hydrazine-fluorine, a choice influenced by an Air Force development contract with Bell Aircraft for a small engine using this combination. In August 1958, the Advanced Research Projects Agency, responsible for planning and coordinating military space missions, ordered the Army to devise a development and funding plan for a large launch vehicle with a first stage using a cluster of existing ICBM engines; this was later to become Saturn I. NASA's request for the transfer of both the large vehicle and the Army's development team met with strong opposition; an agreement in December 1958 left the Army team intact but responsive to NASA needs.

12

Saturn, 1959

The authorization of a large rocket vehicle by the Advanced Research Projects Agency in August 1958 and assignment of its development to the Army Ballistic Missile Agency marked the beginning of a series of successful large launch vehicles. In October 1958, the National Aeronautics and Space Agency asked to absorb ABMA's space group, but an agreement was reached two months later for the group to remain with the Army while being responsive to NASA. This decision was to be reconsidered and reversed within a year.

The competition in large launch vehicles between NASA, ARPA, Air Force, and Army, begun in 1958, continued into 1959. The government settled the issue by selecting Saturn as the single large vehicle to serve all needs. Left unresolved until the closing days of 1959, however, was the configuration of Saturn's upper stages.

The competitive actions between government agencies with respect to launch vehicles, the emergence of Saturn, and the bold decision to use liquid hydrogen–oxygen in Saturn's upper stages are related in this chapter.

First National Space Vehicle Plan

With both military and civilian space managers planning launch vehicles during 1958, it became obvious that a single national plan was needed to avoid costly and needless duplication. The task of preparing a unified plan fell to NASA, and by 15 December 1958 Milton Rosen had prepared a draft plan. Although ABMA had briefed NASA on its plans, the review of Rosen's draft revealed that more information was needed. This led to the formation of a "Joint ARPA-NASA Committee on Large Clustered Booster Capabilities" with Rosen and Richard Canright as cochairmen.* The committee listened to seven presentations during the first week of January 1959+ and on the 8th submitted a two-page report, concise and to the point.

Acknowledging that ABMA had "done the most work, [had] explored the problems of clustering more fully, and, in this case, [was] best qualified from an engineering

---

*Other members: Richard Cesaro and David Young from ARPA; Eldon Hall and Abraham Hyatt from NASA.
+Aerojet-General, Rocketdyne, Convair-Astronautics, Douglas Aircraft, Martin, ABMA, and the Air Force.

standpoint," the report was somewhat critical of the Army organization. Admitting that ABMA's Juno V was feasible, the committee indicated that the time and funds required to solve certain critical problems had been underestimated. Of several possibilities, the committee believed the most practical large vehicle was "a cluster of three Atlases as the first stage and a cluster of three 10-foot diameter stages (liquid oxygen–kerosene at first, liquid oxygen–liquid hydrogen later) as the second stage." The 3-meter oxygen-hydrogen stage referred to was the Centaur. Such a cluster, the committee argued, would be quicker and cheaper to develop than Juno V; but since the latter had a 9-month lead, it should be continued, with limited effort on the cluster initiated as a backup. ARPA and NASA expressed different reasons for having a backup: ARPA wanted limited development started with a second team to broaden national capability; NASA wanted the design of the cluster to proceed to the point of manufacture and be stopped only "if the ABMA configuration is well advanced and shows reasonable promise of success." Both positions reflected a concern over ABMA's ability to perform as promised.

Rosen transmitted the report to Glennan through channels. His boss, Hyatt, added his endorsement, commented favorably on Juno V, and recommended that NASA reopen negotiations to acquire Juno V "as a NASA-sponsored program with ABMA as the developing agency." Silverstein passed the report along without comment, and Glennan took no direct action on it.[1]

Following their work on the committee, Rosen and Eldon Hall prepared the first "National Space Vehicle Program" on 27 January 1959, and it was presented to the National Aeronautics and Space Council the following day. The report was critical of the current launch vehicles—Vanguard, Jupiter C, Juno II, and Thor-Able—calling them hurriedly assembled, not very reliable, and lacking growth potential to meet future needs. A series of general purpose vehicles capable of multiple missions and useful for four or five years was proposed as a means for achieving greater reliability and an orderly progression of payload capability (table 10). Of seven in this series, Atlas-Centaur and Atlas-Hustler (predecessor of Atlas-Agena) were in early stages of development; Scout and Vega were started later in the year, but Vega was cancelled within a few months as being duplicative of Centaur.

Atlas-Centaur had been started by ARPA in August 1958 and was being managed by the Air Force. At the time of the report, NASA was seeking its transfer, and the Air Force was resisting (pp. 200–01). The Centaur stage used two hydrogen-fueled Pratt & Whitney engines for a total thrust of 134 kilonewtons (30 000 lb). With an estimated payload of 1800 kilograms, Atlas-Centaur was seen as useful from 1962 through 1966. (In the event, it has proved more useful than anticipated and is expected to continue serving space needs until replaced by the shuttle at the end of the 1970s, or later.)

Juno V was shown in two configurations in the report, differing only in the third stage. The first version would use kerosene–oxygen and the second, hydrogen-oxygen. The 356 kilonewton (80 000 lb thrust) engines for the latter were never built.

The largest vehicle described in the report was NASA's Nova, which went through a number of different configurations in various proposals. As envisioned in January 1959, Nova would use four Rocketdyne F-1 engines in the first stage for a total thrust of 27 meganewtons (6 million lb) and one F-1 engine in the second stage. The third and fourth stages would use liquid hydrogen–oxygen and the same proposed 356

TABLE 10.—*Characteristics of Proposed New Launch Vehicles, 1959*

Vehicle	Stage 1 Propellants Thrust kN/MN (Thrust lb)	Stage 2 Propellants Thrust kN/MN (Thrust lb)	Stage 3 Propellants Thrust kN/MN (Thrust lb)	Stage 4 Propellants Thrust kN/MN (Thrust lb)	Stage 5 Propellants Thrust kN/MN (Thrust lb)
Scout	Solid 534 kN (120 000 lb)	Solid 258 kN (58 000 lb)	Solid 58 kN (13 000 lb)	Solid 13.3 kN (3000 lb)	
Atlas-Hustler	RP-O₂ 1600 kN (360 000 lb)	N₂H₄-HNO₃ 53 kN (12 000 lb)	Storable 27 kN (6000 lb)	Solid 2.3 kN (500 lb)	
Atlas-Vega	ditto	RP-O₂ 147 kN (33 000 lb)	ditto	ditto	
Atlas-Centaur	ditto	H₂-O₂ 134 kN (30 000 lb)	ditto	ditto	
Juno V-A	RP-O₂ 6.7 MN (1 500 000 lb)	RP-O₂ 890 kN (200 000 lb)	RP-O₂ 356 kN (80 000 lb)	Storable 89 kN (20 000 lb)	Solid 4.5 kN (1000 lb)
Juno V-B	ditto	ditto	H₂-O₂ 356 kN (80 000 lb)	Storable 89 or 27 kN (20 000 or 6000 lb)	ditto
Nova	RP-O₂ or Storable 27 MN (6 000 000 lb)	RP-O₂ or Storable 7.6 MN (1 700 000 lb)	H₂-O₂ 1424 kN (320 000 lb)	H₂-O₂ 356 kN (80 000 lb)	Storable 89 kN (20 000 lb)

Source: "National Space Vehicle Program," 1959

kilonewton engine as the second version of Juno V. Nova would be about 79 meters high, and NASA saw its application as "transporting a man to the surface of the moon and returning him safely to earth." Four additional stages beyond the five shown would be needed for such a mission with a crew of two or three men. Nova's capability, expressed in terms of earth-orbit payload for comparisons, was 68 metric tons.[2]

Although the first national space vehicle plan was little more than a compilation of Department of Defense and NASA plans, it was the first step towards an integrated program. Juno V evolved into Saturn I, and the very large vehicle NASA called Nova evolved into Saturn V, the vehicle used in the Apollo missions of the 1960s and 1970s.

Saturn Runs into Trouble

Juno V, ABMA's "clustered booster" concept for the first stage of a large launch vehicle, weathered the 28 January 1959 review by the National Aeronautics and Space Council in a show of unanimity. Five days later, the Army proposed to change the name to Saturn. (The Army was naming its major vehicles from Greek mythology, and Saturn followed Jupiter on the list.) The Advanced Research Projects Agency approved the name change the following day. The Army, however, had greater ambitions than a simple name change. On 13 February, Medaris submitted a budget request to meet the schedule of a captive firing by December 1959 and first flight by October 1960. His proposal included live second stages for flights 3 and 4 and a live Centaur stage on the 5th flight.

Medaris's estimates called for increases in FY 1959, 1960, and 1961 funding. He gave two alternatives: one, which he labeled as "dead end," consisted of four vehicles; the second consisted of 16 multistage vehicles. Funding estimates were:

	FY 59	FY 60	FY 61
Plan 1	58.3	75.2	41.1
Plan 2	63.5	120.4	128.0

Medaris, of course, wanted Plan 2, which called for about twice as much funding as previous estimates. Again the Medaris–von Braun team was rolling fast, putting on the pressure for a much greater program than ARPA had envisioned. Saturn was going to serve the needs of both the military and NASA. For the former, the justification was a large communications satellite in a "stationary" (24-hour) orbit.

By June, Pentagon budget planning for Medaris's 16-vehicle program had reduced the FY 1960 amount about 10 percent but almost doubled the FY 1961 amount. These, however, were cut in subsequent reviews, with FY 1960 set at $80 million.

The optimistic budget proposals for Saturn swirling about in the Pentagon indicated a bright future for the vehicle, but storm clouds were gathering. Opposition appeared on 17 March when ABMA presented a systems study for Saturn. Roy Johnson, director of ARPA, wanted the program thoroughly reviewed and appointed an ad hoc committee for the purpose. He also asked for a recommendation on the upper stage for Saturn. The committee worked through April and half of May and studied three candidates for upper stages: Atlas, a one-engine Titan, and a two-engine Titan.

Johnson's committee activity was followed with interest by the Air Force because of anticipated needs for the vehicle. On 13 April, in the midst of committee deliberations, Richard Horner, the Air Force's assistant secretary for research and development, proposed to Johnson that Saturn be used for the Dynasoar space glider and that the Air Force be given project responsibility.* Apparently nothing resulted from this move.

On 19 May, Johnson's committee recommended the two-engine Titan as the second stage for Saturn. A Centaur was proposed as the third stage for NASA missions. Johnson approved these recommendations and notified Medaris.

The proposal to use the Martin-built Titan brought the Air Force back into the picture, for the development of a modified Titan as a Saturn second stage would affect the Air Force Titan program, which was in full swing. Johnson sought to avoid the potential conflict by directing Medaris to coordinate with the Air Force on actions involving the Glenn L. Martin Company. This didn't suit the Army which wanted to contract directly with Martin for the second stage of the Saturn. In July, the Air Force counterproposed that all procurement and technical requirements be channeled through its Ballistic Missile Division, with the Air Force being responsible for systems engineering of the second stage for Saturn. Matters remained at an impasse until ARPA, on 9 July, authorized the Army to contract directly for the second stage. ARPA also stressed the need for Army coordination with the Air Force.[3]

While the storm between the Army and Air Force over responsibility for Saturn's second stage was brewing in June, an even greater threat to the Army and Saturn was in the making. Herbert York was promoted from chief scientist of ARPA to director of defense research and engineering—the Pentagon's top position for R&D. He was responsible for all military R&D and for avoiding unnecessary duplication. It was not long before York fixed a critical eye on the escalating plans for Saturn. He was aware that in late 1958 Deputy Secretary of Defense Donald Quarles and NASA Administrator T. Keith Glennan had urged transfer of both the Jet Propulsion Laboratory and the Army Ballistic Missile Agency's space team to NASA, but "had been shot down in flames by the Army."[4] Medaris, commander of the Army Ordnance Missile Command, von Braun, technical director of ABMA, and Wilbur Brucker, Secretary of the Army, were determined to make the Army the leader for large launch vehicles. The trio were tough opponents; even President Eisenhower believed that the transfer of ABMA to NASA should be made, but he did not interfere with the negotiations. A compromise had been reached in December 1958 to transfer JPL to NASA, but leave ABMA with the Army.

York, a nuclear physicist, professor, and former director of the Livermore Laboratory of the University of California, was accustomed to making his own analyses of roles, missions, and needed systems. In his new job, York decided to try again to get the ABMA vehicle team transferred to NASA. He argued that space exploration, including all manned flight, was the responsibility of NASA; the responsibility for all large launch vehicles for space exploration should be NASA's;

---

*Dynasoar was first planned as an airplane boosted to a suborbital altitude followed by skip-glide maneuvers in and out of the atmosphere for maximum range. Later models were to achieve orbit.

von Braun and most of his ABMA team should be transferred to NASA; and the cluster of engines and tanks of Saturn was not the best configuration.

In expanding on his points, York cited the Space Act, his understanding of the President's intentions, and his own belief that "nothing yet suggested by the military, even after trying hard for several years, indicated any genuine need for man in space." York believed that the commitment of the von Braun team to big vehicle development "had been seriously interfering with the ability of the Army to accomplish its primary mission. Whenever the Army was given another dollar, Secretary Brucker put it into space rather than supporting the Army's capability for ground warfare."[5] York's criticism of the Saturn I configuration was based on his analysis which indicated that advanced Titan configurations were superior. He also was convinced that a larger vehicle than Saturn I was needed. These considerations led him to argue that Saturn I, as conceived in mid-1959, was unnecessary and should be cancelled. With these convictions, York made his move in June 1959. ARPA had requested funds for Centaur and Saturn; on 9 June, York informed Johnson that he approved the requested funds for Centaur but not those for Saturn and cited more urgent needs as the reason. He suggested that Johnson might consider shifting funds from other projects for Saturn or, failing this, let the development slip.[6]

With this opening move in his campaign to trim military ambitions in space, York next focused on the Air Force's Titan C proposed by the Glenn L. Martin Company as a launch vehicle for Dynasoar. The first-stage was 4 meters in diameter and was powered by four Aerojet ICBM engines of 667 kilonewtons (150000 lb thrust) each. The second stage was powered by two of the same engines but equipped with larger nozzles for high-altitude operation.[7] On 27 July, York suggested that the Air Force and ARPA should study a common vehicle to meet the requirements of both space missions and Dynasoar. He asked for a report before firm development commitments were made. Two days later, ARPA directed the Army to stop work immediately on using Titan for Saturn's second stage, but soon modified the directive to allow general second stage studies to continue.

Convinced that Saturn was "much bigger than any purely military oriented requirements demanded,"[8] York found that a similar view was held by George Kistiakowsky, the President's science advisor, and others who had reviewed military satellite requirements in particular and had concluded that more small "stationary" communications satellites were better than a few large ones. This was a blow to the main military justification for Saturn. York discussed his analysis and conclusions with Secretary of Defense Neil McElroy and then sent Roy Johnson at ARPA this message:

> I have decided to cancel the Saturn program on the grounds that there is no military justification therefor, on the grounds that any military requirement can be accommodated by Titan C as proposed by the Air Force, and on the grounds that by the cancellation the Defense Department will be in a position to terminate the costly operation being conducted by ABMA.[9]

The big questions facing Johnson, if he were to rebut York's arguments, were: Could Titan C accomplish the missions in the military's ten year plan, be ready as soon as Saturn, and be built at lower cost than Saturn? If the answers to these were affirmative, he would have to agree with York. Johnson went into a huddle with his staff. Meanwhile, he informed McElroy that if York's assertions were correct, he would not oppose the cancellation of Saturn. He also proposed that Saturn and ABMA be transferred to the Air Force.[10] This was apparently a last-ditch effort to get the Air Force's help in saving Saturn.

Secretary of the Army Brucker, who had successfully fought Quarles and Glennan to keep ABMA in late 1958, was outraged. Years later York recalled being summoned by Brucker and threatened with dire consequences, but remained firm.[11]

Transfer of Saturn and ABMA to NASA

York's questioning the military need for Saturn forced the issue and the Air Force, Army, ARPA, and NASA had to reconsider and defend their needs for large launch vehicles. He appointed a committee to review the three vehicles under consideration— Titan C, Saturn, and Nova—with himself and Hugh Dryden, NASA deputy administrator, as co-chairmen.* At the outset, the committee agreed on one point: only one large vehicle should be developed by the government. The presentations on Saturn were made by Canright and House of ARPA and Hardeman of ABMA. Some of the committee members recommended further studies to better define the Saturn upper stages. From committee deliberations, Saturn I emerged as the winner. Titan C was shelved, and Nova was too far in the future to be considered competitive to Saturn I.[12]

York concurred with his committee's recommendation to continue Saturn development. Soon after the committee meeting, he began negotiating with NASA Administrator Glennan for transferring ABMA to NASA. He had Secretary of Defense McElroy's support on this, because McElroy wanted to relieve the Army of the big vehicle program.[13]

In September 1959, there were two issues with respect to Saturn: the second stage configuration and the transfer of the ABMA Saturn development team to NASA. ARPA's stop order on second stage contracting, issued at the end of July, was still in effect, and ARPA had been allocating FY 1960 funds to ABMA on a monthly basis since July, pending resolution of the fate of Saturn.

On 23 September 1959, ARPA responded to the York-Dryden committee suggestions to restudy the second stage by requesting ABMA to make such a study. In the meantime, the transfer of the ABMA Saturn team had come to a head. The top officials of the Department of Defense and NASA were in agreement by October; what remained was convincing von Braun. A meeting had been set with the President on 21 October to formalize the transfer, and the night before, Glennan and Horner met with von Braun in a Washington hotel room.[14] Even at that late hour, von Braun had some grave misgivings about the whole plan. His reluctance to transfer to NASA was not caused by any dislike for the new civilian space agency, the creation of which he had favored. However, several earlier discussions with Glennan had led him to doubt

---

*Other members were Richard Horner, NASA associate administrator; Abe Silverstein, NASA director of space flight development; Richard Morse, director of Army research and development; and Joseph V. Charyk, assistant secretary of the Air Force for research and development.

whether the fledgling agency was ready and able to absorb the entire ABMA team of several thousand people. Von Braun believed that a transfer to NASA of only a portion of his team would seriously jeopardize the continuing development of the Saturn rocket, as well as the orderly completion of unfinished work for Jupiter and the Army's new Pershing missile.[15]

Glennan and Horner, however, convinced von Braun that NASA would support him all the way. The next day the transfer of von Braun's team to NASA was formalized. DoD and NASA officials met with President Eisenhower, who approved the transfer by executive order, subject to the approval of Congress. The transfer became effective on 15 March 1960.

The Gathering Storm over Saturn Configurations

The agreement of 21 October 1959 transferring Saturn and its development team to NASA left the upper stage configuration as the major unresolved issue. The proposal to use a Titan as the second stage had been delayed by the ARPA directive in July. At the time of the transfer agreement, ABMA was restudying Saturn upper stages. This study was assigned to a vehicle analysis group headed by H.H. Koelle and assisted by Francis L. Williams.

Koelle, like von Braun and Ehricke, had become interested in rockets at an early age. A German pilot during World War II, he was shot down by American antiaircraft fire in early 1945. Continuing his interests in rockets after the war, he founded the German Space Society in 1948. Von Braun brought him to the United States in 1955; Koelle specialized in analysis, planning, and designing advanced space vehicles. His large group was assisted by aerospace contractors who also had sizable staffs of advanced design specialists.

In 1959, Koelle's group participated in the Army's bid for a role in manned spaceflight. In March, the Army high command authorized the study of a space project called "Project Horizon," the establishment of an Army lunar outpost which proponents referred to as "high ground."[16] The Project Horizon study, completed in June, is an example of very advanced planning; it is of interest here because of the Saturn configurations proposed and some of ABMA's mid-1959 thinking about the use of liquid hydrogen. Launch vehicles for the lunar mission were designated Saturns I and II. The study also considered a much larger vehicle of 53 meganewtons (12 million lb of thrust) using eight F-1 engines and hydrogen-oxygen upper stages, but concluded that this giant vehicle was not needed for the basic mission.

Saturn I for Project Horizon (fig. 55) had three stages. The first was a clustered tank and engine stage using eight Rocketdyne H-1 engines of 837 kilonewtons (188 000 lb of thrust) each, with kerosene and oxygen as propellants. The second stage was essentially a first-stage Titan I, 3 meters in diameter and powered by two Aerojet LR-89 engines with a thrust of 841 kilonewtons (190 000 lb) each, also using kerosene and oxygen. The third stage was a Centaur powered by two Pratt & Whitney RL-10 engines of 67 kilonewtons (15 000 lb thrust) each, using liquid hydrogen–oxygen. Saturn I was essentially the same configuration ABMA was advocating when work was stopped by the July ARPA directive.

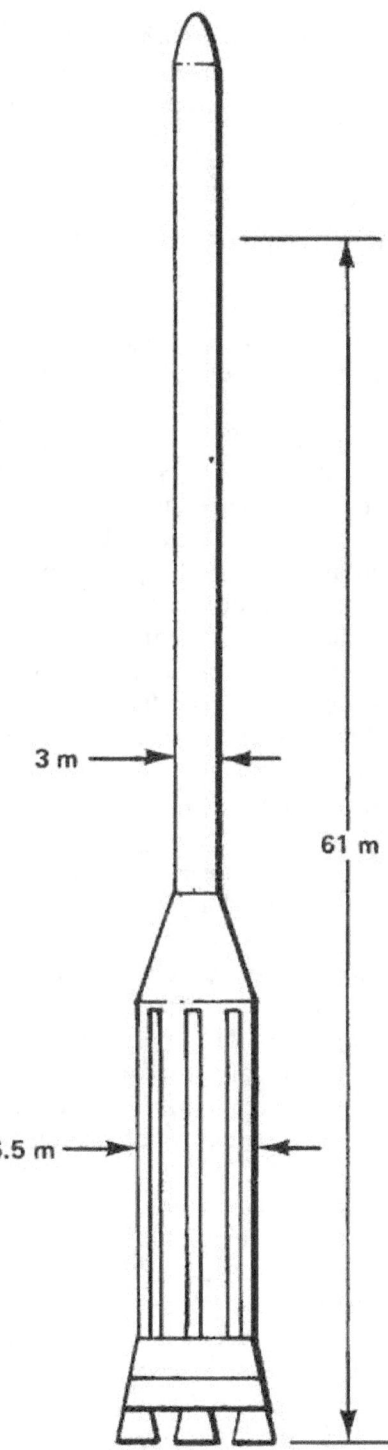

3 m

61 m

6.5 m

Fig. 55. Saturn I as sketched for Project Horizon by the Army Ballistic Missile Agency, 9 June 1959. The first stage used a cluster of tanks and eight engines with a total thrust of 6.7 MN (1.5 million lb); the second stage used two of the same engines burning RP (kerosene) and oxygen. The third stage was a Centaur powered by two hydrogen-oxygen engines.

Saturn II was a second generation vehicle with four stages. The first stage was similar to Saturn I but with uprated engines to provide a total thrust of 9 meganewtons (2 million lb). The second stage was powered by two new proposed engines of 2.2 meganewtons (500 000 lb thrust) each, using liquid hydrogen–oxygen. The third stage was powered by two other new engines of 445 kilonewtons (100 000 lb thrust) each, also using liquid hydrogen–oxygen. The fourth stage was powered by a single engine of the same type as in the third stage. Saturn II, therefore, would use liquid hydrogen–oxygen in all of its upper stages.[17]

In subsequent months, ABMA became much more conservative in its thinking about Saturn upper stages and, ironically, resisted NASA proposals to proceed with liquid hydrogen–oxygen. The reasoning of both parties and the confrontation that was barely avoided complete our story.

NASA had participated in a May 1959 decision recommending the Titan-based second stage for Saturn, which had been stopped by ARPA in July. After the September meeting of the York-Dryden committee on large vehicles, Eldon Hall, Francis Schwenk, and Alfred Nelson began to study Saturn and upper stage configurations. Hall was a leading analyst of flight propulsion and vehicles during his 15 years at the NACA Lewis laboratory. Schwenk was also a propulsion systems analyst who had worked at the Lewis laboratory for eight years before coming to NASA headquarters in 1958. Nelson had been a propulsion analyst at Wright Field for 17 years before joining the group in March 1959.

Two days after the October agreement to transfer Saturn to NASA, Hall sent Silverstein the results of the analysis his group had been making. Among his conclusions: Saturn was basically a good vehicle and could be uprated by using a 4.5 meganewton F-1 engine to replace four of the eight H-1 engines; and by suitable choice of upper stages, development cost could be minimized. Hall recommended that Saturn development be continued and included a phased program (table 11).

Hall's analysis included the new proposed hydrogen-oxygen engine of 668 kilonewtons (150 000 lb thrust) under study by a NASA-DoD group during the year. By agreement with ABMA, the engine was changed to 890 kilonewtons (200 000 lb); it evolved into the J-2 engine by 1960. The NASA B-1 configuration (table 11) was essentially the same as ABMA had proposed as Saturn I of Project Horizon the previous May. Hall was aware of the configuration studies of Koelle and Williams and informed Silverstein that the C-1 configuration in his analysis was similar to the advanced Saturn proposed by ABMA.[18]

The analyses of Saturn by Hall, Schwenk, and Nelson reflected a tradition of NACA and Air Force laboratories. Independent analyses of propulsion systems were not only a means for advancing new concepts, but also for verifying or challenging claims by others. Analyses form the basic framework for interpreting experimental results; it was as routine as tying one's shoe for Silverstein and Hall to do their own analyses of Saturn configurations. This, however, was something new in the experience of ABMA in dealing with headquarters people.

At the end of October, Hall had participated with Abe Hyatt and Adelbert Tischler in a technical survey of ABMA vehicles and attended a meeting in the Pentagon on Saturn configurations. From these meetings, he prepared a table of various Saturn configurations proposed by ABMA, which is reproduced as table 12.[19] The next day

TABLE 11.—*NASA Saturn Configurations, 23 October 1959*

Config-uration	Stage	Name/dia., m	Propellants	No./Type Engine	Stage thrust kN, MN (k, M lb)
B-1	1	Saturn	RP-O₂	8/H-1	6.7 MN (1.5 M)
	2	Titan with thicker skin/3	RP-O₂	2/LR-87	1.8 MN (400 k)
	3	Centaur as proposed for Atlas	H₂-O₂	2/RL-10	134 kN (30 k)
B-2	1	Same as B-1			
	2	High-energy stage	H₂-O₂	2/new 667-kN engines	1.3 MN (300 k)
	3	Enlarged Centaur	H₂-O₂	2/uprated RL-10	greater than 6.7 MN (1.5 M)
C-1	1	Uprated Saturn	RP-O₂	1/F-1 & 4/H-1	not specified
	2	High-energy stage	H₂-O₂	4/uprated 667-kN engines	greater than 2.7 MN (600 k)
	3	Same as Stage 2 of B-2			
	4	Same as Stage 3 of B-2			

he fixed a critical eye on the ABMA configurations and compared them with his own.[20] In the months since June, ABMA had abandoned the use of a Titan I for the second stage because its 3-meter diameter made the vehicle too long and slender, which increased bending loads from aerodynamic forces. Instead, a diameter of 5.6 meters was favored for the second and third stages in the first two models of proposed Saturns (B-1 and B-2, table 12); the fourth stage of a later model (C) went to the same diameter. A feature of the first three ABMA configurations (B-1, B-2, B-3) was the use of either existing engines or engines under development. This would supposedly shorten development time, as engines traditionally took longer to develop than airframes. For this advantage, ABMA was willing to pay a penalty in size and payload. The second stage of the first three configurations used kerosene and liquid oxygen as propellants. NASA, on the other hand, wanted to start development of a 668 kilonewton (150 000 lb thrust) hydrogen-oxygen engine immediately and use it in the second stage of their second model (NASA B-2). ABMA was concerned about bending problems and the need to develop a new, large, hydrogen-oxygen engine for the second stage. NASA was concerned that ABMA's first configuration (ABMA B-1) would cost so much that the development of the large hydrogen-oxygen engine would be seriously delayed and the advanced configurations might never be attained. Hall noted that the second stage of ABMA's first configuration (B-1) was in the Titan C class, yet the payload capability was less than NASA's B-2, for an equal number of stages. Hall became convinced that the ABMA approach was much less than optimum.

TABLE 12.—*Summary of ABMA Saturn Configurations, November 1959*

Config-uration	Stage	Name/dia., m	Propellants	No./Type Engine	Stage thrust kN. MN (k. M lb)
Initial	1	Saturn	RP-O_2	8/H-1	6.7 MN (1.5 M)
	2	Titan/3	RP-O_2	2/LR-87	1.8 MN (400 k)
	3	Centaur/3	H_2-O_2	2/RL-10	134 kN (30 k)
B	1	Saturn	RP-O_2	8/H-1	6.7 MN (1.5 M)
	2	Titan/4.1	RP-O_2	2/LR-87	1.6 MN (360 k)
	3	Centaur/3	H_2-O_2	2/RL-10	134 kN (30 k)
B-1	1	Saturn	RP-O_2	8/H-1	6.7 MN (1.5 M)
	2	/5.6	RP-O_2	4/LR-87	3.9 MN (880 k)
	3	/5.6	H_2-O_2	4/RL-10	356 kN (80 k)
	4	/3	H_2-O_2	2/RL-10	178 kN (40 k)
B-2	1	Saturn	RP-O_2	8/H-1	8.9 MN (2 M)
	2	/5.6	RP-O_2	4/LR-87	3.9 MN (880 k)
	3	/5.6	H_2-O_2	6/RL-10	534 kN (120 k)
	4	/3	H_2-O_2	2/RL-10	178 kN (40 k)
B-3	1	Saturn	RP-O_2	8/H-1 or 4/H-1 + 1/F-1	8.9 MN (2 M)
	2	/5.6	RP-O_2	4/LR-87	3.9 MN (880 k)
	3	/5.6	H_2-O_2	2/new	1.3 MN (300 k)
	4	/5.6	H_2-O_2	4/RL-10	356 kN (80 k)
B-4	1	Saturn	RP-O_2	8/H-1	6.7 MN (1.5 M)
	2	/5.6	H_2-O_2	4/new	2.7 MN (600 k)
	3	/5.6	H_2-O_2	2/new	1.3 MN (300 k)
	4	/5.6	H_2-O_2	4/RL-10	356 kN (80 k)
C	1	Saturn	RP-O_2	8/H-1 or 4/H-1 + 1/F-1	8.9 MN (2 M)
	2	/6.5	H_2-O_2	6/new	4.0 MN (900 k)
	3	/5.6	H_2-O_2	2/new	1.3 MN (300 k)
	4	/5.6	H_2-O_2	4/RL-10	356 kN (80 k)

Source: "Report on Technical Survey of ABMA Activities." Eldon W. Hall, NASA headquarters, 2 Nov. 1959.

Hall got strong support for his views on hydrogen-oxygen for upper stages in a separate but concurrent action. In October 1959, Homer Joe Stewart, NASA's director of program planning and evaluation, wrote a classic memorandum comparing the performance of Atlas-Vega, Atlas-Agena B, and Atlas-Centaur. Differing only in upper stage configurations, Vega used kerosene-oxygen in its upper stage; Agena B, UDMH and nitric acid; Centaur, hydrogen-oxygen. Stewart concluded that since the payloads of Atlas-Vega and Atlas-Agena B were the same, one should be cancelled; subsequently, Vega was. Regarding hydrogen-oxygen, Stewart stated:

Each oxygen-hydrogen stage that is substituted for a conventional propellant stage in a multistage vehicle will increase the payload for a deep space mission two or more times. The figure may be about six times for a marginal conventional propellant system (ratio of payload to first-stage gross weight 0.002). A figure of two to three times is a reasonable generalization. Therefore, substituting oxygen-hydrogen for conventional propellants in two stages of a multistage booster vehicle would increase the payload four to nine times.[21]

While Hall was studying Saturn configurations, Richard Horner, NASA's general manager, initiated an action on ABMA's transfer that provided the mechanism for resolving the issue of Saturn's upper stage configuration.

Saturn Vehicle Team

Following the decision to transfer Saturn to NASA, Richard Horner and Herbert York worked out an agreement for NASA to exercise technical guidance of the project until the formal transfer took place. The agreement provided for a Saturn committee, consisting of NASA and DoD members with a NASA chairman, to provide "advice and assistance" in technical matters. The first and most pressing technical decision was on the upper stages, and Horner requested Silverstein to establish a Saturn Vehicle Team "to prepare recommendations for the guidance of the development and, specifically, for selection of upper stage configurations." Horner made his request on 17 November and wanted the recommendations within thirty days. Silverstein lost no time in getting his team organized.[22] It consisted of:

Abe Silverstein, Chairman	NASA
Col. Norman C. Appold	USAF
Abraham Hyatt	NASA
Thomas C. Muse	DDR&E
G. P. Sutton	ARPA
Wernher von Braun	ABMA
Eldon W. Hall, Secretary	NASA

A brief review of the member's attitudes towards hydrogen is in order.

Silverstein's strong advocacy of hydrogen as a high-energy fuel for aircraft and rockets was well known. Research on hydrogen as a rocket fuel at the NACA Lewis laboratory had been under his direction since 1950. He had initiated a large program on hydrogen for high-altitude aircraft in 1955 and strongly supported more work on hydrogen for rockets. He was familiar with Hall's Saturn studies showing the advantages of using hydrogen-oxygen in the upper stages and was convinced this was the way to go.

Colonel Appold had been the Air Force's manager of the Suntan project using hydrogen for a high-altitude aircraft. In the spring of 1958, he had supported proposals that led later to the initiation of Pratt & Whitney's development of a hydrogen-oxygen rocket engine for Centaur. A large amount of money had been spent on Suntan; and after its cancellation, Appold remained interested in obtaining tangible returns on that investment in technology and facilities. As the only Air Force member of the team, however, Appold had other concerns. The Air Force believed that the Glenn L. Martin Company had its hands full with the Titan ICBM program and took a very dim view of ABMA vehicle proposals using modified Titans, which could interfere with Martin's work on ICBMs.[23] On the other hand, the Air Force was mildly interested in a two stage Saturn as a possible launch vehicle for an advanced Dynasoar—an application that did not need high-energy propellants. Appold, therefore, represented somewhat conflicting views within the Air Force.

Abe Hyatt came from Russia as a small boy, served in the marines during World War II, and rose to chief scientist of the Navy's Bureau of Aeronautics before joining NASA as a flight vehicle and propulsion expert in 1958. He headed launch vehicle and propulsion at NASA headquarters and reported to Silverstein; Eldon Hall worked for him. The three were in agreement on the need to use hydrogen in the upper stages of Saturn from the outset.

Thomas C. Muse worked eleven years as an aeronautical engineer at NACA's Langley laboratory and Douglas Aircraft before joining the Secretary of Defense's staff as an aeronautics expert in 1950. He was neutral with respect to high-energy fuel preferences but recognized their value.[24]

George P. Sutton, chief scientist of ARPA, was the author of the standard rocket propulsion textbook widely used in the United States since it first appeared in 1949. He came to ARPA from Rocketdyne and, like Muse, was neutral on the subject of liquid hydrogen. He was, however, a strong advocate for ARPA interests.

At the time the Saturn Vehicle Team was organized, Wernher von Braun was cold to the idea of using liquid hydrogen. While it is true that his organization proposed Saturn configurations using liquid hydrogen, the early versions would use hydrogen only in the third stage; this was the Centaur and it was being developed by someone else. Of more immediate concern to von Braun was getting confirmation of his plans for the first stage from his new boss, NASA, and settling the long-delayed decision on the second stage. Having been convinced that a cluster of existing engines made sense for early development of the Saturn first stage, he was now equally convinced that a smaller cluster of the same engines made sense for the second stage as well. He could concentrate on building and flight-testing the first two stages, useful for earth-orbital missions, while General Dynamics–Astronautics developed the hydrogen-fueled Centaur as a potential third stage for Saturn. During the Centaur development, already a year old, there would be time to "work out the bugs" in using hydrogen before von Braun had to face the task of adapting it to a third stage.[25] His plan was logical but flawed, as we shall see.

Von Braun's negative attitude towards hydrogen extended far into his background. About 1937, he had observed attempts by Walter Thiel to operate a small rocket engine with liquid hydrogen at Kummersdorf, and the greatest impression he retained was of the numerous line leaks and difficulties of handling liquid hydrogen. It left him with a healthy respect for the safety and fire hazards involved. This attitude would be helpful later in the successful development of the Saturn V, but at the moment was a major roadblock to his acceptance of liquid hydrogen for Saturn I's second stage. At Fort Bliss in the 1940s, von Braun's group had considered a variety of propellants for possible use in the V-2 for a high-altitude sounding mission. The V-2 structure and engine were so heavy that substituting a very low-density fuel like hydrogen would have resulted in poor performance. Krafft Ehricke, who worked for von Braun at Fort Bliss and later at Huntsville, recalls von Braun's objections to low-density propellants.

So does Richard Canright, who wrote a paper on the importance of exhaust velocity and density during that period.[26]

Eldon Hall, the team's secretary, was the sharp analyst who had worked closely with Silverstein since 1955 on the application of liquid hydrogen for high-altitude aircraft and was intimately acquainted with its problems. He had studied very light structures. He had extensive analytical experience in both aircraft and rocket performance. Like Silverstein, he was familiar with liquid-hydrogen research at the Lewis laboratory and had confidence in its practicality. Hall's earlier analyses of Saturn configurations had convinced him that to keep vehicle mass within reasonable limits, the upper stages should use high-energy propellants; and of all the candidates, the combination of liquid hydrogen–oxygen was the closest to practical application.[27] He and Silverstein shared a common understanding and view, and Hyatt—sandwiched between them at NASA—had been persuaded to their view.

Silverstein, therefore, had three working group members favoring his view: Appold, Hyatt, and Hall. Von Braun was the chief opponent—the man who had to be convinced. Silverstein knew that winning von Braun to his view was essential to his and NASA's plans. Von Braun probably was unaware of the extent of NASA's Saturn studies or the intensity of their views on its upper stages. Certainly von Braun wanted to establish good working relationships with his new organization, and he wanted to get on with the job of building large launch vehicles. Although the stage was set for a confrontation, nobody wanted it. Silverstein drew upon all of his skill as chairman to guide the discussions, and he counted on Hyatt and Hall to be strong advocates for his own views. The three met during the course of the team's work to discuss how best to persuade von Braun to their view.[28]

The vehicle team met for the first time on Friday, 27 November. It met four more times and concluded its work on 15 December, with oral and written reports to the NASA administrator.[29]

The first meeting was devoted entirely to briefings: C. Beyer on management aspects of Saturn, E. M. Cortright on NASA missions for Saturn, R. Smith on the Dynasoar program, Wernher von Braun and H. H. Koelle on the technical aspects of the ABMA Saturn systems study, F. L. Williams on the development and funding of the same study, and J. C. Goodwyn on ARPA's evaluation of the study. Upper stages were discussed the next day. Von Braun stressed the importance of an immediate decision and the need to use second stages of 5.6 meters in diameter to lessen bending loads. ABMA was now opposed to using the 3-meter-diameter Titan I as the Saturn second stage, but still favored a modified Titan of larger diameter using RP-oxygen engines.

By the second meeting, the team had agreed on a report outline and assignments of members to write the first five sections, two of which were critical. One of these, about possible Saturn configurations and their performance, was assigned to Koelle of ABMA and Hall and Schwenk of NASA headquarters. The other, on evaluation of Saturn configurations, was assigned to Goodwyn of ARPA, Williams of ABMA, and Hall of NASA. Conclusions and recommendations remained the responsibility of the entire team. The subgroups assigned to prepare the five sections began their work while also participating in meetings of the vehicle team as a whole.

By 3 December a consensus had emerged on one point: to recommend the Saturn first stage under development at ABMA. Attention then shifted to upper stage configurations. A short pitch for solids got little support; von Braun was strongly opposed, because that would combine the handling difficulties of both liquids and solids. Muse argued against a program involving many vehicle changes in favor of going directly to the final desired configuration. Hall noted in the minutes that hydrogen's energy was needed in the upper stages for most missions—although not for Dynasoar—so hydrogen problems had to be faced and solved. Since Dynasoar had an alternative launch vehicle under study, why not go directly to a hydrogen upper stage for Saturn? The problem was really the second stage engine. One solution was to use a stage powered by a cluster of four Pratt & Whitney Centaur engines uprated to a thrust of 89–111 kilonewtons (20 000–25 000 lb) each. At the meeting the next day, von Braun was still not convinced about using hydrogen-oxygen in the second stage. He pointed out that no brand new rocket engine had ever been developed in less than four years and that the development of a liquid hydrogen–liquid oxygen engine more than ten times larger than the Pratt & Whitney engine might take even longer. For this reason, he was not willing to abandon conventional fuels. He also wanted to determine in greater detail the problems with hydrogen-oxygen.

Von Braun expressed concern over aerodynamic heating of liquid hydrogen which required encapsulation of the Centaur stage during flight through the atmosphere—a problem he felt had not been adequately studied for a hydrogen second stage for Saturn. Tank loading and venting problems on the launch pad, with their attendant fire hazards, were other concerns.

By 10 December, Hall had prepared a working draft of the report which contained a recommendation that the second stage be powered by a cluster of four Pratt & Whitney RL-10 (hydrogen-oxygen) engines uprated to 89 kilonewtons (20 000 lb of thrust) each. The stage diameter was 5.5 meters and length, 10.7. There was also a recommendation for Centaur as the third stage and initiation of development of a hydrogen-oxygen engine of 667 to 890 kilonewtons (150 000 to 200 000 lb of thrust) for later Saturn stages.

It was inevitable that at some point during the work of Silverstein's team and its subgroups, the ABMA and NASA representatives would clash head on. Frank Williams, in Koelle's ABMA group on future projects, recalled that the ABMA team was initially so opposed to the use of hydrogen that plans were made "to confront Silverstein with not *no* but *hell no!*" Williams worked hard assembling a four-hour presentation containing great technical detail including cost, probability of success, and impact on Saturn I development schedule, and came to Washington all charged up "to shoot Silverstein out of the saddle." Silverstein was the first to speak and, according to Williams, gave a generalized argument for hydrogen with no technical details: this is the challenge for the long haul; hydrogen is the best fuel; sure it has problems but we can solve them if we dedicate ourselves. Williams considered it a talk along philosophical rather than technical lines and was eager to spring up in rebuttal after von Braun introduced him. To Williams's open-mouthed astonishment, von Braun said, in effect: Abe has a good point. Williams never got the chance to present his arguments. He, not Silverstein, had been "shot out of the saddle"—by his own boss.[30]

Eldon Hall's group at NASA headquarters tangled with Koelle's advanced design group at ABMA on another occasion, which proved to be decisive. The NASA headquarters analysts were using their slide rules to calculate vehicle performance whereas ABMA analysts used a complex program requiring large computer runs. It was not an equal match, but the NASA headquarters analytical group (Hall, Schwenk, and Nelson) had a great deal of experience and judgment. They had noticed that all the Saturn configurations showing promise had at least one upper stage using hydrogen-oxygen. Configurations that used only "conventional" (lower performance) propellants had total masses up to twice as great as those using hydrogen-oxygen stages. The configuration favored by ABMA at one point used four ICBM engines burning RP-oxygen to power the second stage and a modified hydrogen-oxygen Centaur as the third stage. Hall calculated that, by simply replacing the RP-oxygen second stage with the Centaur alone, the resulting two-stage vehicle would lift nearly as much payload to earth orbit as the three-stage ABMA configuration. Hall so argued at one meeting and von Braun considered it incredible. He telephoned Huntsville, where the computer was kept busy all night. The following morning, ABMA telephoned von Braun that Hall was right—the payload without the RP-oxygen stage was indeed close to that with it![31] It was a powerful and convincing argument for the use of high-energy upper stages. This, and the persuasive arguments of Silverstein, convinced von Braun that hydrogen-oxygen for all the upper stages of Saturn was the way to go.

The meeting of 14 December was spent on the proposed report. Sutton questioned the payload figures and wanted to wait for the final "official values" from ABMA, but time was running out. He also argued unsuccessfully for considerably more study before making specific recommendations and questioned the wisdom of omitting a large diameter Titan I as a possbile second stage. By then, however, the von Braun team not only opposed the modified Titan I, because of its high bending stresses, but now strongly supported hydrogen-oxygen in all upper stages. Oswald Lange, representing von Braun, successfully argued that the large diameter Titan I with RP-oxygen was a "dead end" course, and the report so indicated. On 15 December the Saturn Vehicle Team endorsed the recommendation that all upper stages of Saturn be fueled with hydrogen and oxygen. Silverstein, with help from Hall, quickly prepared the final report which bears the same date. The unanimous decision for hydrogen in Saturn's upper stages was a victory for the skillful chairman and his quiet but sharp secretary.

Saturn Development Plan

The report of the Saturn Vehicle Team on 15 December 1959 gave lunar and deep space missions (4500 kg payload) first importance; "stationary" 24-hour equatorial orbit missions second priority; and manned spacecraft missions in low earth orbit (e.g., Dynasoar) third. Five recommendations were made regarding launch vehicles for these missions. A plan was needed for the orderly development of a series of vehicles of increasing payload capability, with emphasis on reliability. All upper stages would use liquid hydrogen–oxygen. The first of the vehicle series should be configuration C-1 (table 13). Fourth, development of a new hydrogen-oxygen rocket engine with a thrust of 668–890 kilonewtons (150000–200000 lb) should begin immediately, along with design studies of stages using it. Finally, a funding plan as prepared by ABMA for vehicle development was recommended.[32]

TABLE 13.—*Possible Saturn Configurations, December 1959*

	Stage			
Vehicle	1	2	3	4
A-1	RP-O₂ 8 H-1 cluster	RP-O₂ Titan 3 m dia.	H₂-O₂ Centaur 3 m dia. 2 × 67 kN (15000 lb)	
A-2		Cluster of IRBM engines	H₂-O₂ Centaur 3 m dia. 2 × 67 kN	
B-1		RP-O₂ 5.6 m dia. 4 H-1 type	H₂-O₂ 5.6 m dia. 4 × 67–89 kN (15000–20000 lb)	H₂-O₂ Centaur 3 m dia. 2 × 67 kN
C-1		H₂-O₂ 5.6 m dia. 4 × 67–89 kN	H₂-O₂ Centaur 3 m dia. 2 × 67 kN	
C-2		H₂-O₂ 5.6 m dia. 2 × 668–890 kN (150000–200000 lb)	H₂-O₂ 5.6 m dia. 4 × 67–89 kN	H₂-O₂ Centaur 3 m dia. 2 × 67 kN
C-3	RP-O₂ 8.9 MN cluster (2 million lb)	H₂-O₂ 5.6 m dia. 4 × 668–890 kN	H₂-O₂ 5.6 m dia. 2 × 668–890 kN	H₂-O₂ 5.6 m dia. 4 × 67–89 kN

Source: Saturn Development Team. "Report to the Administrator on Saturn Development Plan," 15 Dec. 1959.

Of six vehicle configurations considered, only three were recommended (those with "C" designations, table 13). Combinations of only three hydrogen-oxygen upper stages would serve for all three vehicles in a "building block" approach proposed by Hall (figs. 56, 57).[33] The two engines proposed for these upper stages were the Pratt & Whitney RL-10, part of the legacy of the Suntan project, and a new and larger engine which later became the J-2.

The Saturn Vehicle Team presented the results in a meeting with T. Keith Glennan, NASA administrator; Hugh L. Dryden, deputy administrator; and Richard Horner, associate administrator; and its recommendations were approved. On 29 December, Horner discharged the vehicle team and replaced it with a new Saturn committee that he and Herbert York had agreed would be useful in technical guidance for Saturn during the interim period, before ABMA and the Saturn were formally transferred to NASA the following March.[34] By this time, NASA had split its space effort into two parts with Silverstein heading the office of spaceflight programs, concerned chiefly with

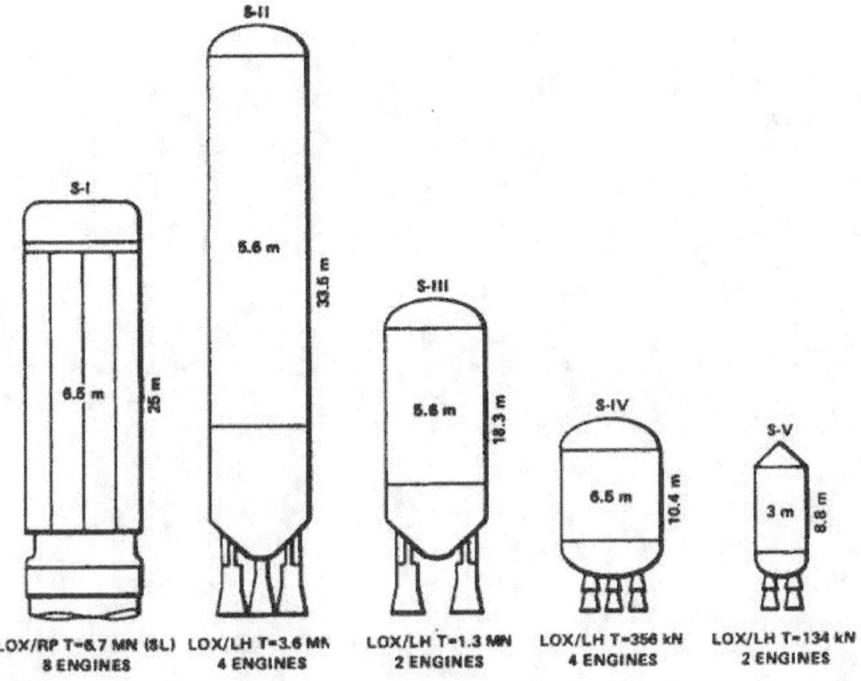

Fig. 56. Saturn "building block" stages recommended for a series of launch vehicles by the Saturn vehicle team, December 1959.

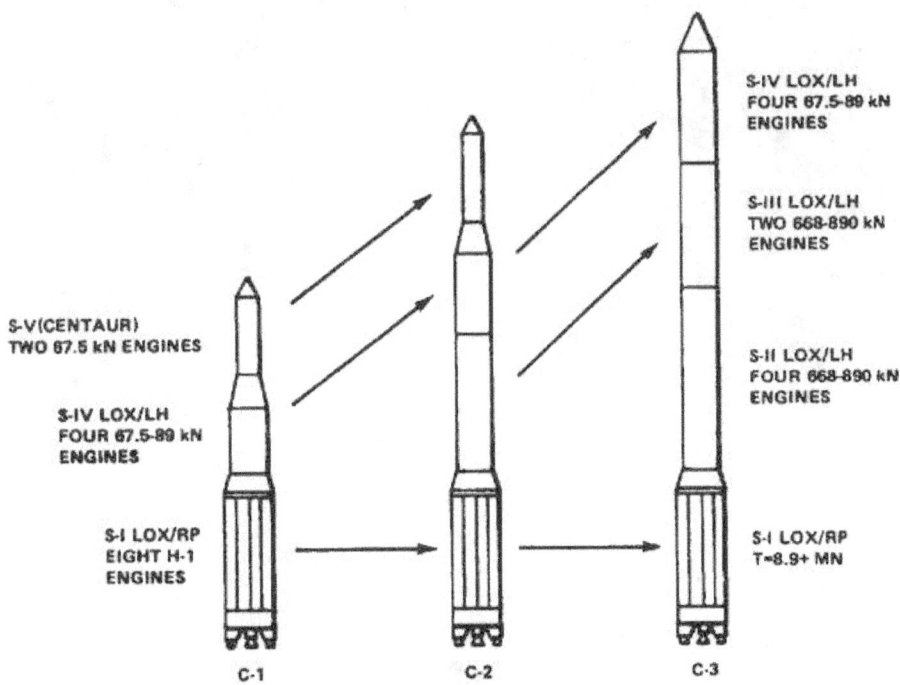

Fig. 57. Saturn configurations recommended by the Saturn vehicle team, December 1959.

Fig. 58. Saturn V, the Apollo launch vehicle, 10 m in diameter and 111 m tall with the Apollo spacecraft. First launched in 1967, Saturn V used liquid hydrogen–oxygen in its two upper stages.

space missions and payloads, and Maj. Gen. Don R. Ostrander heading the newly created office of launch vehicle programs, responsible for the launch vehicles and propulsion development and operations.

During the next two years, Saturn configurations were restudied as part of the national commitment in 1961 for a manned lunar landing, but one basic concept established by Silverstein did not change: the use of hydrogen in all Saturn upper stages. The work of Silverstein, Hall, and the other members of the Saturn Vehicle Team in taking a bold stand in choosing to use liquid hydrogen for Saturn was one of the major decisions that made the great manned spaceflight events of the 1960s and 1970s possible.

SUMMARY, PART III

Russian achievements with satellites caused a rapid change in United States policy on space; from a second-priority scientific investigation, space became a major national effort. During the 1950s, missile development and research and development projects on hydrogen provided the basis for substantially increasing U.S. launch vehicle capability. The Air Force project on a hydrogen-fueled airplane, started in 1956, did not reach fruition; but its managers, technology, liquefiers, transport dewars, and other equipment transferred directly to the development of an upper stage for Atlas-Centaur—which became the first hydrogen-fueled rocket that flew.

Research sponsored by the Air Force on the feasibility of large rocket engines, beginning in 1955, provided the basis for moving quickly during 1958 to start the development of a large rocket engine of 6.7 meganewtons—ten times larger than the largest current engine. Concepts pushed by the Advanced Research Projects Agency led to the authorization of the first large launch vehicle, using a cluster of existing ICBM engines in the first stage.

Competition between the Air Force, Army, ARPA, and the National Aeronautics and Space Administration for planning and developing large launch vehicles led to decisions for a single large launch vehicle, Saturn I, with NASA as the responsible agency. Disagreement over the upper stage configuration of the vehicle, and particularly whether to use conventional fuel or liquid hydrogen, led to a bold decision at the end of 1959 to use hydrogen. This decision was one of the keys to the success of the Apollo moon landing missions of the 1960s and 1970s.

Appendixes

Appendix A

Hydrogen Technology through World War II

Hydrogen is the simplest element, a molecule of two atoms, a gas at normal conditions, colorless, odorless, nontoxic but asphyxiating, and non-corrosive but reactive. It occurs in many substances, the most common of which is water, and ranks ninth in abundance of the chemical elements on the earth. Active chemically, hydrogen exists free only in minute quantities. There is plenty of hydrogen in stars, and Jupiter is believed to be made up entirely of hydrogen in several forms under intense pressure.

Only helium is more difficult to liquefy than hydrogen. At atmospheric pressure, hydrogen boils at 20.3 K. It is the lightest element; its liquid density is 1/14 that of water.

Hydrogen ignites very easily and burns over a wider range of mixtures with air or oxygen than any other fuel. It releases more than twice as much energy on burning as gasoline on a mass basis, but because of its low density rates low on a volume basis. Hydrogen's low density and liquid temperature, coupled with other characteristics, are major obstacles to its more widespread use as a fuel.

The first experimental investigation of liquid hydrogen for flight propulsion in the United States began in 1945. It started at the same time that documents from Germany on technical details of jet aircraft and rockets became available. Nevertheless, research indicates that U.S. interest in hydrogen for aircraft and rockets was not directly linked to German work.

Hydrogen had interested scientists and engineers for centuries, and since its liquefaction at the end of the nineteenth, had been considered as a fuel by three rocket pioneers. Providing some background for a better understanding of post-World War II hydrogen developments in the United States is the purpose of this appendix.

Appendix A-1

Hydrogen through the Nineteenth Century

Man became aware of hydrogen as a flammable gas when he began mixing iron and sulfuric acid near a flame, an event that could have occurred as early as the eighth century and certainly no later than the sixteenth.= The use of hydrogen became such an integral part of the history of chemistry from the seventeenth through the nineteenth century that only selected highlights, covering properties pertinent to its use as a fuel, will be briefly summarized here.

Gaseous Hydrogen

Robert Boyle, one of the founders of modern chemistry, published the first description of the flammability of hydrogen in 1673 and Henry Cavendish described its flammability limits in air in 1766.÷ Cavendish's results were close to the modern limits of 4 to 75 percent hydrogen by volume. In contrast, the limits for gasoline vapor are 1 to 7 percent and for natural gas, 4 to 15 percent by volume. The very wide flammability limits of hydrogen make it easy to burn over a wide range of conditions, a great asset in a fuel. The same property, however, makes hydrogen hazardous to handle.

Cavendish, who called hydrogen "inflammable air," was also the first to measure its density. He reported in his 1766 paper that hydrogen is 7 to 11 times lighter than air (the modern value is 14.4). Cavendish's results not only opened up a new chapter in the history of gases, but also attracted attention to hydrogen as an alternate to hot air as a buoyant gas. Jacques Charles was first to take advantage of this soon after the first public demonstration of a hot-air balloon. After four days of struggling with his iron-acid hydrogen generation equipment, Charles launched his 4-meter balloon on 27 August 1783. Just over three months later, he and one of his balloon builders, Aine Robert, became the first men to ascend in a hydrogen balloon.×

With all the enthusiasm over balloning that began with the Mongolfier brothers and Charles in 1783, it was inevitable that the good and bad properties of hydrogen would meet. The worst happened on 15 June 1785 when Pliatre de Rozier and an assistant, P. A. Ronaon, attempted to cross the English channel in a hydrogen balloon carrying a small hot-air balloon for altitude control. Thirty minutes into the flight the hydrogen ignited and the two men perished. Hydrogen's flammability was the underlying cause of the first air tragedy.[4] Nevertheless, the attractiveness of hydrogen as a readily available buoyant gas was to outweigh the danger of flammability for 150 years of lighter-than-air flight. This application of hydrogen is the reverse of hydrogen's later role as a fuel, where its flammability is a major advantage and its low density the disadvantage that inhibits its use for flight in the atmosphere.

The increased demand for hydrogen for balloon flight following Charles's successes brought an early improvement in the technology of its generation. Although iron and sulfuric acid were readily available, their use in generating hydrogen was difficult to apply in the filling of balloons. In the winter of 1783–1784, a scientist and an inventor collaborated to bring a great improvement in hydrogen generation. The great French chemist, Antoine Lavoisier, and Charles Meusnier, army officer and inventor, generated hydrogen by passing steam through the red-hot barrel of an iron cannon. The Lavoisier-Meusnier process, with refinements, became the most effective and economical way to obtain hydrogen during the first part of the nineteenth century. Although still in use, the steam-iron method was largely replaced at the start of the twentieth century by two other methods: passing superheated steam over incandescent coke and electrolysis of a dilute solution of caustic soda.[5]

For use as a fuel, a property of hydrogen that is of even greater importance than flammability is the large amount of energy released during combustion. Lavoisier and Pierre Laplace measured the heat of combustion of hydrogen in 1783–1784 using an ice calorimeter. The experiment took $11\frac{1}{2}$ hours and the amount of ice melted was equivalent to about $97 \cdot 10^6$ joules per kilogram of hydrogen. This was much higher than values obtained for other substances, and whether for this reason or other uncertainties, the results were not published until 1793. During the nineteenth century, the heat of combustion was measured many times. The modern value is $120 \cdot 10^6$ J/kg for gaseous reactants to gaseous products, so the Lavoisier-Laplace value was not too far off. In comparison, the heat of combustion of gasoline is $48 \cdot 10^6$ J/kg, less than half that of hydrogen.[6]

The density of gaseous hydrogen is so low that its heat of combustion on a volume basis does not compare favorably with denser fuels. Hydrogen was used in heating torches, but has largely been replaced by acetylene. Gaseous hydrogen was used in 1820 as fuel for one of the earliest internal combustion engines, but it was quickly replaced by coal gas which was much more readily available and had a higher heating value per unit volume.[7]

During the nineteenth century, other hydrogen properties, useful in fuel applications, were determined. The explosive limits of hydrogen-oxygen mixtures were found to be from 5.5 to 95 percent hydrogen by volume—much wider than its limits in air. The flame temperature of two parts hydrogen and one of oxygen was measured as 3117 K in 1867, not far from the modern value of 2760 K. In 1881, flame speed at the same mixture was measured as 2810 m/s.[8]

So far, only gaseous hydrogen has been discussed. Important to the application of hydrogen as a fuel are its properties as a liquid, which were not known until near the end of the nineteenth century.

Liquefaction of Gases through the Nineteenth Century

"The production of cold is a thing very worth of the inquisition both for the use and disclosure of causes," wrote Francis Bacon, the first systematic investigator of low temperature phenomena, in 1627.[9] Starting in the late eighteenth century, investigators sought to liquefy gases and reach lower and lower temperatures. By mid-nineteenth century, all but six of the known gases had been liquefied, and temperatures as low as 163 K had been attained by evaporating a mixture of ether and solid carbon dioxide. The six remaining gases were oxygen, nitrogen, nitric acid, carbon monoxide, methane, and hydrogen. (Helium, observed in the sun's gases in 1868, was not discovered on earth until 1895.)

In 1883, a Polish professor of physics, Zygmunt von Wroblewski, achieved the static liquefaction of oxygen and air. Thereafter, with the capability to cool compressed hydrogen to 73 K, efforts to liquefy it intensified. Wroblewski's attempt resulted in only a transient vapor. However, in 1885 he published some remarkably accurate physical data He gave hydrogen's critical temperature as 33 K (modern value, 33.3 K); critical pressure, 13.3 atmospheres (modern value, 12.8 atm); and boiling point, 23 K (modern value, 20.3 K).

Gas liquefaction techniques up to 1895 involved three basic steps: 1) compressing the gas to a high pressure, usually 50 atmospheres or more, 2) chilling the compressed gas to as low a temperature as possible using various cooling methods, and 3) expanding the chilled, compressed gas slowly from a high to a lower pressure by means of a needle valve. The cooling methods included, for example, evaporating an ether–solid carbon dioxide mixture or evaporating liquid ethylene, which Wroblewski used in liquefying oxygen. The third step made use of the Joule-Thomson effect for gases, based on experiments by Joule in 1845 and later refined by Thomson. They found that a gas, in slowly expanding from a high to a lower pressure, undergoes a change in temperature. The gas may be either heated or cooled by the expansion, depending upon the initial temperature and the particular gas. For most gases at room temperature, expansion results in cooling, as anyone who has used an aerosol can or operated a carbon dioxide fire extinguisher has experienced. Compressed air initially at 273 K, for example, will drop about ¼ K for each atmosphere drop in pressure, while carbon dioxide will drop 1½ K for each atmosphere drop in pressure. The temperature below which an expansion produces cooling is called the inversion temperature; it is high for most gases, but for hydrogen, it is about 193 K. Hydrogen, therefore, must be cooled below this temperature before it is expanded.

In 1895, a breakthrough occurred in gas liquefaction techniques, although it is not clear whether scientist or engineer first used it. The technique was to employ regenerative cooling in the liquefaction process, a simple concept in retrospect. Regenerative cooling means using a fluid as the coolant in a process in which the fluid is itself involved. In the liquefaction of gases, it means that the gas that is cooled by the Joule-Thomson expansion process is later used to cool the incoming compressed gas before expansion.

The regenerative cooling concept was an old idea, first introduced by Siemens in 1857 and used by Kirk, Coleman, Solvay, Linde, and others in refrigeration apparatus. In 1895, and within two weeks of each other, William Hampson in England and Carl von Linde in Germany obtained patents for equipment to liquefy air using the Joule-Thomson expansion process and regenerative cooling.[10] Linde described his apparatus to physicists and chemists at Munich in 1895. A number of publications appeared that same year, among them one by James Dewar, English physicist and chemist, who described his apparatus for liquefying air using regenerative cooling.

Hampson's process for liquefying air was simple. He compressed air to 200 atmospheres, expanded it to one atmosphere, and passed the expanded, cooled air through a baffled heat exchanger to cool the incoming compressed air. His method was not very efficient, using power at a rate equivalent to 3.7 kilowatts to produce 1 liter per hour of liquid air. Hampson had his apparatus working at Brin's Oxygen Works by April 1896.[11]

Linde's approach was more complex than Hampson's, but it was also more efficient and suitable for large-scale production of liquid oxygen, nitrogen, and air. He used two stages of gas compression, precooled the air with a separate ammonia refrigeration system, and employed a coiled-tube heat exchanger having three concentric tubes for regenerative cooling. The heat exchanger was insulated by a wood case filled with wool.

Regenerative cooling proved to be the technological link needed to liquefy hydrogen. On 10 May 1898, James Dewar used it to become the first to statically liquefy hydrogen. Using liquid nitrogen he precooled gaseous hydrogen, under 180 atmospheres, then expanded it through a valve in an insulated vessel, also cooled by liquid nitrogen. The expanding hydrogen produced about 20 cubic centimeters of liquid hydrogen, about 1 percent of the hydrogen used.[12]

Dewar measured the density of liquid hydrogen at 0.07 kilogram per liter, the modern value, which is 1/14 the density of water and about 1/12 the density of kerosene or gasoline.

The insulated vessel Dewar used was the vacuum container flask he developed earlier which became known as "Dewar flasks," now simply dewars. His design was a very significant contribution to the storage and transportation of very cold liquefied gases such as oxygen, nitrogen, air, hydrogen, fluorine, and helium. Dewars are double-walled vessels with a vacuum in the annular space to minimize heat transfer by conduction and convection. The walls are silvered to reflect radiant heat. Following liquefaction of hydrogen, Dewar became very confident about storing and transporting it in his vacuum vessels, predicting that it could be handled as easily as liquid air. Dewar vessels, with engineering refinements, are used today to transport liquid hydrogen with very low loss rates.

By 1900, then, many of the major properties of gaseous and liquid hydrogen were known. Liquid air, oxygen, and nitrogen were being produced in quantity. Hydrogen had been liquefied, and dewar flasks made its storage and transportation feasible. Suggestions for using liquid hydrogen were not long in coming. The use of gaseous hydrogen for ballooning and its loss from venting brought a suggestion for using liquid hydrogen by storing and evaporating it in a double-walled bag.[13] Of greatest interest to our story, however, was a suggestion made by an obscure Russian schoolteacher, barely five years after Dewar's accomplishment, to use liquid hydrogen to fuel a space rocket.

Appendix A-2
Rocket Pioneers

Konstantin Eduardovich Tsiolkovskiy (1857–1935) was sixteen years old when he was struck by a fascinating idea: Why not use centrifugal force to launch a spacecraft from earth? He became so excited about his idea that he could not sleep and wandered about the streets of Moscow all night, thinking about it. By morning, however, he saw the flaw in the concept. The experience had a profound effect on his later activities. Throughout his life, he continued to dream of flying to the stars; when he did so, he felt the same excitement as he had on that memorable night.[1] Tsiolkovskiy became the first to develop the theory of rocket flight and the first to consider hydrogen-oxygen to propel rockets.

Born of poor parents, Tsiolkovskiy lost his hearing at the age of nine and with it went the normal social relationships and education of children in his village, 900 kilometers east of Moscow. He became an avid reader and daydreamer, educating himself and qualifying as a schoolteacher in spite of his deafness, little guidance, and lack of books. Tsiolkovskiy's father, a forester, encouraged his son to build models, do physical labor, and be self-reliant. His mother showed him a collodian balloon filled with hydrogen when he was eight.* At fourteen, Tsiolkovskiy attempted to make a paper balloon filled with hydrogen but failed. He later developed the idea of a metal dirigible and published papers on the notion. He became interested in winged flight also and built a model which his father proudly showed to guests. In his twenties, he became involved with steam engines, fans, and pumps but after building a few he realized that his talents lay more in the direction of analyses and theoretical studies.[2] All of Tsiolkovskiy's rocket contributions are theoretical; he did not attempt experiments.

Tsiolkovskiy was a prolific writer on rockets and other subjects, displaying a remarkable insight into physical phenomena. He did not possess advanced academic credentials, yet he was recognized and accepted by eminent scientists for his contributions. In 1891 he sent a paper on the theory of gases to the Petersburg Physico-Chemical Society where it was well received by the members, including Dmitri Mendeleyev, famed Russian chemist. After a later contribution on the mechanics of

---

*Flexible collodian, made by dissolving guncotton in alcohol and ether and adding balsam and castor oil, was commonly used for making small balloons in the latter half of the nineteenth century. Tsiolkovskiy's mother may have bought a hydrogen-filled collodian balloon from an itinerant peddler.

252

animal organisms, Tsiolkovskiy was unanimously elected a member of the Society.

Of Tsiolkovskiy's many contributions to rocket technology, the best known is the theory of rocket flight which he developed from the laws of motion.[3] In its simplest form, the velocity of a rocket can be expressed as:

$$V = V_j \ln(M_o / M_e)$$

where V is the maximum velocity of the rocket in gravity-free, drag-free flight, V_j is the rocket exhaust jet velocity, ln is the natural logarithm, M_o is the initial or full rocket mass, and M_e is the empty rocket mass. This equation is called the Tsiolkovskiy equation in fitting tribute to the great pioneer.

The Tsiolkovskiy equation states a key relationship for understanding the advantages and disadvantages of liquid hydrogen as a fuel. The equation shows that rocket vehicle velocity is directly proportional to the rocket exhaust jet velocity. The latter is essentially constant for a given rocket design, propellants, and operating conditions. It depends upon the amount of heat energy released during combustion, the combustion pressure, the combustion products, and the nozzle for expanding the gases.

The second term of the Tsiolkovskiy equation, $\ln(M_o / M_e)$, involves two masses differing only in the amount of propellant (fuel and oxidizer) expended. The initial or full mass includes everything—payload, vehicle structure, tanks, engines, controls, guidance, and propellants. The empty mass is the initial mass minus the propellants that have been expended. During operation, the continuous burning of propellant and expelling of exhaust means that the total mass of the vehicle is continuously decreasing, starting with M_o and ending with M_e. Tsiolkovskiy understood this well, deriving the equation of flight based on the conservation of momentum, integrating, and using the initial and final conditions of the rocket to obtain his equation.*

Tsiolkovskiy explained the mass ratio term by showing that if the mass ratio is written as a geometric progression, the corresponding relative velocity ratios are an arithmetic progression.[4] As an example, he chose a progression of mass ratios to the base 2 and computed velocity ratios relative to the velocity ratio for a mass ratio of 2. This gave the geometric and arithmetic progressions shown in the first two rows of figure 59. The other relationships in the figure show the actual values of mass ratios and velocity ratios and also the former in terms of propellant mass fraction and the useful range for rockets.†

Examination of the values of figure 59 shows why there is an incentive for rocket designers to make the vehicle's structure, engines, guidance, and controls as light as possible, for they and the payload constitute the empty mass. Every kilogram shaved

---

* Applying the conservation of momentum to rocket flight without drag or gravity loss gives $dm\, V_j = m\, dv$. where m is vehicle mass, d the derivative, V_j exhaust jet velocity (assumed constant), and v is flight velocity at the time the mass is m. Separating the variables and integrating gives $v/V_j = -\ln m + C$, where C is the constant of integration. Using initial conditions ($v=0$, $m=M_o$) to solve for C gives $C = \ln M_o$; the equation becomes $v/V_j = \ln(M_o/m)$, the vehicle velocity v when its mass is m. The rocket reaches its maximum velocity, V, when all propellant is expelled; at this time, $m = M_e$ and $v = V$. Substituting these gives Tsiolkovskiy's equation.

†Since the mass of propellant expended (M_p) is the difference between initial mass (M_o) and final mass (M_e), mass ratios can be expressed in terms of any two. Tsiolkovskiy later used the ratio M_p/M_e, now known as the Tsiolkovskiy ratio. In the United States, the propellant mass fraction M_p/M_o is often used; it is easy to visualize physically.

TSIOLKOVSKIY EQUATION, $V = V_j \ln \dfrac{M_o}{M_e}$, WHERE V = ROCKET VELOCITY, V_j = EXHAUST VELOCITY,

In = NATURAL LOGARITHM, M_o = FULL MASS, AND M_e = EMPTY MASS

	2^1	2^2	2^3	2^4	2^5	2^6	2^7
MASS RATIO, M_o/M_e (AS GEOMETRIC PROGRESSION)							
VELOCITY RATIO, V/V_j (RELATIVE TO V/V_j FOR M_o/M_e = 2)	1	2	3	4	5	6	7
MASS RATIO, M_o/M_e (SAME AS FIRST ROW)	2	4	8	16	32	64	128
ACTUAL VELOCITY RATIO, V/V_j	.693	1.39	2.08	2.77	3.47	4.16	4.85
FRACTION OF TOTAL MASS THAT IS PROPELLANT (FUEL + OXIDIZER)	.500	.750	.875	.938	.967	.984	.992

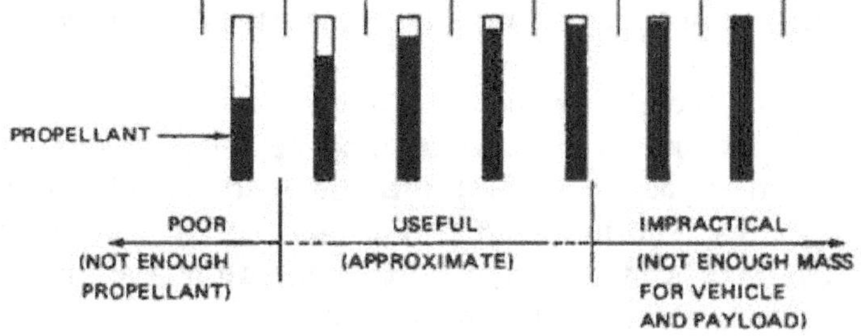

PROPELLANT →

POOR   USEFUL   IMPRACTICAL

(NOT ENOUGH PROPELLANT)   (APPROXIMATE)   (NOT ENOUGH MASS FOR VEHICLE AND PAYLOAD)

Fig. 59. Tsiolkovskiy's equation of rocket flight, neglecting drag and gravity. He illustrated the relationship by the geometric progression of the first row and the arithmetic progression of the second. The other rows illustrate the same relationship in other forms.

off the vehicle hardware means either a kilogram of payload gained or an increase in mass ratio and hence, vehicle velocity. By this reasoning, it can also be seen that given a choice, vehicle designers prefer dense propellants, for a greater propellant load can be put in a given tank size and mass. This is one of the reasons many investigators avoided liquid hydrogen, which has the lowest density of any fuel.

The progressions of figure 59 also show that in the higher mass ratios, gains in velocity come slowly. This diminishing return effect focuses attention on the other term of Tsiolkovskiy's equation, exhaust jet velocity. How can it be increased? By using fuels yielding higher heat energy, which translates into higher exhaust jet velocity. Tsiolkovskiy recognized the rocket as a heat engine and was aware of Joule's measurement of the mechanical equivalent of heat. It follows that the more heat that could be generated per kilogram of burning propellants, the greater the amount of work that could be done. He therefore searched the chemistry texts, particularly Mendeleyev's *Principles of Chemistry*, to find the fuels giving the highest heat per unit mass of reactants.[5] From what was known about heats of combustion by the start of the twentieth century, it is not at all surprising that Tsiolkovskiy became the first to propose the use of liquid hydrogen and oxygen to propel a rocket, which he did in his classic "Exploration of the Universe with Reaction Machines," first published in 1903.[6] Tsiolkovskiy used a heat of formation of water of 16 · 10[6] joules per kilogram (from reacting hydrogen and oxygen) and recognized that some heat would be expended in converting liquid hydrogen and liquid oxygen to their gaseous states.* He converted this heat of formation into mechanical energy and obtained a potential energy of 1633 kilogram-meters. Using Newton's relationship between potential and kinetic energy, Tsiolkovskiy calculated the exhaust jet velocity of a liquid hydrogen–liquid oxygen rocket as 5700 meters per second.† Using this value, he calculated vehicle velocities for a range of mass ratios. For example, at a mass ratio M_o/M_e of 5, vehicle velocity was 9170 meters per second. Tsiolkovskiy preferred a mass ratio of 5, which means that 80 percent of the vehicle mass is propellant, because he calculated this ratio gave the greatest utilization of propellants.‡

In his 1903 paper, Tsiolkovskiy described a manned rocket (fig. 60):

Visualize . . . an elongated metal chamber . . . designed to protect not only the various physical instruments but also a human pilot . . . . The chamber is partly occupied by a large store of substances which, on being mixed, immediately form an explosive mass. This mixture, on exploding in a controlled and fairly uniform manner at a chosen point, flows in the form of hot gases through tubes with flared ends, shaped like a cornucopia or a trumpet. These tubes are arranged lengthwise along the walls of the chamber. At the narrow end of the tube the explosives are mixed: this is where the dense, burning gases are obtained. After undergoing intensive rarefaction and cooling, the gases explode outward into space at a tremendous relative velocity at the other, flared end of the tube. Clearly, under definite conditions, such a projectile will ascend like a rocket . . . . The two liquid gases are separated by a partition. The place where the gases are mixed and exploded is shown, as is the flared outlet for the intensely rarefied and cooled vapors. The tube is surrounded by a jacket with a rapidly circulating liquid metal [mercury]. The control surfaces serving to steer the rocket are also visible.[7]

The fuel in this rocket is labeled "hydrocarbon," although in the article Tsiolkovskiy discussed hydrogen-oxygen more than any other fuel. In an article in 1911, summarizing work to that time, the caption to a drawing of his 1903 rocket showed liquid hydrogen as the fuel.[8]

The hydrogen-oxygen combination greatly appealed to Tsiolkovskiy because the thermal energy released in its reaction was the highest he knew. In his initial enthusiasm, he brushed aside the difficulty of liquefaction; barely five years after Dewar's initial success, he stated: "At the present time the transfer of hydrogen and oxygen into their liquid states poses no special problem."[9] He does not refer to Dewar's work, but he was aware of the liquefaction of air and the effect of low temperature on

---

*His value for water converts into a heat of combustion of $144 \cdot 10^6$ J/kg for hydrogen, almost the same as obtained by Dulong in 1838.

†Using $\frac{1}{2}V^2 = gh$, where V is velocity, g acceleration of gravity (9.81 m/s²), and h height in m. Solving for velocity gives Tsiolkovskiy's result. He did not allow for the thermal cycle efficiency of a rocket engine, which is on the order of 40 to 50 percent, although he did in a later (1911) paper.

‡He computed utilization as the kinetic energy of the empty rocket at burnout velocity divided by the kinetic energy of the propellant mass at its jet velocity. The mass ratio of 5 is a good value for modern rockets; the German V-2, however, had a mass ratio of about 3.

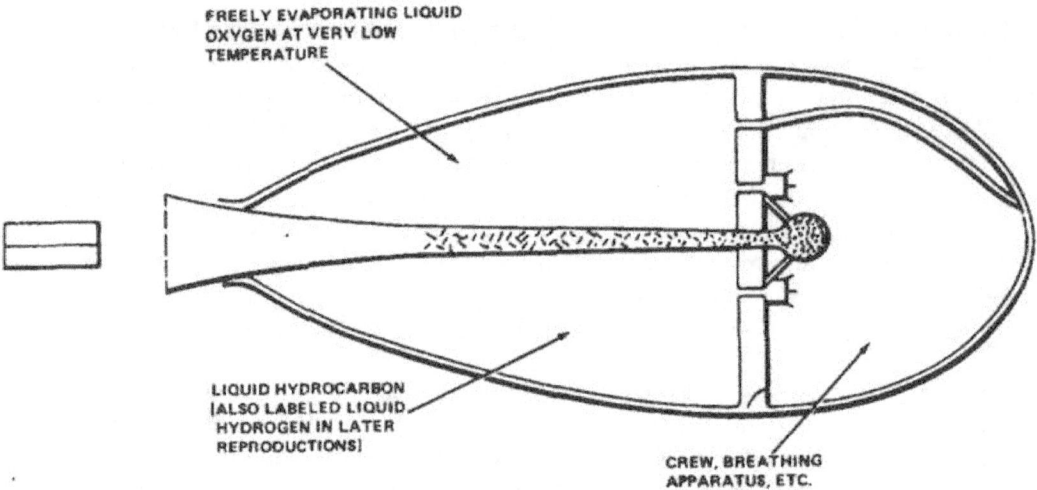

Fig. 60. Tsiolkovskiy's manned rocket, 1903. The rocket thrust chamber and a partition separate the fuel and oxidizer. The rectangles in the rear are rudders, part of the control equipment.

metals. He may have heard or read about Dewar's success in liquefying hydrogen—which would have been a feat for a secondary schoolteacher in a small Russian village.

Although liking hydrogen-oxygen, Tsiolkovskiy hedged his selection of it by observing: "The hydrogen may be replaced by a liquid or condensed hydrocarbon; for example, acetylene or petroleum."[10] The consideration of several candidate fuels characterized many of Tsiolkovskiy's later papers; as he became increasingly aware of the difficulties of using hydrogen, he gradually shifted away from it. In 1911, he observed that liquid oxygen could be obtained cheaply and so could gaseous hydrogen, but added: "The liquefaction of hydrogen is difficult (as of now), but it can be replaced with equal or even greater advantage by liquid or liquefied hydrocarbons such as ethylene, acetylene, etc."[11] Fifteen years later, he was even cooler towards the use of hydrogen:

> Liquefied pure hydrogen contains less potential energy, since it is cold and absorbs energy on turning into a gas, and its chemical effect is weaker. It is difficult to liquefy and store, since, unless special precautions are taken, it will rapidly evaporate. Liquid or easily liquefiable hydrocarbons are more favorable.[12]

In 1929–1930, Tsiolkovskiy published papers on rocket-propelled aircraft and argued that the way to the stars was to gradually increase the capability of aircraft from atmospheric to interplanetary flight. He discussed hydrogen, carbon, and benzene as fuels and felt it was a pity that liquid hydrogen was scarcely available.[13] In a still later paper, he considered hydrogen and light hydrocarbons as aircraft fuels, but expressed another problem when using hydrogen for aircraft:

> The fuels must be dense, so that they do not occupy much space. In this respect, liquid hydrogen is not suitable because it is 14 times lighter than water.[14]

A year before his death in 1935, Tsiolkovskiy submitted a paper in which he summarized his thinking about fuels for rockets.[15] He listed six properties a fuel should have: (1) maximum work per unit mass on combustion; (2) gaseous combustion products; (3) low combustion temperature to prevent chamber burnout; (4) high density; (5) liquid that readily mixes; and (6) if gaseous, must have high critical temperature and low critical pressure for use in liquefied form. He added that costly compounds should be avoided and found that hydrogen-oxygen satisfied all conditions except (4) and (6). Other fuels (methane, benzene, acetylene, and ethylene) were discussed, with a preference indicated for benzene and ethylene. Methanol, ethanol, ether, and turpentine also came under scrutiny; their high heats of combustion impressed him. Among oxidizers, he considered ozone, oxygen, nitrous oxide, nitrogen dioxide, and nitrogen tetroxide; of these, he liked nitrogen dioxide, an oxidizer much used in later rockets in the United States and elsewhere. Tsiolkovskiy also considered solid propellants, but rejected them on the basis of their low energy and danger of unexpected explosion. In his summary, he concluded: "Hydrogen is unsuitable because of its low density and storage difficulties when in the liquid form."[16] He might well have added his earlier comment about very limited availability. Many others who followed were to experience the same attraction for hydrogen-oxygen and abandon it for the same reasons.

While Tsiolkovskiy dreamed of spaceflight and published theoretical papers of very limited distribution, another man was dreaming of spaceflight, but with a difference. This man, also an educator, took rockets a giant step forward by adding practical experimentation to his ideas and theoretical calculations.

Robert Hutchings Goddard (1882–1945) suffered ill health during his youth in Massachusetts, but it did not deter his great love of experimental science.[17] His first experience with hydrogen came at sixteen when he was heating a tube containing hydrogen with an alcohol flame. Air must have entered the tube for it exploded violently, hurling glass fragments into the ceiling and through the door of his attic laboratory. His parents convinced him to redirect his interests, and he soon was experimenting with a pillow-shaped aluminum balloon filled with hydrogen; it was too heavy to rise.

At sixteen, Goddard began describing his experiments and calculations in a notebook—a lifetime practice that has provided a rich source of information. Like Tsiolkovskiy, Goddard had a dream about spaceflight that had a profound effect upon his life. His occurred in a cherry tree on his grandmother's farm on 19 October 1899, and from that moment spaceflight became his greatest single goal and the tree, the symbol of his resolve.[18]

In his student years (1904–1913) at Worcester Polytechnic Institute, Clark University, and a year of postdoctoral research at Princeton University, Goddard studied physics and chose thesis subjects that were more acceptable to academicians than rockets.[19] During this time, however, he continued studying rockets in his spare time. His initial experiments with powder rockets came during his undergraduate days. In 1907, he calculated the lifting of scientific equipment to a great height by a combination of balloon and rocket. Like Tsiolkovskiy, he was attracted to the hydrogen-oxygen combination because of its high energy. In 1909, he calculated that the energy from 45 kilograms of hydrogen-oxygen was sufficient to send a kilogram payload to infinity. He also found that it took nearly 50 times as much gunpowder as hydrogen-oxygen for the same mission. Convinced of the potential of hydrogen-oxygen, he wrote in 1910 about a method for producing hydrogen and oxygen on the moon. Later he took out a patent for producing hydrogen and oxygen where there was ice and snow at low temperatures, such as high on a mountain. He proposed to generate the gases by electrolysis of water using solar energy.[20]

Goddard began his rocket experiments with gunpowder, by far the most convenient and available explosive, but he was soon attracted to the more powerful guncotton. In 1913, he compared it with hydrogen-oxygen:

> . . . it should be a comparatively simple matter to construct an apparatus using guncotton, having a net efficiency of 70 percent; which would require 500 lb initially to send 1 lb off the earth. . . . Hydrogen and oxygen, of course, give the greatest energy on burning, but the hydrogen would have to be liquid and the oxygen solid for the sake of lightness, and this would introduce difficulties which would more than offset the trouble of using a larger mass, with guncotton.[21]

By that time, he saw a much smaller performance gap between a practical solid propellant and hydrogen-oxygen and recognized the problems of obtaining and working with liquefied gases.

From 1915 to 1916, Goddard conducted a series of remarkable experiments, demonstrating for the first time that rocket efficiency increases in a vacuum. Elated with this success but sorely in need of funds, Goddard described his results and ideas for reaching extreme altitudes in an impressive report and sent it to the Smithsonian Institution with a request for financial support. He got the support and three years later his report, with added notes, was published by the Smithsonian. Unfortunately, he had much less success with his next series of experiments. It took four years of hard effort and many failures before he realized that he was on the wrong track. In all this time he had been firing solid propellant in a series of discrete explosions. He sought a mechanical method that would detonate the succession of charges in much the same way as a machine gun.[22] Such designs are not only mechanically difficult but also heavy.

In a report to the Smithsonian in 1920, Goddard pointed out that liquid hydrogen and solid oxygen had greater capability than smokeless powder for lunar and planetary missions.[23] He suggested that the liquid hydrogen–solid oxygen mixture might be enclosed in a capsule of solid smokeless powder. As previously mentioned, however, Goddard had difficulty with smokeless powder charges and the liquid hydrogen–solid oxygen mixture was not tried.

Goddard's failure to make a practical rocket using discrete charges of smokeless powder was a blessing in disguise, for it turned him to liquid propellants and continuous burning—the approach that led to his greatest successes. He abandoned discrete charges, and in July 1921 began experiments with continuous burning, working first with the familiar smokeless powder. By September, he had switched to liquids. He first considered alcohol as a fuel but never used it. Instead he chose ether because it had a lower boiling point and heat of vaporization—advantages for mixing and combustion—and a higher heating value. He soon switched from ether to gasoline, and thereafter he stuck with the gasoline–liquid oxygen combination in building his flight rockets. On 1 November 1923, he static-fired a rocket using this combination pumped to the combustion chamber.[24]

In spite of Goddard's early interest in hydrogen-oxygen he did not operate a rocket with either gaseous or liquid hydrogen as the fuel. This was probably due more to his practical nature than to lack of interest. He had enough problems getting liquid oxygen for his initial experiments; liquid hydrogen would have been much harder to obtain, as well as expensive and difficult to handle.[25] He apparently was satisfied with his propellant combination and absorbed in the myriad other problems of building and flying a complete rocket vehicle—a highly complex task.

On 3 May 1922, nine months after Goddard began experiments with ether and oxygen in continuous combustion, he received a letter from another man obsessed with spaceflight, a man whose publication and priority claims a year later were to upset Goddard.

Hermann Oberth (1894–   ) became interested in space travel at the age of eleven when his mother gave him a copy of Jules Verne's *From the Earth to the Moon*. In the next two years he analyzed and discarded several methods for achieving spaceflight, including Verne's cannon, the pull of a powerful magnet on skids in a long tunnel lined with ice, and a wheel utilizing centrifugal force. He came to the conclusion that reaction propulsion was the only feasible method, but was dismayed by the problems of fuel consumption, handling of liquid fuels, hazards of solids, and the high cost of chemicals. These problems discouraged him from attempting experiments in his youth. Instead he turned to the theoretical analyses that are his lasting contributions to rocketry.

Oberth's educational experiences in Germany were unhappy. In his secondary schooling the emphasis was on the classics, with exercises requiring good memory and flawless writing, and he found it difficult to conform. He wrote a thesis on rocket propulsion for his doctor's degree at Heidelburg University, but it was rejected.

Oberth's career is marked with difficulties and lack of recognition for his contributions. In 1917, he submitted a proposal to the German War Department to build a long-range liquid propellant missile, but it was rejected as impractical. After his thesis was rejected, Oberth paid to have it published in 1923 as a pamphlet of less than one hundred pages with the title *Die Rakete zu den Planetenraumen* (The Rocket into Interplanetary Space). This was the publication that disturbed Goddard; but according to Oberth, it was largely ignored in Europe in the 1920s. Oberth became a secondary schoolteacher in 1923, but left a year later to accept a banker's offer to finance the building of his space rocket. After six months of indecision, the banker backed out, leaving Oberth disillusioned and broke. In 1929, a third edition of his book was published under the new title *Wege zur Raumschiffahrt* (Ways to Spaceflight). In 1929–1930, he served as a scientific advisor for the film *Frau im Mond* (Girl in the Moon) produced by Fritz Lang—a film that greatly influenced Krafft Ehricke when he was eleven (p. 191). The making of the film brought the opportunity to obtain funding for rocket experimentation. Oberth undertook to build a liquid-fueled rocket to be launched on the day of the film's premiere to publicize it, but time was too short and the task too difficult. The film was a great success, but Oberth returned to relative obscurity.

In 1938, Oberth began working on military rocket development for the Germans. He went to Peenemünde about the time the A-4 (V-2) development was completed, but worked on solid rather than liquid rockets until the end of the war. The Allies, gathering up rocket experts, ignored Oberth, who returned to his family in Feucht. He came to the United States in 1955, where he worked quietly in Huntsville on rocket research, but returned to Feucht four years later.[26]

Oberth's 1923 book was based on his conviction that existing technology made possible rocket flight beyond the earth's atmosphere and that, with refinements, vehicles could carry man beyond the earth's gravitational field. He developed and improved the theory of rocket flight and performance. He proposed a space rocket having an alcohol-water mixture and liquid oxygen as the first stage propellants and hydrogen-oxygen as the upper stage propellants. With a remarkable insight into physical phenomena, his theoretical analyses were a great contribution to the general advancement of rocket technology. A number of his proposals were later adopted and put into practice. The slow acceptance of his theoretical work, the criticisms, and the controversies may have been due to his lack of academic credentials and the boldness of his proposals. His lost opportunities and experimental failures were probably due to his naiveté in politics and business as well as his lack of practical engineering experience. He deserves much credit for providing the theoretical basis for European rocket development and, later, space boosters. His first public recognition came from France, when the Société astronomique de France gave him the REP-Hirsch award in 1929, for the third edition of his book.*

Oberth recognized and showed the advantages of using multistage (step) rockets, an old concept, to achieve the very high velocities necessary for spaceflight. A step rocket is one rocket riding piggyback on top of another so that their velocity increases during burning are additive, a concept recognized as early as the seventeenth century.[27] The Tsiolkovskiy equation (p. 253) can be used to approximate the final velocity of a multistage rocket by considering each stage in succession. In addition to designing step rockets, Oberth also recognized that a rocket had greater thrust when operating outside the earth's atmosphere than within it and cited Goddard's experiments as proof.

One of Oberth's contributions is very helpful to a better understanding of rocket performance and the reason hydrogen makes such a good fuel. He applied the theory of gas flow through nozzles to rockets and cited Zeuner's *Turbinen* (Turbines) as his source.[28] Oberth was the first to publish the relationship for the case of a rocket, showing that the rocket exhaust velocity is a function of four variables: the pressure and the specific volume of the gases in the combustion chamber, the ratio of the specific heats (constant pressure to constant volume) of the gases (assumed to remain constant

---

*The award, with a 5000-franc prize, was established that year by the famed French aviator and space pioneer Robert Esnault-Pelterie and banker André Hirsch for the experimenter who had done the most for spaceflight. Oberth was the first to win it. As a special compliment to him the prize was doubled. Ley, *Rockets, Missiles & Space Travel* (New York: Viking, 1961), pp. 23–24.

during the gas expansion), and the pressure of the expanded gases at the nozzle exit.*
Using the perfect gas equation, rocket exhaust velocity can also be expressed as a
function of three ratios: specific heat, nozzle exit pressure to combustion pressure, and
combustion gas temperature to the mean molecular mass of the gases. Of these, the
pressure ratio is determined by the rocket design; the other two depend upon the
particular fuel and oxidant plus their relative proportions. Since the specific heat ratio
varies over a relatively small range, the rocket engineer focuses his attention on the
ratio of combustion temperature to mean molecular mass in comparing the
performance potential of various propellant combinations. The exhaust velocity is
proportional to the square root of this ratio—so the higher, the better. The rocket
engineer also is concerned with cooling the engine and would like to keep the
combustion temperature as low as possible without great sacrifice in exhaust velocity.
He would, therefore, want to keep the mean molecular masses of the combustion gases
as low as possible. This is where hydrogen excels, for it is the lightest of the elements.
Exhaust gases rich in hydrogen exceed the performance of heavier gases at the same
temperature.

Oberth liked the hydrogen-oxygen combination because it had the highest jet
velocity of any combination he could find. The second best combination, he believed,
was ethyl alcohol and oxygen. His choice of hydrogen-oxygen for the second stage of
his space rocket and alcohol-oxygen for the first stage did not change as he refined and
expanded his calculations from the first through the third editions, over a period of six
years. The reasons for his choices, other than energy content, and his explanations—
scattered in his book—are somewhat obscure and confusing. They involved
considerations of propellant density, aerodynamic forces, choice of propellant mixture
ratio, and the effect of these on mass ratio and performance. Oberth was well aware of
the low density of hydrogen, its effect on structural volume and weight, and the effect
of these, in turn, on aerodynamic drag and bending forces in flight through the
atmosphere. For this reason, dense propellants were favored for the first stage, which,
of course, had to start from the ground and fly through the atmosphere.

Oberth was also concerned about dissociation losses in the hydrogen-oxygen
reaction. Dissociation is the breakup of molecules (in this case, water) into its elements,
which occurs at high pressures and temperatures. It is an endothermic process—one
that absorbs heat—so part of the heat released in the combustion process is absorbed
by dissociation and therefore is unavailable for conversion into work. From structural
and aerodynamic considerations, Oberth wanted to use the stoichiometric (exact
proportion) mass mixture ratio of 1 part hydrogen to 8 of oxygen to have a high
propellant density of the hydrogen-oxygen combination (of 1.02 kilograms/liter). The
dissociation consideration, however, convinced Oberth that a hydrogen-rich mixture,
1 part hydrogen to 2 of oxygen, would be necessary for a first stage. Since this low

---

*Oberth's equation for exhaust velocity was

$$\sqrt{C\left(\frac{\gamma}{\gamma-1}\right) P_o \, V_o \left[1 - \frac{P_e}{P_c}\right]^{\frac{\gamma-1}{\gamma}}}$$

where C is a constant, γ the ratio of specific heats, P_o the combustion pressure, V_o the specific volume, P_c the combustion pressure, and P_e the exit pressure. This equation can be obtained by algebraic combinations of the basic thermodynamic relationships of the equation of state for a perfect gas, conservation of energy, reversible adiabatic (isentropic) gas expansion, and specific-heat relationships.

propellant density meant structural and aerodynamic disadvantages, Oberth rejected hydrogen-oxygen for his first stage.

The situation with respect to dissociation for the upper stage rocket was different. A lower pressure could be used, which reduced the problem of dissociation. Oberth first believed he could use the stoichiometric mixture ratio to keep overall density at a minimum, but later, cooling problems led him to choose a hydrogen-rich mixture—1 part hydrogen to 1.43 of oxygen.[29]

He designed the second stage very light and fragile and protected it during the first part of the flight by enclosing it within the first stage. Oberth's analysis indicated that the alcohol-oxygen combination was superior to hydrogen-oxygen at lower altitudes, but at an altitude between 5500 and 8400 meters (1/2 to 1/3 atm), the hydrogen-oxygen combination became and remained superior.[30] He was right in choosing hydrogen-oxygen for upper stages, as later events showed.

In his writings, Oberth was fiercely defensive and made numerous references to apparent critics. He appeared particularly sensitive about the practicality of his designs. He sought to give just enough information to show that his rocket "could be built under all circumstances" and to show that he knew something about practical design, yet not give enough information so that someone else could build his rocket. He described his "Model B" rocket as more complicated than his "Model C," which he would not detail. Model B was a high-altitude meteorological rocket consisting of an alcohol-oxygen first stage and a hydrogen-oxygen upper stage. An "auxiliary" stage boosted it, so it really was a three-stage rocket. The hydrogen-oxygen stage was very small—a decided disadvantage in using low-density hydrogen. It weighed 6.9 kilograms and carried 3.3 kilograms of propellant—1.36 kilograms of which was liquid hydrogen. The hydrogen tank, made of a copper-lead material 0.0144 millimeter thick, had a capacity of about 19 liters; Oberth specified its weight as 33 grams. He provided a hydrogen pump to generate 5 atmospheres pressure and specified its weight, plus the oxygen tank and reinforcements, as 0.5 kilogram.[31] Oberth's sketches and description give few details. These specifications would severely tax the skills of an engineer even with today's advanced technology. In fairness to Oberth, however, his tanks were light because he conceived the idea of using internal pressure for reinforcement (like a metal balloon). A quarter of a century later, the same basic idea appeared in several U.S. designs and is used today in the Atlas and Centaur vehicles.

Although often citing experiments by others to support his statements, Oberth seldom mentioned his own experiments or revealed details about them. He apparently experimented with a gaseous hydrogen–oxygen burner. In responding to a critic who claimed that jet speeds beyond 2000 meters per second were impossible, Oberth stated he had achieved velocities of 3800–4000 meters per second with a gas burner. Later, in discussing a design for a mail rocket, he observed: "For example, with a correctly built oxygen-hydrogen nozzle I achieved a burn lasting 21 minutes." In describing another design in 1929, however, he stated: "Unfortunately, I have not yet been able to experiment with liquid hydrogen."[32]

Regardless of the practicality of his design or the lack of experimentation, Oberth's contributions represented the most comprehensive theoretical analyses of rockets available; Tsiolkovskiy's earlier publications were practically unknown outside Russia, and Goddard's principal rocket publication up to that time was the 1919

Smithsonian paper. Oberth's publications did much to stimulate others to work on rockets.

Like Tsiolkovskiy and Goddard, Oberth wrote imaginative versions of space travel to popularize the subject. In his 1929 book he spun a yarn about a rocket flight around the moon. The rocket, which used hydrogen-oxygen, was to ascend from the Indian Gulf. He relates an incident that could apply to present-day interest in hydrogen:

When I arrived . . . I was suprised to see the many automobiles which caused neither fumes nor noise and, in spite of their sometimes considerable speed, seemed to have extremely small and light motors.

Well remember . . . we have liquid hydrogen and oxygen factories . . . All these automobiles have hydrogen motors. . . .

Yes, but is not all the hydrogen produced by the plants needed for the rocket?

At first . . . no large rockets were launched for months. To prevent our hydrogen plants from being completely idle in the meantime, we sought to utilize at least part of the liquid hydrogen in industry. . . . Today we can hardly fill the demand. We are obliged to enlarge the plants almost every month.[33]

After Oberth wrote to Goddard on 3 May 1922, Goddard replied and enclosed a copy of his 1919 Smithsonian paper. When Oberth's book appeared in 1923, he sent Goddard a copy. An addendum to the book deals with Oberth's becoming aware of Goddard's work and stating the independence of his work. His book disturbed Goddard, who never overcame his adverse reaction to Oberth.[34] It stimulated Goddard to prepare an addendum to his 1922 Annual Report to Clark University, and he sent a copy of it to the Smithsonian. It was a response to Oberth's work and contained details of experimental work not mentioned in his own previous reports. He described his first liquid-propellant experiments with ether as the fuel. Although he mentions Oberth's proposal to use hydrogen-oxygen and his own considerations of this combination, he did not mention any work, past or contemplated, using either gaseous or liquid hydrogen.[35]

In 1929, Goddard submitted supplementary notes to his 1920 Smithsonian report indicating that he experimented with continuous burning of liquid propellants in 1920. He added that "continuous combustion by the use of hydrogen and oxygen was first considered by the writer several years ago, in June 1907, and was patented in 1914." As in 1920 and 1922, however, he made no mention of experimental work with hydrogen-oxygen.

To sum up, Tsiolkovskiy, Goddard, and Oberth—the pioneers in the history of rocket technology—represent both the theoretical and the practical side of rocketry. They not only had their consuming interest in rockets in common, but Tsiolkovskiy and Goddard had inspirational dreams about space travel, and all three wrote popularized accounts of space travel. All three were educators; two were secondary schoolteachers, one a college professor. All three were attracted to hydrogen because of its high energy, but theoretician and experimenter alike saw its disadvantages of low availability, low density, and difficulty in handling. Goddard, the only one of the three to experiment extensively with liquid propellants in continuous combustion, passed over hydrogen in favor of gasoline—most likely because of gasoline's availability and ease in handling. Hydrogen, then, successively attracted and disillusioned rocket investigators. This pattern of attraction-repulsion is repeated many times in the history of hydrogen as a flight propulsion fuel.

Although Tsiolkovskiy, Goddard, and Oberth were perceptive in seeking hydrogen's advantages as a rocket fuel, advancements in hydrogen technology useful for flight applications came from other concurrent activities, one of which was development of the dirigible.

Appendix A-3

Hydrogen Technology, 1900–1945

In addition to the considerations of liquid hydrogen for rockets by Tsiolkovskiy, Goddard, and Oberth, other concurrent activities contributed to hydrogen technology. The largest and best known of these were the development and operation of the large dirigibles in Germany, Italy, Great Britain, and the United States from 1900 to 1937. Much has been written about these giants of the sky that need not be repeated here.[1] Thousands of passengers were carried safely in hydrogen-filled dirigibles, yet there were many accidents, which finally killed the giants. The most spectacular—and final—accident involved the *Hindenburg*, filled with 200 000 cubic meters of hydrogen, which burst into flames at Lakehurst, New Jersey, on 6 May 1937, killing 13 passengers and 22 members of its crew as well as one ground crewman.[2] When the ease with which hydrogen-air mixtures can be ignited is examined, the wonder is that there were not more accidents. For example, the Germans found in 1912–1913 that faint, finger-length flames of hydrogen-air mixtures could be produced merely by rubbing rubberized surfaces together, and the same fabric generated static electricity if torn. One modern safety manual points out that the minimum spark ignition energy to ignite a hydrogen-air mixture at atmospheric pressure is 0.000019 joule.[3] If that means nothing to you, the manual warns against certain actions that can generate static electricity to ignite hydrogen in test areas: combing the hair, wearing clothes made of nylon or other synthetics or wool, and allowing furred animals in the area. The legacy of dirigible operations, where millions of cubic meters of hydrogen were generated, stored, transferred, and flown, is not so much the safety procedures, but fears of using hydrogen that the accidents instilled in the minds of so many people.

One aspect of dirigible activity not so well known was attempts to use hydrogen as a fuel. Dirigibles had to vent buoyancy gas and Paul Haenlein obtained a U.S. patent in 1872 to use that otherwise wasted gas in the dirigible's engines. Haenlein, however, used coal gas and apparently did not get around to using hydrogen.[4] Using hydrogen in dirigible engines surfaced again after World War I in Italy, Great Britain, Germany, and the United States. In 1920, two British investigators estimated that dirigible range could be increased 20 percent by burning the hydrogen usually vented. They found that an engine could operate on hydrogen as an additive or on hydrogen alone, but in the latter case there was a tendency to knock.[5] Similar results were found elsewhere but the idea never gained widespread use. Experiments in Germany and the United States on using hydrogen in diesels met with some success in 1935, but by then the use of hydrogen in dirigibles was close to the end.[6]

In addition to dirigible developments, another great stimulus to the development of hydrogen technology during the first four decades of the twentieth century was scientific investigation. Unlike dirigible applications, which were centered on gaseous hydrogen, the scientific investigations that advanced hydrogen technology were primarily concerned with low temperature phenomena.

Hydrogen Technology from Science, 1900–1940

Progress in developing equipment for liquefying gases during the last decade of the nineteenth century was matched by equally impressive gains during the first decade of the twentieth. Carl von Linde's air liquefaction equipment, capable of liquefying 8 liters per hour, was exhibited in Paris in 1900 and purchased by the College of France.[7] In 1902, a process for separating oxygen from air, developed by Georges Claude, was in commercial operation in France. Two years later the British Oxygen Company exhibited a hydrogen liquefier, designed by James Dewar, at the Louisiana Purchase Exhibition in St. Louis. The National Bureau of Standards purchased it for $2400 for low-temperature thermometry.[8] By 1905, Linde liquefaction plants were operating in both Germany and France, and in 1907 the Linde Air Products Company began operations in the United States.

In 1906 interest in low temperature phenomena was stimulated when a German chemist, Walther Nernst, postulated the third law of thermodynamics—that the total and free energies become equal as absolute zero is approached. Heike Kamerlingh Onnes, founder of the cryogenic laboratory at Leyden in 1894, reached 4.2 K when he first liquefied helium in 1908. By evaporating helium, scientists were soon able to reach within one kelvin of absolute zero.

In 1924–1926, a new era began in theoretical physics—wave (quantum) mechanics. It was introduced by a 32-year-old scientist, Louis Victor de Broglie, in his doctoral thesis. He postulated a relationship between the velocity or momentum of electrons and wave lengths of radiation. His work stimulated many other physicists, and among those who carried the theoretical work further were Clinton Davisson, George Thomson, Erwin Schrodinger, and Weiner Heisenberg. All won Nobel prizes in physics for their contributions.

Heisenberg, a 24-year-old German physicist, believed that the theory should include only observable elements. His new wave mechanics theory expressed wave length frequencies and intensities of radiation emitted by the atoms in matrix mathematics. He used his theory to postulate in 1926 that the hydrogen molecule existed in two forms, which subsequently were called orthohydrogen and parahydrogen. Heisenberg's 1932 Nobel prize in physics was awarded "for the creation of quantum mechanics, the application of which has, among other things, led to the discovery of the allotropic forms of hydrogen."[9]

In orthohydrogen, the two hydrogen nuclei in the molecule spin in the same direction; in parahydrogen, the two nuclei spin in opposite directions. The two molecules have different physical properties but their chemical properties are the same.

In 1927 a British physicist, D. M. Dennison, used Heisenberg's postulate to make one of his own. Earlier observations of the specific heat of hydrogen had indicated an anomaly; the rotational specific heat decreased with time and temperature. Dennison postulated that this was caused by the two kinds of hydrogen not being in equilibrium at the lower temperature. In 1928, William Giauque and Herrick L. Johnston at the University of California at Berkeley attempted to test Dennison's postulate by keeping a sample of hydrogen at a low temperature for six months, but the observed changes were so small that their experiment was inconclusive. The following year another team of investigators, K. F. Bonhoeffer and P. Harteck, used a catalyst to obtain equilibrium at low temperature and obtained almost pure parahydrogen. They showed the differences between orthohydrogen and parahydrogen in terms of specific heat and thermal conductivity of the gases.[10]

At room temperature and above, ordinary hydrogen is 75 percent orthohydrogen and 25 percent parahydrogen. At 77.4 K (temperature of liquid nitrogen used for cooling) the hydrogen mixture at equilibrium is 52 percent orthohydrogen and 48 percent parahydrogen. At the boiling point of liquid hydrogen, 20.3 K, the equilibrium composition is 99.8 percent parahydrogen.* When gaseous hydrogen is liquefied, it will slowly and spontaneously seek equilibrium, with orthohydrogen changing to parahydrogen. At 20.3 K, the conversion releases more heat (532 joules per gram) than is required to vaporize the liquid (453 joules per gram), so that liquefied normal hydrogen evaporates completely on conversion to parahydrogen—even in a perfectly insulated container—a situation Dewar did not foresee. The vaporization loss during the conversion at 20.3 K amounts to about 1 percent of the stored liquid hydrogen per hour, a loss much too high to be tolerated in practical applications.[11]

Another line of scientific investigation that led to new information about hydrogen and provided a powerful stimulus for developing liquid hydrogen technology began at the University of California at Berkeley with the research of Gilbert Lewis and William Giauque in testing the validity of the third law of thermodynamics. In 1926 Giauque devised a method for attaining very low temperatures by an adiabatic demagnetization technique. It was now possible to get within a few thousandths of a degree of absolute zero. At these temperatures, thermal motion of atoms almost ceases and Giauque was able to measure energy changes associated with the transition in the states of the atoms. In 1929, Giauque and an associate, Herrick L. Johnston, published the results of a discovery that set in motion a train of events leading to the discovery of heavy hydrogen in 1931. In studying the spectrum of oxygen, they found that in addition to atoms of atomic mass 16, there were others with masses of 17 and 18. The three types of oxygen atoms existed in the atmosphere in the proportions of 3150:1:5, respectively, and gave an average atomic mass for oxygen of 16.0035.[12] This startled physicists and chemists, for the whole scale of atomic mass was based on oxygen with an atomic mass of 16.0; now the base and all masses related to it had to be changed. Giauque was awarded the 1949 Nobel chemistry prize for this and other contributions to low temperature physics.

---

*Sources differ as to the boiling point of liquid hydrogen at 1 atm with some quoting 20.3 K and others 20.4 K. Some of the confusion comes from the fact that liquid hydrogen can be "normal" hydrogen (75% ortho, 25% para), "equilibrium" hydrogen (21% ortho, 79% para) or parahydrogen (99.8% para). Two National Bureau of Standards authors, Richard B. Steward and Hans M. Roder, in chap. 11, "Properties of Normal and Para Hydrogen," in *Technology and Uses of Liquid Hydrogen*, ed. R. B. Scott, W. H. Denton, and C. M. Nichols (New York: Macmillan, 1964), p. 380, give the boiling point at 1 atm for normal hydrogen as 20.380 K and for parahydrogen, 20.268, citing the work of Woolley, Scott, and Brickwedde for the former and Roder, Diller, and Weber for the latter.

Prior to the Giauque-Johnston discovery, Francis Aston developed a highly accurate (1:20000) spectrographic measurement technique and investigated a number of elements, including hydrogen. He measured hydrogen's mass as 1.00778, based on an oxygen mass of 16.0, which compared well with chemical determinations of hydrogen's mass of 1.00777.[13] The Giauque-Johnston change in oxygen's mass meant a greater difference between the spectropic and chemical measurements of hydrogen's mass. In 1931, R. T. Birge and D. H. Menzel concluded this difference to be too great for experimental error and postulated that among the hydrogen atoms of atomic mass 1 must be some of atomic mass 2 in the proportion of about 1 in 4500.[14] This was an exciting challenge to physicists and chemists and the race began to determine whether the Birge-Menzel postulate was correct.

The winner of the race was Harold Urey, who had studied at the University of California and was influenced by the work of Lewis and Giauque. Urey first had to concentrate the isotope to identify it. He calculated that the difference in vapor pressures would provide the means for concentrating deuterium by distillation of solid hydrogen at the triple point. He postulated that the same differences in vapor pressure might also apply to the liquid state. He turned to the National Bureau of Standards where F. G. Brickwedde agreed to help. Brickwedde evaporated 4000 cubic centimeters of liquid hydrogen near the triple point, ending up with only one cubic centimeter. In the fall of 1931, Urey and his assistant, G. M. Murphey, placed Brickwedde's sample in a spectrograph and established the presence of deuterium, beyond all doubt.[15]

Urey won the 1934 Nobel chemistry prize for his achievement. Eight months after Urey's discovery, E. W. Washburn discovered that hydrogen and deuterium could be separated by electrolysis. When water is electrolyzed and hydrogen gas escapes, the residual water contains a greater proportion of deuterium oxide (heavy water). This discovery led Norway to undertake large-scale production of heavy water at a hydroelectric plant at Rjukan. Since heavy water is a good moderator for atomic reactors, the Allies raided the Norwegian plant during World War II to prevent Germany's obtaining a supply of the isotope. Deuterium can also be concentrated by a diffusion process.[16]

In 1935, the third hydrogen isotope, tritium, was prepared by Lord Rutherford, Marcus Oliphant, and Paul Harteck by bombardment of deuterophosphoric acid with fast deuterons.[17]

In summary, the scientific interest in low temperature phenomena provided a powerful driving force for advancing the technology of liquid hydrogen. The spontaneous conversion of orthohydrogen to parahydrogen, the release of enough heat in the conversion process to vaporize the liquid hydrogen, and the use of a catalyst to speed the conversion process were discoveries essential to later developments of technology for the storage and transportation of liquid hydrogen in quantity.

Rocket Experiments with Liquid Oxygen and Liquid Hydrogen, 1937-1940

The first to experiment with a low temperature liquefied gas in a rocket was Robert Goddard, who began using liquid oxygen in 1921. By 1923, Goddard had successfully operated a gasoline–liquid oxygen rocket, incorporating pumps for both, on a test stand. Three years later, on 16 March 1926, Goddard launched the world's first liquid-fueled rocket at Auburn, Massachusetts.

The first to profit from Goddard's experience were the Germans during the 1930s. The German A-4 (V-2) using alcohol–liquid oxygen was the first practical application of a liquid-fueled rocket and the first to be mass produced. The V-2 established beyond all doubt the practicality of using a low temperature liquefied gas as a rocket propellant.

With all the German experience with gaseous hydrogen in dirigible operations, plus their experience with liquid oxygen for rockets, it was inevitable that they would consider liquid hydrogen for rockets. They did, but according to Wernher von Braun, the experience was brief and the results not very satisfactory.

In 1932, Walter Dornberger, a Germany army officer, organized a small rocket research station on the artillery proving grounds at Kummersdorf.[18] Among the engineers brought there were von Braun, Walter Riedel, and Walter Thiel. By 1936, the Kummersdorf group had the basic concept for the A series of rockets, and Dornberger started construction of a new rocket station at Peenemünde the same year. In April 1937, von Braun left Kummersdorf to become the technical director at the new station.[19] Thiel stayed at Kummersdorf and continued research on novel injection methods, more effective cooling, and higher combustion chamber pressures using alcohol–liquid oxygen as propellants for experimental rocket engines. Thiel also tried other propellant combinations including gasoline–liquid oxygen, methane–liquid oxygen, hydrazine–nitric acid, liquid hydrogen–liquid oxygen, and liquid hydrogen–liquid oxygen–fluorine mixtures. The experiments covered combustion characteristics, cooling, and general handling aspects of the fuels and oxidizers. The small rocket engine (less than 200 newtons, or 44 lb thrust) could be regeneratively cooled with one or both propellants or by water in a separate system. Von Braun observed an experiment with liquid hydrogen:

> As to Thiel's liquid hydrogen tests with this set-up, I remember seeing liquefied (outside) air dripping from the supercold liquid hydrogen line. In discussing liquid hydrogen's potential, Thiel fully endorsed Oberth's earlier optimism, but pointed out that tightness of plumbing connections was a critical problem and the ever-present explosion hazard caused by accumulation of leaked-out hydrogen gas in an unvented structural pocket would require extreme care in the design of a liquid hydrogen-powered rocket or rocket stage.[20]

Von Braun remembered the hydrogen experiments as occurring between 1937 and 1940. The exploratory work was not followed up as the Germans concentrated on developing rockets using alcohol–liquid oxygen.

Appendix A-4
Hydrogen as a Rocket Fuel

By the end of World War II, the major properties of liquid hydrogen were well established. The technique for its liquefaction, first developed by James Dewar in the nineteenth century, was refined but remained basically the same. Hydrogen liquefaction equipment and capacity remained small because the only demand was for research investigations in government laboratories and universities.

Liquid hydrogen was one of the first liquid rocket fuels proposed, but it was abandoned because of its low density, low availability, and handling hazards. Gaseous hydrogen technology, including its use as a fuel, was developed in association with dirigibles, but those airships were abandoned. Walter Thiel experimented with liquid hydrogen–liquid oxygen in a rocket engine in Germany in the late 1930s, but he experienced difficulties with leakage. In general, the experiences in using either gaseous or liquid hydrogen in flight applications were not favorable. To understand why, let us note the desirable characteristics of rocket fuels and evaluate hydrogen against each criterion.

High exhaust velocity. This is the single most important performance goal and is related to the heat of combustion of the fuel and the molecular mass of the combustion products. The importance of having a high exhaust velocity was first expressed by Tsiolkovskiy in the early 1900s. Hydrogen surpasses all other chemical fuels in exhaust velocity, and were this the only consideration, it would have been chosen and used in preference to other fuels long ago.

High fuel density. The second most important characteristic is a high fuel density, for this increases the mass ratio of a vehicle and increases its range or payload capability. Higher density also reduces drag during flight through the atmosphere, by allowing smaller and lighter tanks. Unfortunately, hydrogen has the lowest density of all fuels, a characteristic most responsible for Tsiolkovskiy's—and others following him—abandoning the consideration of hydrogen as a flight fuel. Oberth, however, believed correctly that this handicap could be overcome by very light construction techniques and by using hydrogen only in the upper stages of a multistage rocket.

Desirable cooling characteristics. These include a relatively low combustion temperature to lessen the heat flow into the engine walls, and fuel characteristics of high thermal stability and specific heat so that it can be effectively used as a regenerative coolant. In addition, a low vapor pressure or low critical pressure keeps the fuel-coolant from boiling or existing as both liquid and gas in coolant passages—an undesirable situation. Hydrogen scores well on these characteristics (except low vapor pressure) but there is no evidence that anyone considered or experimented with liquid hydrogen as a coolant prior to 1945—probably because nobody got beyond hydrogen's undesirable characteristics.*

High reaction rate. Rapid reaction of fuel and oxidizer over a wide range of conditions is advantageous in converting the energy content of a chemical fuel to heat in a minimum volume. Hydrogen's high flame speed, low ignition energy, and wide flammability limits—all advantages—have been known a long time but were not fully appreciated until after 1945.

Desirable handling and storage characteristics. An ideal fuel for handling and storage has a low vapor pressure, low freezing point, high shock stability, high ignition energy, and is nontoxic and noncorrosive. Hydrogen scores poorly on these desirable characteristics with its very low temperature (high vapor pressure), low ignition energy, and wide flammability-explosion limits.

Available in quantity. Hydrogen scored low on availability before World War II primarily because the only demand was for small quantities for scientific research. Gaseous hydrogen and the technology for its liquefaction were available, however.

From these six general considerations of fuels, it can be seen that hydrogen's properties represented the extremes in both desirable and undesirable characteristics and offered a fitting challenge to those interested in exploring the potential of new fuels.

---

*Robert Goddard has been credited with the idea of regenerative cooling with liquid hydrogen, but the author questions this based on research for this book.

Appendix B
Propulsion Primer, Performance Parameters, and Units

Since man first began to fly, he has striven to go faster. His progress in pushing speed records upwards was followed by an upward trend for commercial aircraft (fig. 61). As speed of aircraft increases, so does drag—the frictional resistance on its surfaces. It follows that higher speeds require higher thrust to overcome drag (fig. 62).

The Propulsion Principle

The thrust of all flight propulsion systems comes from the same principle—reaction—as expressed by Newton's second law of motion:

$$\text{force} = \text{mass} \cdot \text{acceleration}$$

In the case of airplanes flying through air, boats moving across the water, or rockets flying in the atmosphere or outside it, the reaction principle can be expressed in the equivalent forms:

$$\text{force} = \text{rate of change of momentum, or}$$
$$\text{force} = \text{mass flow rate} \cdot \text{change in fluid velocity.}$$

The airscrew or propeller, the boat propeller, the jet engine that swallows atmospheric air in operation, and the rocket engine that carries its oxidizer as well as fuel, all use the reaction principle. If you have stood at the stern of a boat under power, for example, and observed the wake, you have seen that the wake moves faster than the adjacent water, which is a visual indication of the change in fluid velocity.

Flight propulsion systems vary in the relative amounts of mass flow rates and changes in fluid velocity, as illustrated diagrammatically by figure 63. Assume, for comparison purposes, that the three types of vehicles are moving horizontally through the atmosphere at the same flight speed, designated by V_o, and that all are producing the same propulsive thrust, designated as F. Let us now examine, on a relative basis, the amount of fluid mass affected and the velocity given to it. On this relative basis, the propeller affects the largest air mass but gives it the lowest increase in velocity. The

273

MAN'S DESIRE TO GO FASTER

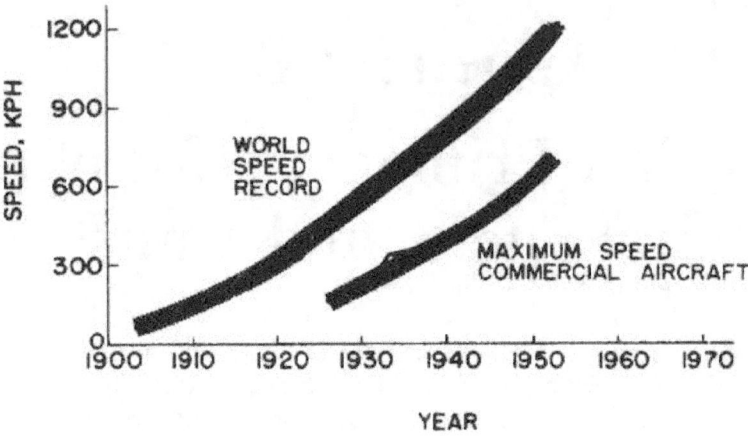

Fig. 61. Growth of aircraft speed. (Adapted from NACA figure. 1953.)

AIRCRAFT THRUST REQUIREMENTS

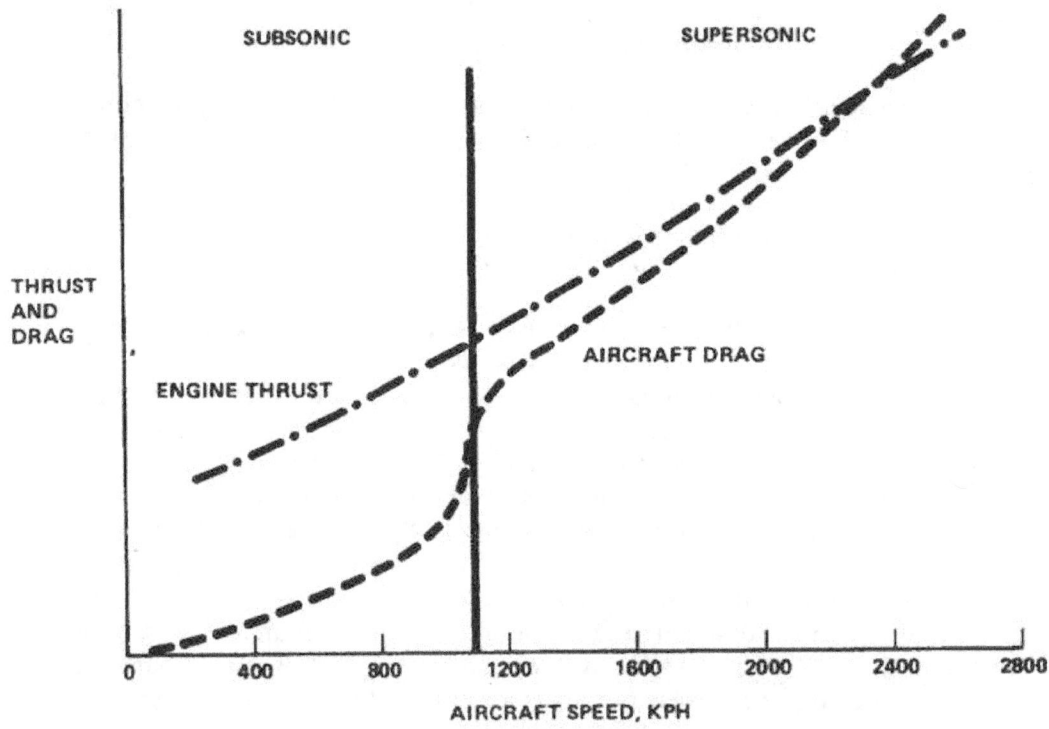

Fig. 62. Aircraft thrust requirements increase with speed. (Adapted from NACA figure. 1953.)

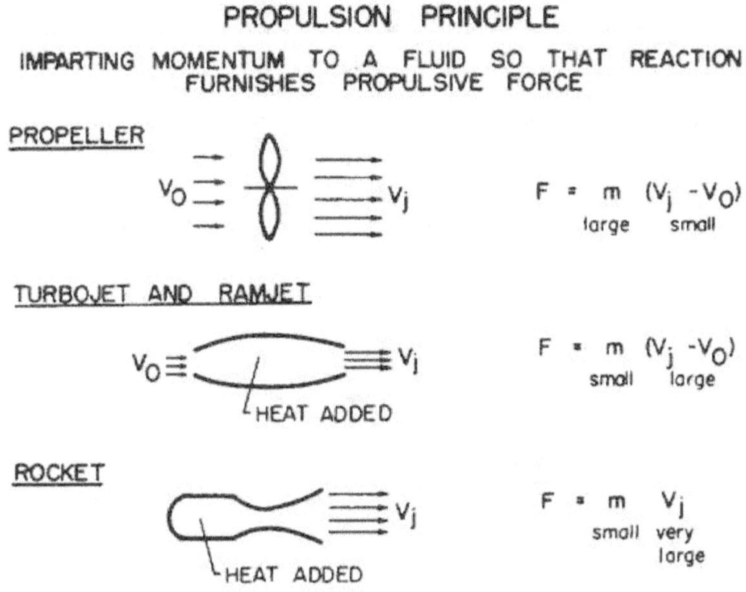

PROPULSION PRINCIPLE

IMPARTING MOMENTUM TO A FLUID SO THAT REACTION FURNISHES PROPULSIVE FORCE

<u>PROPELLER</u>

$$F = m \ (V_j - V_O)$$
$$\text{large} \quad \text{small}$$

<u>TURBOJET AND RAMJET</u>

HEAT ADDED

$$F = m \ (V_j - V_O)$$
$$\text{small} \quad \text{large}$$

<u>ROCKET</u>

HEAT ADDED

$$F = m \ V_j$$
$$\text{small} \quad \text{very} \\ \text{large}$$

Fig. 63 Propulsion principle is the same for propeller, turbojet, ramjet, and rocket. F = thrust (assumed same for all in comparison); m = mass flow of working fluid (air or exhaust gas); V_j = exhaust gas velocity; V_o = flight velocity. (NACA figure, 1953.)

turbojet and ramjet engines affect a smaller mass of air than the propeller, but give the air a much higher velocity. The rocket, carrying its own working fluid as fuel and oxidizer, does not accelerate air but for the same thrust, ejects the smallest amount of fluid mass at the highest velocity.

Since both the propeller and air-consuming jet engines must accelerate the air they fly through to produce thrust, it follows that their thrust depends upon flight velocity. For the propeller, the limitation begins to show when its tip approaches the speed of sound in the air (340 m/s at sea level; 295 m/s at 10 700 m altitude). In the mid-1930s, an airplane speed of 805 km/hr (224 m/s) was attainable only in power dives and there was uncertainty over whether full power on the engine speeded up the dive or slowed it down. For the air-consuming jet engine, the momentum drag of the incoming air increases with flight speed and when flight speed equals the exhaust jet speed, thrust falls to zero. The rocket, on the other hand, carries all of its working fluid and its jet thrust is independent of flight speed, in or out of the atmosphere. These relationships are illustrated qualitatively by figure 64.

Propulsive Efficiency

To go deeper into the subject of propulsive effectiveness, we can consider the propulsive efficiency. Marine engineers call it the Froude efficiency after William Froude (1810–1879) who first used it. It is defined as the ratio of useful power output to the rate of energy input. For flight, propulsive efficiency is expressed as:

$$\frac{\text{thrust} \cdot \text{flight speed}}{\text{kinetic energy increase of fluid}}$$

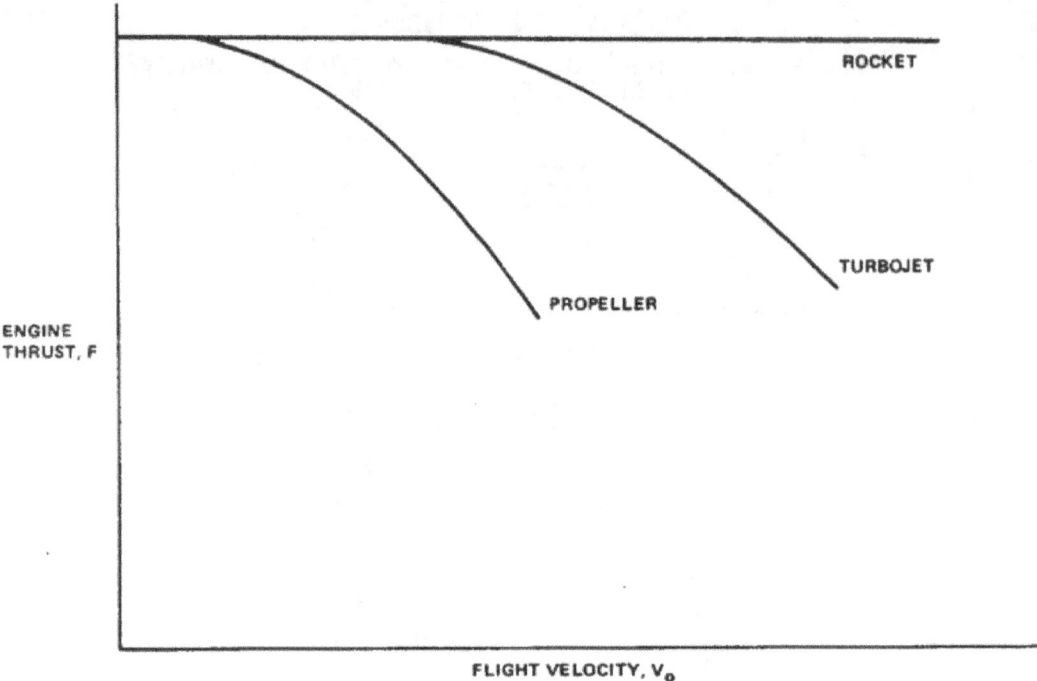

Fig. 64. Rocket thrust is independent of flight velocity, while the turbojet and propeller depend upon it.

Its principal use for flight is to indicate that the various propulsion systems operate best in different speed ranges. This is shown qualitatively by figure 65.* The propeller is the most efficient propulsive method at low speeds, while the jet engine achieves best efficiency only at relatively high flight speeds. The very high exhaust velocities of the rocket make its propulsive efficiency high only at very high flight speeds. Figure 66, from an NACA figure of the 1950s, shows similar information in a slightly different way.

Propulsion System Comparisons

The considerations of thrust and propulsive efficiency are by no means the whole story in comparing various propulsion systems. We must consider, for example, the thermodynamic efficiency of the engine cycle, or how efficiently the chemical energy of the fuel is converted to useful work such as driving the propeller and compressor, or accelerating the exhaust jet. The weight and size of the engine, its complexity, and its service life are other factors to be considered. Figure 67, from an NACA chart before

---

*Since thrust is the mass flow rate   change in fluid velocity, propulsive efficiency is

$$\frac{m\,(V_j - V_o\,V_o)}{1/2\,m\,(V_j - V_o)\,V_o}, \text{ which reduces to } \frac{2}{1 + V_j/\,V_0}$$

where m is mass flow rate, V_j is exhaust jet velocity, and V_o is flight velocity.

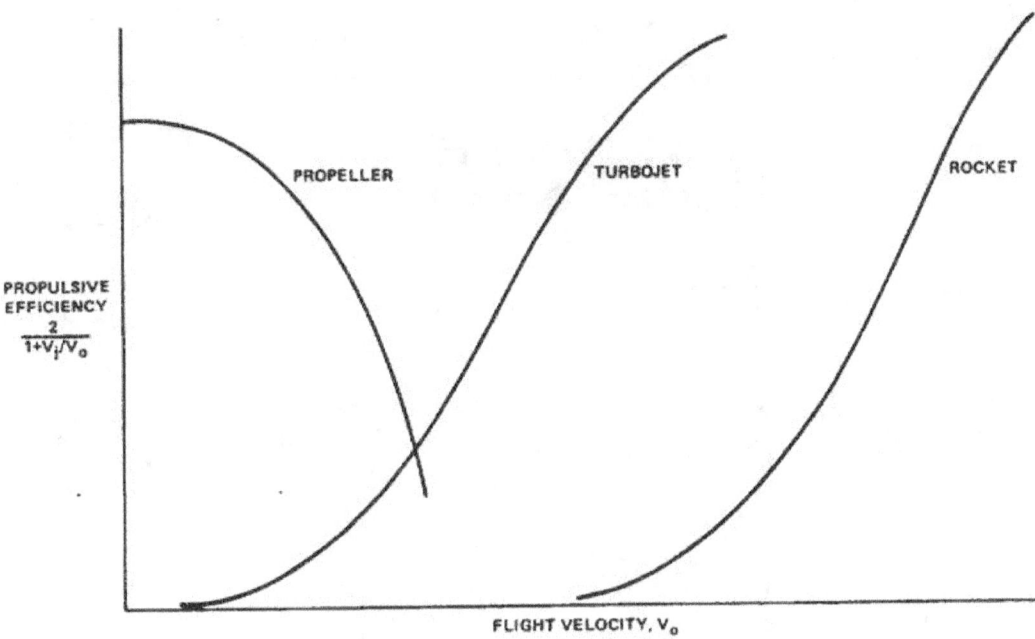

Fig. 65. Propulsive efficiency peaks at different speeds for the propeller, turbojet, and rocket.

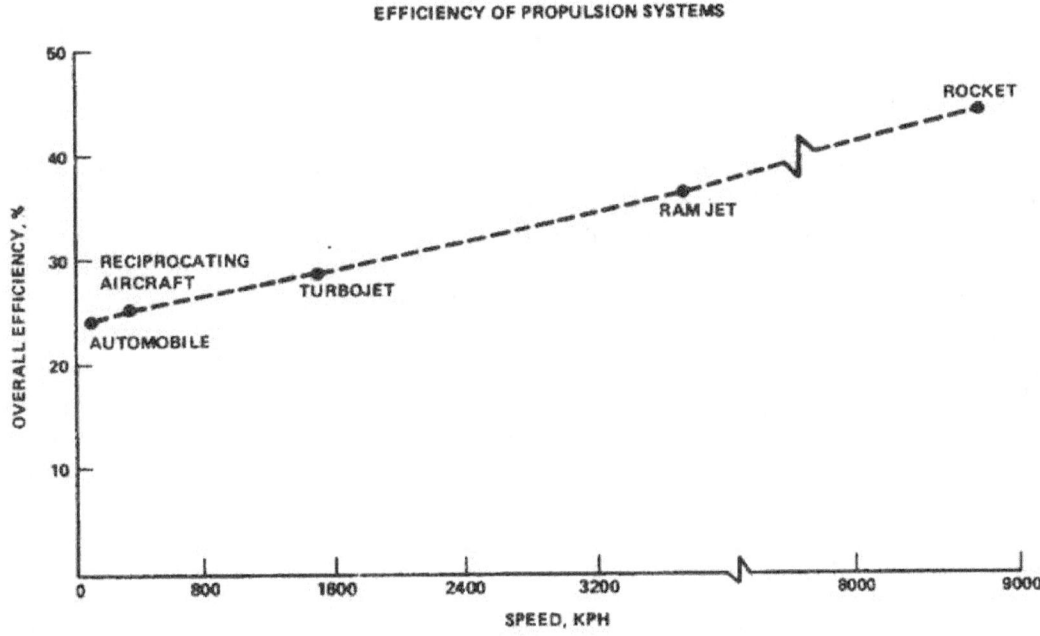

Fig. 66. Efficiency of various propulsion systems peaks at different speeds. (Adapted from NACA figure, ca. 1953.)

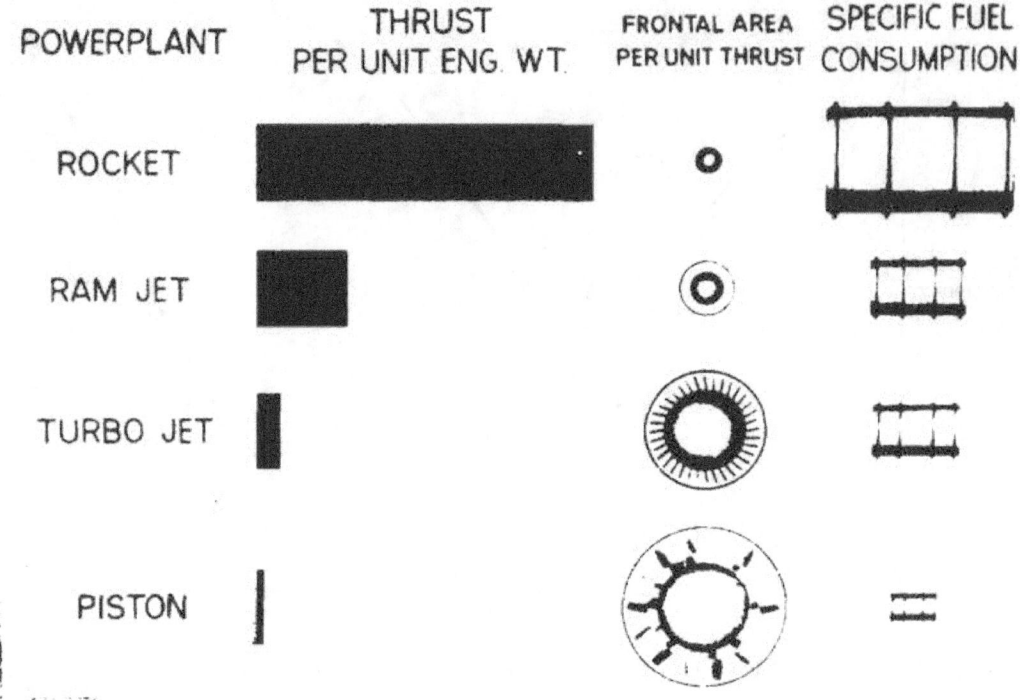

Fig. 67. Comparison of flight propulsion systems; relative values. (NACA figure, ca. 1949.)

1950, compares flight engines on the basis of specific weight, frontal area (which affects drag), and specific fuel consumption.

Gas Turbine Engines

The gas turbine is much older than the piston (reciprocating) engine. Hero used a gas turbine (steam) to drive a toy merry-go-round in 130 B.C. A patent for a gas turbine was granted to John Barber in England in 1791, and many were developed for various applications in the nineteenth century. Jets of hot gas impinge against turbine blades to spin the turbine which, in turn, performs some useful function such as driving a generator, compressor, pump, or other mechanical device. In an aircraft turbine engine, the turbine drives a compressor, fan, and sometimes a propeller, depending on the engine design. The most usual turbine engine is the turbojet where the turbine drives a compressor. Figure 68 is a sectional view of a turbojet, with the various components labeled. Air entering the inlet is compressed by the axial flow compressor having multiple stages. A liquid is sometimes injected in the compressor of high-performance turbojets as a coolant. The main fuel is injected and burned in the combustors in the annular space between the outer shell and the inner "spool." The hot combustion gases drive the turbine, shown in two stages, and in the after section additional fuel may be injected and burned to increase gas temperature and provide more thrust. The hot exhaust gases expand through the nozzle to produce thrust.

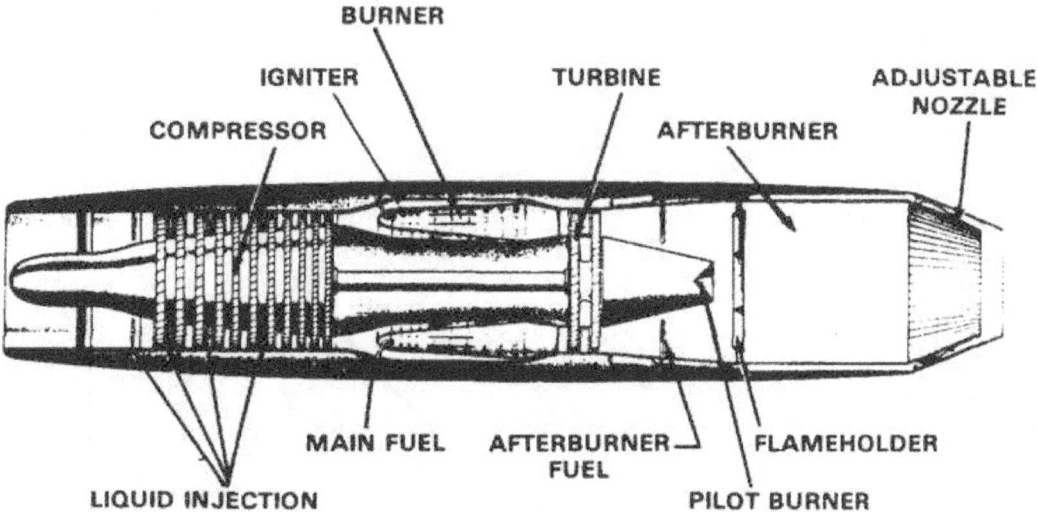

Fig. 68. Turbojet engine with thrust augmentation by an afterburner. (NACA figure. 1952.)

Thrust augmentation with an afterburner is for high performance military turbojets; afterburners are not used on commercial turbojets, where low specific fuel consumption is a major factor. Figure 69 shows the growth of thrust of turbojet engines from the early 1940s until 1952.

Ramjet

The ramjet engine began to receive attention during the second half of the 1940s and reached its peak during the 1950s. The ramjet has been called a flying stovepipe, for the absence of rotating parts that characterize the turbine engine. The ramjet gets its name from the method of air compression; it cannot operate from a standing start but must first be accelerated to a high speed by another means of propulsion. The air enters the spike-shaped inlet and diffuser (fig. 70) which serve the same purpose as the compressor. Fuel is injected and burns with the aid of flameholders that stabilize the flame. The burning fuel imparts thermal energy to the gas, and the expansion through the nozzle at speeds greater than the entering air produces the forward thrust. The ramjet, always needing an auxiliary propulsion system for starting, got squeezed between improved turbine engines and rockets during the 1950s and never recovered.

Rocket

A typical liquid propellant rocket, as diagrammed by NACA in 1952, is shown by figure 71. It consists of a guidance compartment, payload, fuel tank, oxidant tank, and engine compartment. A gas generator, operated from the main propellants or an auxiliary propellant, drives a turbine which drives pumps to supply fuel and oxidant at pressures of 20 to 40 atmospheres to the thrust chamber. The fuel is usually circulated in a cooling jacket surrounding the nozzle and combustion chamber prior to injection

GROWTH OF THRUST OF TURBOJET ENGINES

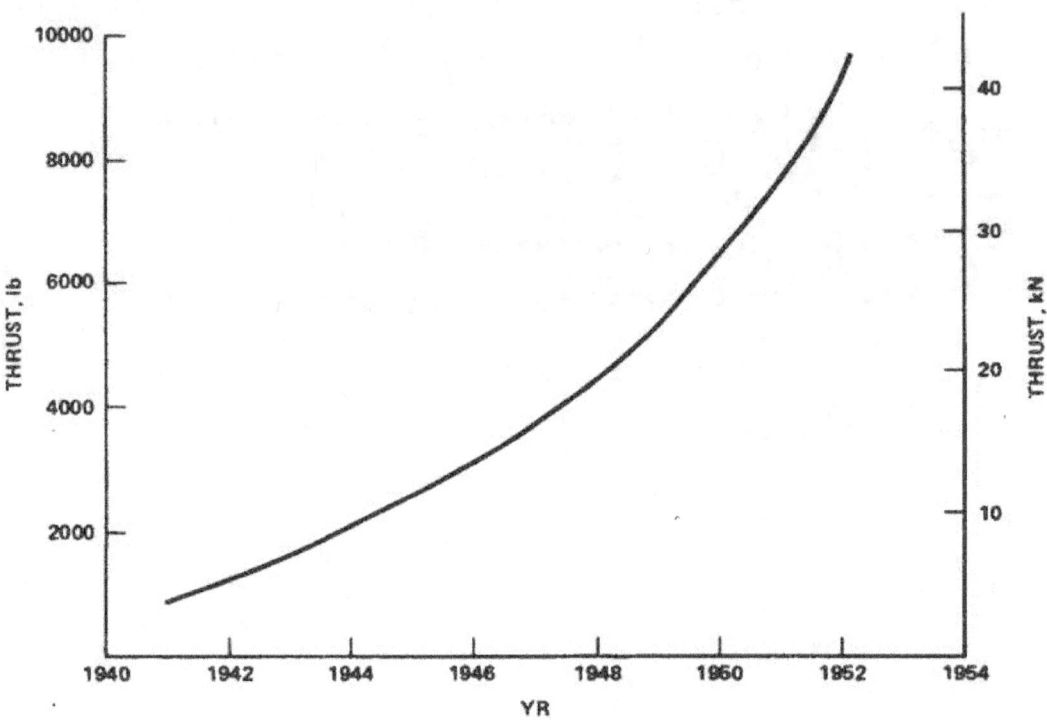

Fig. 69. Growth of thrust of turbojet engines. (Adapted from NACA figure, 1953.)

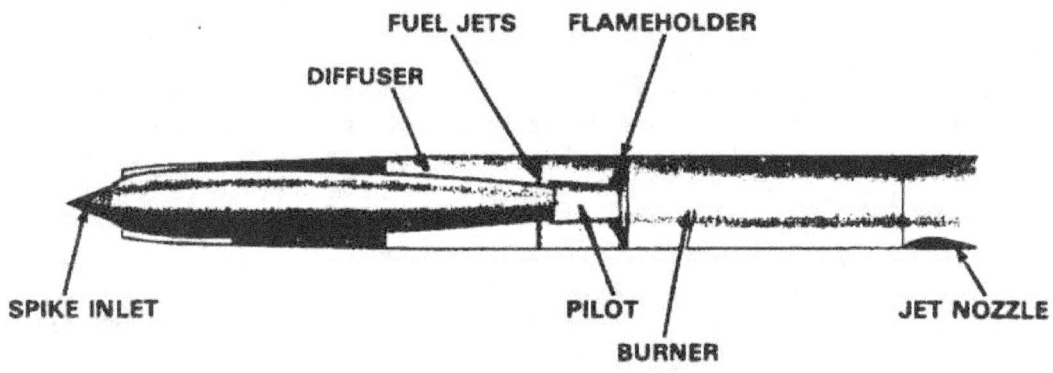

Fig. 70. A typical ramjet engine. (NACA figure, 1952.)

and burning. This is called regenerative cooling, for the heat picked up by the coolant-fuel is returned to the combustion chamber during combustion. Some fuel-oxidant combinations react spontaneously on initial contact—aniline and nitric acid, for example, as well as all combinations that use fluorine as the oxidizer. Other combinations, such as gasoline-oxygen, alcohol-oxygen, and hydrogen-oxygen,

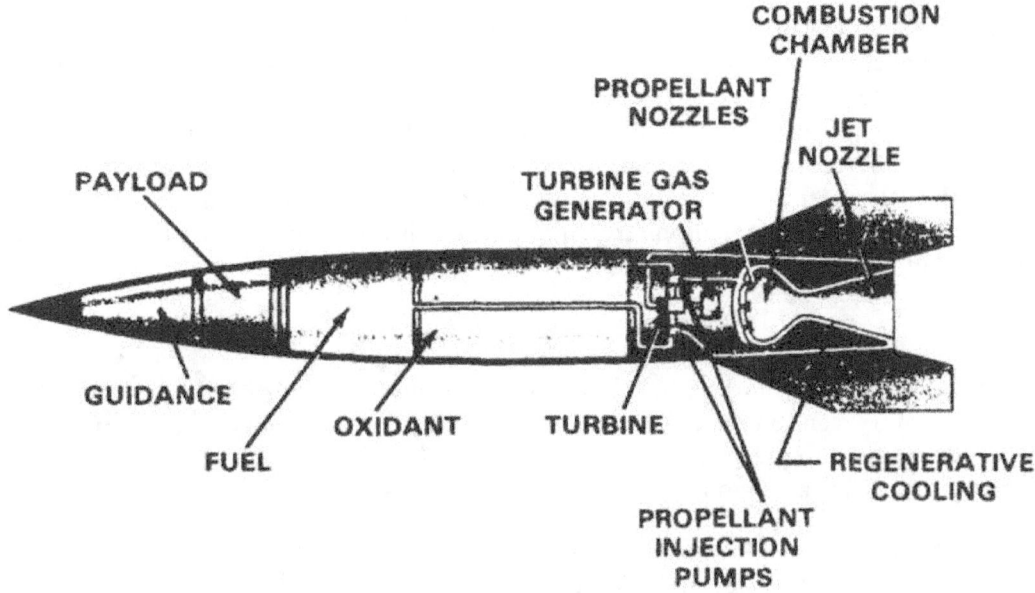

Fig. 71. A typical liquid-propellant rocket. (NACA figure, 1952.)

require an ignition source, which may be electrical or chemical. The fuel and oxidizer burn in the combustion chamber, and the hot gases expand through the nozzle to produce forward thrust. Because they use concentrated oxidizers, rockets are more compact than air-consuming engines where the oxidizer is gaseous, and diluted 80 percent by nitrogen at that.

Rocket Performance Parameters and Units

American rocket research and development during 1945–1959 used the English system of units. As NASA has directed that metric units be used in all publications, including this one, the conventional units have been converted using E. A. Mechtly, "The International System of Units: Physical Constants and Conversion Factors," (NASA SP-7012, 1973). The international system, designated SI in all languages, mandates newtons for force, newtons per square meter for pressure, and joules per second per square meter for heat transfer rates. In the interest of readability and as a concession to those brought up in the English system, a few compromises have been made. These are described in the following discussion of the major performance parameters used in rocketry.

Thrust. Force, expressed in newtons (N), kilonewtons (kN), and meganewtons (MN). A newton is that force that gives 1 kilogram (kg) of mass an acceleration of 1 meter per second per second (m/s²). To convert from 1 pound force to newtons, multiply by 4.448. This has usually been rounded off to 4.45 for conversions in the text. To offset the unfamiliarity of the newton, thrust is normally expressed in pounds in parentheses.

Propellant flow. Expressed in kilograms per second (kg/s). 1 lb = 0.4536 kg.

Mixture ratio. The proportions in which fuel and oxidizer are burned in the combustion chamber. Mixture ratio is expressed in several ways; one of the most common is the ratio of mass flow of oxidizer to mass flow of fuel, abbreviated O/F; sometimes the inverse is used. Another way of expressing mixture ratio, popular in the 1950s, is the percentage of the total mass flow that is fuel; a third is the molar ratio of one to the other. A mole is the mass equivalent of the molecular weight of the fuel and oxidizer. Sometimes a stoichiometric mixture ratio is mentioned; this means the exact proportion of fuel and oxidizer for complete combustion. As an example,

$$H_2 + {}^1/_2 O_2 \rightarrow H_2O$$

is the stoichiometric mixture. In this example, one mole of hydrogen combines with a half mole of oxygen to produce a mole of water. Expressed in mass units, 2.016 grams of hydrogen plus 16 grams of oxygen produce 18.016 grams of water. The O/F is 7.9, percent fuel is 11, and molar oxidizer to fuel ratio is ½.

Specific impulse and exhaust velocity. In 1903 Tsiolkovskiy, and other Europeans after him, expressed rocket engine performance in terms of the velocity of the exhaust emerging from the nozzle in meters per second (m/s). This made sense because the rocket exhaust velocity was a term in the equation expressing the velocity of a rocket-propelled vehicle. In the United States, it became the custom to express rocket performance in terms of the measured quantities: thrust and mass flow of the propellants. The thrust divided by the total mass flow of propellant was defined as the specific impulse. Specific impulse is the inverse of specific fuel consumption used in discussing the performance of other types of propulsion systems. In English units, specific impulse is in pounds force per pounds mass per second (lbf·sec/lbm). On seeing pounds in both numerator and denominator, many succumbed to the temptation to cancel them and express specific impulse incorrectly in units of seconds: the two pounds represent different physical phenomena, force and mass, and are connected by the conversion factor 32.2 lbm·ft/lbm·sec². In SI, specific impulse is expressed in newtons per kilogram per second or N·s/kg. English values of specific impulse are converted to SI by multiplying by 9.807, which can be rounded to 10 for approximations. The numerical value of specific impulse and exhaust velocity in SI are the same; only the units are different. Since exhaust velocity is a simple concept to visualize physically and since specific impulse expressed in newtons per kilogram per second is unfamiliar to many, including the author, all performance values in this text have been converted to exhaust velocity in meters/second (m/s). Typical values of exhaust velocity for liquid propellant rockets range from 2000 to 4500 m/s. The V-2 had an exhaust velocity of about 2200 m/s, very good for 1944. High energy propellants give exhaust velocities in the range of 3000 to 4500 m/s, and the liquid hydrogen–oxygen combination is in the upper part of this range.

Pressure. Expressed as newtons per square meter (N/m²) in SI. One pound per square inch is 6895 N/m². Rocket combustion pressures used during 1945–1959 were generally in the range of 300–600 lb/in² or 2069–4137 kN/m². Rocket combustion pressures were also expressed in atmospheres—multiples of sea-level pressure which is the same in any system of units. Since atmospheres are easier grasped than kN/m^2, combustion and higher pressures in this text are expressed in atmospheres (atm). One atmosphere is slightly over $100 \ kN/m^2$; the 2069–4137 kN/m^2 pressure range above becomes 20.4–40.7 atm.

Nozzle area ratio. The ratio of nozzle exit area to throat area. Area ratio determines the amount of expansion of the exhaust gases through the nozzle and is related to exhaust gas pressures. If a rocket designer is asked to provide a nozzle that is to be operated only on the ground—as in the case of an experimental rocket engine—he chooses an area ratio such that the exhaust gas pressure at the nozzle exit is equal to ambient pressure, usually considered as sea-level pressure or 1 atmosphere. If the combustion pressure is 20 atmospheres, the gases undergo an expansion ratio of 20 and this corresponds roughly to a nozzle area ratio of 4. If, for some reason, the designer provides a nozzle area less than that needed for complete expansion, the exhaust gases emerge from the nozzle exit at greater than ambient pressure. In this case the gases are said to be underexpanded, for they have to expand further to reach ambient pressure. On the other hand, if the designer provides a larger area ratio than that needed for complete expansion, the exhaust gases reach a pressure equal to ambient while still in the nozzle. In some cases the gases will continue to follow the nozzle walls and expand to a pressure lower than ambient. In this case the gases are said to be overexpanded. Sooner or later, overexpanded gas must be reconciled with the ambient pressure and nature provides for this adjustment by means of a shock wave. The ideal nozzle is one that provides for complete gas expansion, neither more nor less, for theory shows that this yields maximum exhaust velocity. This poses a problem to the designer of a launch vehicle, because as soon as the vehicle is launched, the ambient pressure begins to fall, approaching zero at very high altitudes. Since the nozzle is fixed, what area ratio should he choose? If he designs for sea-level conditions, he is getting less performance at altitude than could be realized; if he designs for altitude, he suffers some performance loss from overexpansion at sea-level. For rocket engines in upper stages, he doesn't have this problem for the ambient pressure is close to zero at ignition. In rocket engines for all stages, however, he must balance the gain in performance from providing a larger area ratio nozzle against the added weight and cooling requirements of the larger nozzle and must also stay within special limits of the vehicle. A typical rocket engine for an upper stage will have an area ratio of about 40. In the text, you will seldom encounter area ratio, but performance at sea-level and altitude is used and in each case it is implied that a nozzle is provided for complete, or nearly complete, expansion of the exhaust gases.

Heat transfer rate. Heat flow per unit time per unit area. In SI, joules/second·meter^2 ($J/s \cdot m^2$). One $Btu/s \cdot in^2$ equals 1.637 $J/s \cdot m^2$. Rocket combustion temperatures, from 2500 to 4500 K, and pressures, from 20 to 40 atmospheres, produce very high heat transfer rates—on the order of 20 to 200 times greater than those produced in boilers and superheaters in a steam plant, for example, and much higher than in other types of internal combustion engines. Typical values of rocket heat transfer rates range from 1 to 20 $J/s \cdot m^2$, but can be higher in local areas.

Mass ratio. The ratio of the gross (total) mass of a rocket vehicle to its empty mass (M_o/M_e). The difference between the two is the mass of propellant expended during the operation. The empty mass includes the rocket engine, the structure, controls, and the payload. For the last stage of a multistage launch vehicle, the payload is the spacecraft. For other stages, the payload is the loaded stages above it. Mass ratios range from about 3 to 10. The V-2 had a mass ratio of about 3.

Vehicle velocity. Expressed in meters per second (m/s). Vehicle velocity is a function of the rocket exhaust gas velocity, mass ratio of the vehicle, aerodynamic drag during flight through the atmosphere, gravitational pull expressed in terms of burning time of the rocket, and the trajectory. In 1903 Tsiolkovskiy derived the velocity for a rocket in vertical flight, disregarding drag and gravitational pull. This yielded the fundamental rocket flight equation named after him and so identified in the text. The equation is:

$$V = V_e \ln(M_o/M_e)$$

where V is vehicle velocity, V_e is exhaust gas velocity, M_o is gross mass, M_e is empty mass, and ln is the natural logarithm. The direct relationship between vehicle velocity and exhaust gas velocity accounts for the interest in high energy propellants that yield high exhaust velocities.

Multistage rocket. A vehicle composed of two or more complete rocket systems, each including propellant and associated controls and structure. The last (upper) stage carries the spacecraft as the payload. The payload of the first stage is the upper stages and the spacecraft. Investigators learned a very long time ago that rocket vehicle velocity could be increased by having one rocket unit riding piggy-back on another until the first expends all of its propellant. The dead weight of the first stage is normally jettisoned. The second rocket unit or stage begins operating at the velocity given it by the first stage and adds its own velocity increase, for a much higher final velocity. If the first two stages have identical exhaust velocities and mass ratios, for example, the final velocity will be twice that of either operating singly.

Other parameters and units. Horsepower is given in its equivalent kilowatts (kW), and revolutions per minute (rpm) was retained. The conversion of feet to meters and pounds to kilograms is obvious.

TABLE 14.— *Propulsion Parameters and Units*

Parameter	English	SI	This text
Thrust	lb	N	N (lb)
Propellant flow	lbm/s	kg/s	kg/s
Mixture ratio	various		various
Specific impulse	lbf·s/lbm	N·s/kg	not used
Exhaust velocity	ft/s	m/s	m/s
Pressure	lbf/in^2	N/m^2	kN/m^2 or atm
Nozzle area ratio	A_e/A_t		A_e/A_t
Heat transfer rate	$Btu/s·in^2$	$J/s·m^2$	$J/s·m^2$
Mass ratio	M_o/M_c		M_o/M_c
Vehicle velocity	ft/s	m/s	m/s
Power	hp	W	kW
Length	ft	m	m
Mass	lbm	kg	kg

SI abbreviations

  N = newtons
   s = seconds
 kg = kilograms
  m = meter
kW = kilowatts
  J = joule
  c = centi (10^{-2})
  k = kilo (10^3)
 M = mega (10^6)

Other units and prefixes are given in E.A. Mechtly. *The International System of Units: Physical Constants and Conversion Factors*. NASA SP-7012, 1973.

Source Notes

Chapter 1: Introduction

1. Theodore von Kármán with Lee Edson, *The Wind and Beyond* (Boston: Little Brown, 1967), pp. 267–68. See also H. H. Arnold, *Global Mission* (New York: Harper, 1949), pp. 532–33; Arnold to von Kármán, 7 Nov. 1944, NASA History Office.
2. As quoted by Thomas A. Sturm, *The USAF Scientific Advisory Board: Its First Twenty Years, 1944–1964* (Washington: USAF Historical Division Liaison Office, 1967), p. 2.
3. Arnold, *Global Mission*, p. 530.
4. Eugene M. Emme, *Aeronautics and Astronautics: An American Chronology of Science and Technology in the Exploration of Space, 1915–1960* (Washington: NASA, 1961), p. 46.
5. Robert Schlaifer, "Development of Aircraft Engines" (Boston: Harvard Univ., 1950); reprinted in *Development of Aircraft Engines and Fuels* (Elmsford, NY: Maxwell Reprint, 1970), pp. 321–508. Schlaifer gives an excellent account of aircraft gas turbine developments from origins through World War II.
6. Ibid., pp. 457–61; also Durand committee files, NASA History Office.
7. The original XP–80 was built to use a British jet engine, which Schlaifer, "Development of Aircraft Engines," p. 475, identifies as a de Haviland Goblin; Emme, *Aeronautics and Astronautics: An American Chronology*, p. 47, identifies it as a Halford engine. The time for building the XP–80 came from an interview with C. L. Johnson and B. R. Rich, Lockheed California Co., Burbank, 2 May 1974.
8. Arnold, *Global Mission*, p. 544.
9. Schlaifer, "Development of Aircraft Engines," p. 321.
10. Telephone interview with Robert E. Littell, Falls Church, VA, 20 Aug. 1973.
11. James Phinney Baxter, 3d, *Scientists against Time* (Cambridge: MIT Press, 1946), p. 15; John E. Burchard, ed., *Rockets, Guns and Targets*, in the Office of Scientific Research and Development series, *Science in World War II* (Boston: Little, Brown, 1948), p. 5.
12. Alexis W. Lemmon, Jr., "Fuel Systems for Jet Propulsion" (Washington: Navy, 1945).
13. Ibid., p. 18.
14. Ibid., p. 27.

Chapter 2: Air Force Research on Hydrogen

1. F. Simon, "Liquid Hydrogen as a Fuel for Aircraft," Clarendon Laboratory, Oxford Univ., 16 Apr. 1942, 10 pp., including diagram. Sent to NACA, Washington, with cover letter, 5 May 1942. Reproduced by NACA and sent to members of its committee on power plants for aircraft, 30 June 1942 (NACA Hdqs. library card 5020/262). A copy of the report has not been located. The copy sent to Wright Field went first to Col. E. R. Paige, chief of the power plant laboratory and a member of the NACA power plants committee. Nancy Arms, *A Prophet in Two Countries: The Life of F. E. Simon* (New York: Pergamon, 1966), p. 105, briefly mentions the incident: "He suggested a scheme for using hydrogen as a fuel for long distance planes; it was turned down because of

the practical difficulties and because new developments soon made it possible to carry enough petrol for trans-Atlantic flights." Information on Simon's career also came from this biography.

2. Robert V. Kerley admits to being co-author of the verses, which were circulated to members of NACA's subcommittee on fuels and lubricants and apparently to the power plants committee—the latter by Sam Heron. Kerley to author, 29 Jan. 1974.

3. It is not clear who at Wright Field had the idea to investigate hydrogen as a fuel for aircraft and rockets, but Opie Chenoweth, Weldon Worth, Robert Kerley, John Duckworth, Ewell Phillips, and Marc Dunnam were all involved in the work. Kerley left Wright Field in Feb. 1945 on an overseas assignment but returned in May. During his absence, Duckworth took his place in the fuels and oil branch at Wright Field. Kerley left Wright in September to work for the Ethyl Corp. and Duckworth became the Air Force representative on the NACA fuels and lubricants subcommittee. Opie Chenoweth to author, 23 Jan. 1974; John Duckworth to author, 25 Jan. 1974; interview with Chenoweth and Weldon Worth, Dayton, OH, 7 June 1974; Robert V. Kerley to Monte D. Wright, NASA History Office, 14 May 1976.

4. Earle R. Caley, "History of the Department of Chemistry," Ohio State Univ., Columbus, n.d.; Herrick L. Johnston, "The Cryogenic Laboratory," *Engineering Experiment News* 18 (June 1946): 3–21, in which Johnston describes the origins of the laboratory, the equipment, and plans.

5. Gwynne A. Wright in group interview that included Howard A. Altman and Charles Weisend, CVI Corp., Columbus, 4 June 1974; interview with Prof. Thor A. Rubin. Ohio State Univ., Columbus, 5 June 1974.

6. "Liquid Hydrogen Propellant for Aircraft and Rockets," AFP:431952, Project MX-588, Contract W33-038-ac-11101 (14552), 1 July 1945–30 Apr. 1948; "Liquid Hydrogen. Rocket Motor Development," Contract W33-038-ac-19382 (19126), OSU Research Foundation Project 333. The second contract, on properties, initially carried the old contract title "Liquid Hydrogen Propellant for Aircraft and Rockets," but later became "Properties of Liquid Hydrogen," Contract W33-038-ac-14794 (16243), OSU Research Foundation Project 264.

7. Herrick L. Johnston and William L. Doyle, "Final Report—Development of the Liquid Hydrogen–Liquid Oxygen Propellant Combination for Rocket Motors," TR-7, Columbus: OSU Research Foundation, Dec. 1951.

8. Interview with Wright, Altman, and Weisend, 4 June 1974; interview with Dr. William V. Johnston (H. L. Johnston's son), Gaithersburg, MD, 7 Apr. 1976.

9. Herrick L. Johnston, "Rocket Motor Development—First Summary Report," Columbus: OSU Research Foundation, Apr. 1949; Johnston and C. A. Huntley, "Second Annual Summary Report—Liquid Hydrogen Thrust Chamber and Pump Development," SPR 333-2, Columbus: OSU Research Foundation, Apr. 1950; Johnston and William L. Doyle, "Final Report—Development of the Liquid Hydrogen–Liquid Oxygen Propellant Combination for Rocket Motors," TR-7, Columbus: OSU Research Foundation, Dec. 1951.

10. Interview with Irwin J. Weisenberg, Los Angeles, 1 May 1974.

11. Interview with Dr. Willard P. Berggren, Univ. of Bridgeport, Bridgeport, CT, 21 Mar. 1974.

12. Johnston and Doyle, "Final Report—Development of the Liquid Hydrogen–Liquid Oxygen Propellant Combination for Rocket Motors"; Irwin J. Weisenberg and W. P. Berggren, "Hydrogen in the Critical Region as a Rocket Engine Coolant," 1960 (unpublished).

13. Leroy F. Florant and Harold F. Snider, "Centrifugal Pumping of Liquid Hydrogen," TR 333-4, Columbus: OSU Research Foundation, May 1950.

14. Interview with William L. Doyle, Redondo Beach, CA, 26 Apr. 1974, with addendum by Doyle, 8 Nov. 1974; Doyle, "Experimental Evaluation of the Liquid Hydrogen–

Liquid Fluorine Propellant Combination at Various Chamber Pressures," TR 333-5, Columbus: OSU Research Foundation, Oct. 1951.

Chapter 3: Hydrogen-Oxygen for a Navy Satellite

1. Telephone interviews with Harvey Hall, 31 Oct., 6 Nov. 1973; interview with Robert Gordon, Azusa, CA, 23 Apr. 1974.

2. Telephone interview of Abraham Hyatt by R. Cargill Hall, 21 Jan. 1970; interview with Hyatt, El Segundo, CA, 26 Apr. 1974; F. Zwicky, "Report on Certain Phases of War Research in Germany," Summary Report F-U-3RE, Hq., Air Materiel Command, Wright Field, Jan. 1947.

3. Beacon Desk, Identification Sect. Elect. Mat. Br., to Head, Special Weapons Sect., Elect. Mat. Br., Engineering Div., Bureau of Aeronautics, Navy Dept., Washington, by R. P. Haviland, 10 Aug. 1945, NASA History Office.

4. F. Zwicky et al., "Final Report as of December 31, 1944, to the Bur. of Aero., Navy Dept., on Contract NOa(s)-3055," Report R-50, Aerojet Engineering Corp., Pasadena, 31 Dec. 1944.

5. Interview with Gordon, 23 Apr. 1974.

6. George H. Osborn, Robert Gordon, and Herman L. Coplen, "Liquid Hydrogen Rocket Engine Development, 1944–1950," XXI International Astronautical Congress, Constance, West Germany, 9 Oct. 1970. Hydrogen research at Aerojet before 1945 has not been documented. Interview with George James, NSF, Washington, 19 July 1973.

7. Head, Special Weapons Sect. (R&E Elect. Mat. Br.) to Head, Experiments and Development Br., Bur. of Aero., Navy Dept., by J. A. Chambers, 25 Aug. 1945, NASA History Office; Dep. Dir. of Engineering (R&D) to Head, Experiments and Development Br., et al., Bur. of Aero., Navy Dept., by R. S. Hatcher, 30 Oct. 1945, NASA History Office; Minutes of 1st Meeting of Space Rocket Committee, 8 Oct. 1945, by Lt. (jg) Max, NASA History Office.

8. Minutes of 2d Meeting of Space Rocket Committee, 15 Oct. 1945, by Lt. (jg) Max, NASA History Office.

9. Minutes of 3d Meeting by Lt. (jg) Max, 22 Oct. 1945, with addendum by Harvey Hall, NASA History Office.

10. Minutes of Fourth Meeting, Space Rocket Committee, 29 Oct. 1945, by Lt. (jg) Max, NASA History Office.

11. O. E. Lancaster and J. R. Moore, "Investigation on the Possibility of Establishing a Space Ship in an Orbit above the Surface of the Earth," A.D.R. report R-48, Nov. 1945, Bur. of Aero., Navy Dept.

12. The JPL study, with all reports referenced, is summarized by H. J. Stewart, "A Summary of Performance Studies for a High Altitude Orbiting Missile," report 8-5, JPL-GALCIT, Pasadena, 10 July 1946.

13. F. J. Malina and Martin Summerfield, "The Problem of Escape from the Earth by Rocket," Publication 5, JPL-GALCIT, Pasadena, 23 Aug. 1946. Malina and Summerfield presented a paper with same title at Sixth International Congress for Applied Mechanics (Sec. II), Paris, 22–29 Sept. 1946; also *Journal Aeronautical Sciences* 14 (1947): 471–80.

14. Eugene M. Emme, *Aeronautics and Astronautics: An American Chronology of Science and Technology in the Exploration of Space, 1915–1960* (Washington: NASA, 1961), p. 51.

15. R. Cargill Hall, "Earth Satellites, A First Look by the United States Navy in the 1940s," presented at XXI International Astronautics Congress, Constance, West Germany, 9 Oct. 1970.

16. As quoted by R. Cargill Hall, "Early U.S. Satellite Proposals," *Technology and Culture* 4 (Fall 1963): 411.

17. J. C. Hunsaker, "Jet Propulsion," an address before the National Academy of Sciences, 23 Apr. 1946, NASA History Office.

18. Cargill Hall, "Early U.S. Satellite Proposals," p. 414.

19. Excerpts from "Project RAND, First Quarterly Report," RA-15000, June 1946, NASA History Office.

20. "Preliminary Design of an Experimental World-Circling Spaceship," report SM-11827, Project RAND, Douglas Aircraft Co., Santa Monica, CA, 2 May 1946; "First Quarterly Report," RA-15000, Project RAND, Santa Monica, June 1946.

21. Cargill Hall, "Early U.S. Satellite Proposals," p. 423.

22. R. G. Wilson, "Structural Design Study—High Altitude Test Vehicle," report NA 46-758, North American Aviation, Inglewood, CA, 26 Sept. 1946.

23. "Second Quarterly Report," RA-15004, Project RAND, Santa Monica, Sept. 1946; Cargill Hall, "Early U.S. Satellite Proposals," pp. 22–23.

24. "Fourth Quarterly Report," RA-15033, Project RAND, Santa Monica, Mar. 1947.

25. "Contract NOa(s)-8496: Report of Progress during Period 27 September to 11 October 1946," Aerojet Engineering Corp., 17 Oct. 1946; Osborn, Gordon, and Coplen, "Hydrogen Rocket Engine."

26. Pedro C. Medina, "HATV—Summary Report," Engineering Report 2666, Glenn L. Martin Co., Baltimore, June 1947.

27. Robert Gordon, "Performance, Pressure Distribution, and Heat Transfer in a Water-Convection-Cooled Flared Tube Hydroxygen Motor of 400-lb Thrust," report RTM-34, 17 Mar. 1948.

28. "Contract NOa(s)-8496, Report of Progress during Period 27 Sept. to 11 Oct. 1946," Aerojet Engineering Corp., 17 Oct. 1946; "Report of Progress for the Month of May 1947, Contract NOa(s)-8496, Item 2," report R-78, Aerojet Engineering Corp., Azusa, 15 June 1947; and Robert Gordon, Herman L. Coplen, and David A. Young, "Final Report on Item 2, Contract NOa(s)-8496," report R-79, Aerojet Engineering Corp., Azusa, 15 July 1947.

29. R. Cargill Hall, "A Chronology of Some Events in Early United States Satellite Studies during the 1940s," JPL/HN-7, Jet Propulsion Laboratory, Pasadena, April 1970.

30. Cargill Hall, "Early U.S. Satellite Proposals," p. 427.

31. R. B. Canright, "The Relative Importance of Specific Impulse and Propellent Density for Large Rockets," report 4-29, Jet Propulsion Laboratory, Pasadena, 7 Jan. 1947. The von Braun, Hager, and Tschinkel report is a reference in Canright's report.

32. Under Contract NOa(s)-8496, Item 4 was the development of the rocket engine and Item 5 was the hydrogen liquefier. The Aerojet Engineering Corp. issued a variety of reports and memoranda. The following were used: "Progress for July 1947," R-80, 15 Aug. 1947; "Progress for Aug. 1947," R-81, 15 Sept. 1947; "Progress for Sept. 1947," R-83, 15 Oct. 1947; "Progress for Nov. 1947," R-85, 15 Dec. 1947; "Progress for Dec. 1947," R-87, 15 January 1948; "Semi-Annual Summary, 1 July–31 Dec. 1947," R-88, 30 Jan. 1948; "Progress for Jan. 1948," R-90, 15 Feb. 1948; "Quarterly Summary, Jan.–Mar. 1948," R-94, 30 Apr. 1948; "Semi-Annual Summary Report, 1 Jan.–30 June 1948," R-100, 20 Aug. 1948; "Quarterly Progress Report, 1 July–30 Sept. 1948," R-335, 13 Dec. 1948; "Semi-Annual Summary, 1 July–31 Dec. 1948," R-358, 25 Feb. 1949; George H. Osborn and Wayne D. Stinnett, "Development of a 3000-lb Thrust Liquid Hydrogen–Liquid Oxygen Rocket Motor," research technical memo. 54, 29 Aug. 1949; and David A. Young, "Research and Development of Hydrogen-Oxygen Rocket Engine, Model XLR 16-AJ-2, Final Report," R-397, 28 Sept. 1949. Other Aerojet reports are cited in Osborn, Gordon, and Coplen, "Hydrogen Rocket Development." These references are the sources for Aerojet technical results described in the remainder of the chapter.

33. Interview with Robert Gordon, 23 Apr. 1974.

34. Dwight I. Baker, "Regenerative Cooling Tests of Rocket Motors Using Liquid Hydrogen and Liquid Oxygen," report 4-53, Jet Propulsion Laboratory, Pasadena, 11 Aug. 1949.

35. Cargill Hall, "Early U.S. Satellite Proposals," p. 429.

Chapter 4: Hydrogen Technology from Thermonuclear Research

1. Edward Teller with Allen Brown, *The Legacy of Hiroshima* (Garden City, NY: Doubleday, 1962), chap. 3; Herbert York, *Race to Oblivion; A Participant's View of the Arms Race* (New York: Simon & Schuster, 1970), p. 35; Richard G. Hewlett and Francis Duncan, *Atomic Shield, 1947-1952*, vol. 2 of *A History of the Atomic Energy Commission* (University Park: Penn. State Univ. Press, 1969).
2. Interview with Dr. Willard P. Berggren, Univ. of Bridgeport, Bridgeport, CT, 21 Mar. 1974; interview with William L. Doyle, Redondo Beach, CA, 26 Apr. 1974, and addendum submitted by Doyle, 8 Nov. 1974.
3. Telephone interview with Dr. David White, Philadelphia, 18 June 1974; Doyle interview.
4. Eduard Farber, *Nobel Prize Winners in Chemistry, 1901-1961*, rev. ed. (New York: Abelard-Schuman, 1963), pp. 208-11.
5. Interview with Dr. Bernard Rubin, NASA, Washington, 23 Jan. 1974. Rubin took undergraduate chemistry courses from Prof. Johnston.
6. White interview; telephone interview with Prof. Clyde A. Hutchinson, Chicago, 23 June 1974; interview with Prof. Thor A. Rubin, Ohio State Univ., Columbus, 5 June 1974; Dr. Ed Mack, Jr., to Dean W. Paul Hudson, Graduate School, 5 June 1952, OSU Archives.
7. Johnston file, OSU Archives; Rubin interview.
8. White interview; interview with Arthur A. Brooke, OSU, 5 June 1974; interview with Gwynne A. Wright, Howard A. Altman, and Charles Weisend, CVI Corp., Columbus, 4 June 1974.
9. Brooke interview; Wright, Altman, and Weisend interview.
10. Johnston file, OSU Archives.
11. Wright, Altman, and Weisend interview; Johnston file, OSU Archives.
12. Johnston file, OSU Archives; Wright, Altman, and Weisend interview; York, *Race to Oblivion*, p. 40.
13. Rexmond C. Cochrane, *Measures for Progress: A History of the National Bureau of Standards* (Washington: GPO, 1966), pp. 83-84.
14. F. G. Brickwedde, "A Few Remarks on the Beginning of the NBS-AEC Cryogenic Engineering Laboratory," Cryogenic Engineering Conference, National Bureau of Standards, Boulder, CO, 8-10 Sept. 1954, in K. D. Timmerhaus, ed., *Advances in Cryogenic Engineering* (New York: Plenum Press, 1960), 1:1-4; Cochrane, *Measures for Progress*, p. 446.
15. Brickwedde, "A Few Remarks," pp. 2-3.
16. "Research Facilities of the NBS-AEC Cryogenic Engineering Laboratory," *Advances in Cryogenic Engineering*, 1:5-22.
17. C. B. Hood, Jr., H. W. Altman, M. L. Yeager, N. C. Hallett and L. D. Wagner, "The Herrick L. Johnston Air Tactical Dewar," *Advances in Cryogenic Engineering*, 1:44-48; B. W. Birmingham, E. H. Brown, C. R. Class, and A. F. Schmidt, "Experimental Dewars Developed by the National Bureau of Standards," *Advances in Cryogenic Engineering*, 1:49-61.
18. Wright, Altman, and Weisend interview.
19. T. Stearns, D. J. Sandell, and J. S. Burlew, "The Refrigerated Transport Dewar," *Advances in Cryogenic Engineering*, 1:35-40.
20. H. L. Johnston, C. B. Hood, Jr., H. W. Altman, J. G. Pierce, and C. W. Weisend, "Mobile Liquid Hydrogen Plant," *Advances in Cryogenic Engineering*, 1:324-28.
21. Wright, Altman, and Weisend interview.
22. *Advances in Cryogenic Engineering*, vol. 1.
23. Ibid., vol. 2.
24. Ibid., vol. 3, K. D. Timmerhaus, ed. (New York: Plenum Press, 1960).
25. *Advances in Cryogenic Engineering*, vols. 4-6, K. D. Timmerhaus, ed. (New York: Plenum Press, 1960-1961). The growth of cryogenic technology during the 1950s is

summarized by Russel B. Scott, *Cryogenic Engineering* (Princeton: Van Nostrand, 1959). *Technology and Uses of Liquid Hydrogen* (New York: Macmillan, 1964), edited by Scott, is an excellent collection on liquid hydrogen technology.

26. A textbook, *Nuclear Rocket Propulsion* by R. W. Bussard and R. D. DeLauer, was published in 1958 (New York: McGraw-Hill). Bussard, a nuclear rocket pioneer, and DeLauer do not cover the history of nuclear rockets. That has been ably done by a latter-day James Dewar in a doctoral thesis: James Arthur Dewar, "Project Rover: A Study of the Nuclear Rocket Development Program, 1953–1963," Ph.D. dissertation, History Dept., Kansas State Univ., Feb. 1974.

Chapter 5:   NACA Research on High-Energy Propellants

1. Robert H. Goddard, *The Papers of Robert H. Goddard*, ed. Esther C. Goddard and G. Edward Pendray, 3 vols. (New York: McGraw-Hill, 1970), p. 1237.

2. Ibid., p. 1238.

3. Ibid., p. 1079.

4. Ibid., p. 1083.

5. Telephone interview with Robert E. Littel, Falls Church, VA, 20 Aug. 1973.

6. R. O. Miller and P. M. Ordin, "Theoretical Performance of Rocket Propellants Containing Hydrogen, Nitrogen, and Oxygen," RM E8A30 (NACA, 1948).

7. V. N. Huff and C. S. Calvert, Jr., "Charts for the Computation of Equilibrium Composition of Chemical Reactions in the Carbon-Hydrogen-Oxygen-Nitrogen System at Temperatures from 2000° to 5000°K," TN 1653 (NACA, 1948); V. N. Huff, C. S. Calvert, Jr., and Virginia Erdmann, "Theoretical Performance of Diborane as a Rocket Fuel," RM E8I17a (NACA, 1949); Virginia E. Morrell, "Effect of Combustion-Chamber Pressure and Nozzle Expansion Ratio on the Theoretical Performance of Several Rocket Propellant Systems," RM E50C30 (NACA, 1950); V. N. Huff, Sanford Gordon, and Virginia E. Morrell, "General Method and Thermodynamic Tables for Computation of Equilibrium Composition and Temperature of Chemical Reactions," report 1037 (NACA, 1951; superseded TN 2113 and 2161 reported earlier).

8. P. M. Ordin, R. O. Miller, and J. Diehl, "Preliminary Investigation of Hydrazine as a Rocket Fuel," RM E7H21 (NACA, May 1948); W. H. Rowe, P. M. Ordin, and J. Diehl, "Investigation of the Diborane–Hydrogen Peroxide Propellant Combination," RM E7K07 (NACA, 1948); W. H. Rowe, P. M. Ordin, and J. Diehl, "Experimental Investigation of Liquid Diborane-Liquid Oxygen Propellant Combination in 100-Pound-Thrust Rocket Engines," RM E9C11 (NACA, 1949); P. M. Ordin and R. O. Miller, "Experimental Performance of Chlorine Trifluoride–Hydrazine Propellant Combination in 100-Pound-Thrust Rocket Engines," RM E9F01 (NACA, 1949).

9. J. L. Sloop, "NACA High Energy Rocket Propellant Research in the Fifties," read at Panel on Rocketry in the 1950s, AIAA 8th Annual Meeting, Washington, 28 Oct. 1971, condensed in *Astronautics and Aeronautics* 10 (Oct. 1972):52–57; Sloop, "Memoir: High Energy Rocket Propellant Research at the NACA/NASA Lewis Research Center, 1945–1960," read at the 7th International History of Aeronautics Symposium, 24th International Astronautical Congress, Baku, USSR, 8 Oct. 1973; Walter T. Olson to the author, 21 Jan. 1972; interview with P. M. Ordin, NASA Lewis Research Center, 30 May 1974.

10. J. L. Sloop, P. M. Ordin, and V. N. Huff, "Theoretical and Experimental Investigation of Diborane as a Rocket Fuel," *NACA Conference on Fuels—A Compilation of the Papers Presented by NACA Staff Members*, Flight Propulsion Research Laboratory, Cleveland, 26 May 1948.

11. P. M. Ordin, H. W. Douglass, and W. H. Rowe, "Investigation of the Liquid Fluorine-Liquid Diborane Propellant Combination in a 100-Pound Thrust Rocket Engine," RM E51I04 (NACA, 1951).

NOTES TO PAGES 75 - 83

12. The date—even the year—of the meeting referred to is, unfortunately, in question. The author, a participant in the meeting, believes it was 1950 and so stated in a paper given before the International Astronautical Congress in 1973. On 26 Apr. 1974, another participant, William Lawler Doyle of Redondo Beach, CA, asserted during an interview that the date was 19 May 1950. The month and year are supported by an internal memorandum of Pratt & Whitney Aircraft, East Hartford, CT, entitled "Rocket Motors" by C. Branson Smith on 31 July 1956 (PWA files). In discussing choice of propellants, Smith cited a reference given as "Minutes of Meeting for Selection of Rocket Propellants—NACA, May 1950—best overall discussion." The official NACA-Lewis laboratory file copy of the minutes, prepared by the author, was destroyed in a purge of classified documents in 1969. The NASA-Lewis "shelf list" of documents (Records Management) indicates that a document entitled "Conference on Rocket Propellant Selection held at Lewis on 19 May 1950" was destroyed 15 Sept. 1969. Undoubtedly a rocket propellant selection meeting was held 19 May 1950, but what is not completely proved is whether this was the particular meeting the author is referring to. The recollections of the author, of Doyle, the Smith document, and the NASA-Lewis record all but settle the matter except for one major problem. The caller and chairman of the meeting, Abe Silverstein—a man proud of his memory—insists that the meeting was held as late as 1952. Since the meeting was one of the early indications of positive NACA-Lewis interest in liquid hydrogen for rockets, the author hopes that documents by other participants at the meeting will surface and resolve the issue. Until then, the 1950 date has been adopted.

13. Interview with William Lawler Doyle, Redondo Beach, CA, 26 Apr. 1974.

14. House Committee on Armed Services, *Authorizing New Construction of the National Advisory Committee for Aeronautics*, 82d Cong., 1st sess., 1951, H. Rept. 452, p. 14.

15. *NACA Conference on Supersonic Missile Propulsion: A Compilation of the Papers Presented*, Lewis Flight Propulsion Laboratory, Cleveland, 13 March 1952.

16. "Rocket Engine Research Program at National Advisory Committee for Aeronautics, Lewis Flight Propulsion Laboratory, Active as of June 1952," NASA History Office.

17. B. E. Gammon, memo for the record, "Resolution Passed at the June 26–27 Meeting of the Special Subcommittee on Rocket Engines," Washington, 30 June 1952, NASA History Office.

18. "Minutes of Meeting, Special Subcommittee on Rocket Engines, Committee on Power Plants for Aircraft," 13–14 Nov. 1952, NASA History Office.

19. Abe Silverstein to NACA Hq., Attn: Mr. A. M. Rothrock, "Discussion with Dr. M. J. Zucrow of Proposed Addition to Rocket Research Facilities," Nov. 1952, NASA History Office.

20. Herbert York, *Race to Oblivion: A Participant's View of the Arms Race* (New York: Simon & Schuster, 1970), p. 83.

21. A copy of the von Neumann committee recommendations is in "Documents in the History of NASA: An Anthology," NASA History Office, June 1975, pp. 24–29.

22. Logbook of Cell 22, Rocket Laboratory, NACA Lewis Flight Propulsion Laboratory, from files of Frank J. Kutina, Lewis Research Center, copy in NASA History Office; Sloop to Chief, Fuels and Combustion Research Div., "Hydrogen-Oxygen Combustion at Hydrogen-Rich Mixtures," 6 Jan. 1955, NASA History Office. The mixture was 46 percent hydrogen by weight, pressure was 29 atm., exhaust velocity was 3033 m/s.

23. NACA research memos E53J20, E55D27, and E55D29.

24. "Resolution Concerning Increased NACA Effort Adopted by the Subcommittee on Rocket Engines," 28 Nov. 1955, NASA History Office.

25. *Forty-First Annual Report of the National Advisory Committee for Aeronautics* (Washington, 1957), p. ix.

26. Harold G. Price, "Hydrogen Work at Rocket Lab," a summary sheet, 23 Apr. 1946, NASA History Office. The Cell 22 Logbook notes a 15-second successful run.

27. Dwight D. Eisenhower, *The White House Years: Waging Peace, 1956–1961* (Garden City, NY: Doubleday, 1965), p. 208.

28. *Forty-Second Annual Report of the National Advisory Committee for Aeronautics* (Washington, 1957), p. ix.

29. Hugh L. Dryden, NACA, to Thon.as E. Myers, Chairman, NACA Subcommittee on Rocket Engines, 3 July 1956, NASA History Office.

30. Thomas E. Myers to Hugh L. Dryden, 17 July 1956, NASA History Office.

31. "Member Suggestions to Agenda Items for Advanced Concepts of Rocket Propulsion," ca. Fall 1956, NASA History Office; "Research Recommended by the NACA Subcommittee on Rocket Engines as Requested by the Department of Defense," appendix II, transmitted to the subcommittee members with a cover letter by Benson E. Gammon, subcommittee secretary, on 15 May 1957, NASA History Office.

32. Richard B. Canright, "Rocket Propellants: Views of an Airframe Manufacturer," 31 July 1957, transmitted to the members of the NACA Subcommittee on Rocket Engines with a cover letter by Benson E. Gammon, 21 Aug. 1957.

33. Ibid.

34. Cell 22 Logbook.

35. Carmon M. Auble, "A Study of Injection Processes for Liquid Oxygen and Gaseous Hydrogen in a 200-Pound-Thrust Rocket Engine," RM E56I25a (NACA, 1957). The report was written 26 Sept. 1956.

36. M. F. Hiedmann and Louis Baker, Jr., "Combustor Performance with Various Hydrogen-Oxygen Injector Methods in a 200-Pound-Thrust Rocket Engine," RM E58E21 (NACA, 1958).

37. J. E. Dalgleish and A. O. Tischler, "Experimental Investigation of a Light-Weight Rocket Chamber," RM E52L19a (NACA, 1953).

38. A. O. Tischler, "Rocket Branch: Engines and Injectors, 100 Hr. Ests. Only," a status data sheet, 2 Mar. 1956, NASA History Office; Sloop to E. J. Wasielewski, "Fabrication of Rocket Engines," 20 Mar. 1956, NASA History Office; Edward F. Baehr, "Construction Method for Lightweight Rocket Thrust Chambers," appendix A of "Experimental Performance of Liquid Hydrogen and Liquid Fluorine in Regeneratively Cooled Rocket Engines," by Howard W. Douglass, Glen Hennings, and Harold G. Price, Jr., TM X-87 (NASA, 1959); also "Making High Thrust Rocket Chambers," *Astronautics* 5 (Mar. 1960): 34.

39. Work diary of Dr. W. T. Olson, NASA Lewis Research Center, Cleveland, vol. 8, for the period Mar. 1957–July 1958.

40. *NACA 1957 Flight Propulsion Conference*. Lewis Flight Propulsion Laboratory, Cleveland, 21–22 Nov. 1957.

41. Sloop, "NACA High Energy Rocket Propellant Research in the Fifties"; Sloop, "Memoir: High Energy Rocket Propellant Research at the NACA/NASA Lewis Research Center, 1945–60"; interview with Howard W. Douglass, NASA Lewis Research Center, 28 May 1975; logbook of Cell 22.

Chapter 6:   NACA Research on Hydrogen for High-Altitude Aircraft

1. *NACA Conference on Fuels: A Compilation of the Papers Presented by NACA Staff Members*, Flight Propulsion Laboratory, Cleveland, 26 May 1948.

2. Walter T. Olson, J. Howard Childs, and Edmund R. Jonash, "Turbojet Combustor Efficiency at High Altitudes," RM E50I07 (NACA, 27 Oct. 1950), p. 2.

3. Walter T. Olson and Louis C. Gibbons, "Status of Combustion Research on High-Energy Fuels for Ram Jets," RM E51D23 (NACA, 3 Oct. 1951), p. 33.

4. Benson E. Gammon, "Preliminary Evaluation of the Air and Fuel Specific-Impulse Characteristics of Several Potential Ram-Jet Fuels; IV: Hydrogen, Methylnapthalene, and Carbon," RM E51F05 (NACA, 1951).

5. Hugh M. Henneberry, "Effect of Fuel Density and Heating Value on Ram-Jet Airplane Range," RM E51L21 (NACA, 25 Feb. 1952), p. 1.
6. House Committee on Science and Astronautics, *Boron High Energy Fuels*, 86th Cong., 1st sess., 1959, H. Rept. 1191, pp. 2, 5–6.
7. "Minutes of Meeting of Fuels and Propulsion Panel of the USAF Scientific Advisory Board," 16–18 Apr. 1952; and USAF response by Maj. Gen. James E. Briggs, 12 Aug. 1952. Author's notes, declassified by the USAF, are in the History Office.
8. 4–5 Feb. at WADC, 21–22 Mar. at WADC, and 24 Mar. 1954 at Langley AFB. Author's declassified notes in NASA History Office.
9. Addison M. Rothrock, "Turbojet Propulsion-System Research and the Resulting Effects on Airplane Performance," RM 54H23 (NACA, 17 Mar. 1955); "Minutes of Meeting of Fuels and Propulsion Panel of the USAF Scientific Advisory Board," 29 Sept. 1954. Author's declassified notes in NASA History Office.
10. Edmund R. Jonash, Arthur L. Smith, and Vincent F. Hlavin, "Low-Pressure Performance of a Tubular Combustor with Gaseous Hydrogen," RM E54L30a (NACA, 9 May 1955), p. 1.
11. Interview with Abe Silverstein, Cleveland, 29 May 1974.
12. "Minutes of Meeting of Fuels and Propulsion Panel of the USAF Scientific Advisory Board," Mar. 1955, declassified notes in NASA History Office.
13. Abe Silverstein and Eldon W. Hall, "Liquid Hydrogen as a Jet Fuel for High-Altitude Aircraft," RM E55C28a (NACA, 15 Apr. 1955), pp. 1–2.
14. Ibid., pp. 18–19.
15. T. W. Reynolds, "Aircraft-Fuel-Tank Design for Liquid Hydrogen," RM E55F22 (NACA, 9 Aug. 1955).
16. Jerrold D. Wear and Arthur L. Smith, "Performance of a Single Fuel-Vaporizing Chamber with Six Injectors Adapted for Gaseous Hydrogen," RM E55I14 (NACA, 1955).
17. Interviews with Sol Weiss, William Rowe, Tom Gelder, Thaine Reynolds, and Loren Acker, NASA Lewis Research Center, Cleveland, 29 May; Paul Ordin, Lewis, 30 May; and Abe Silverstein, Cleveland, 29 May 1974.
18. David B. Fenn, Willis M. Braithwaite, and Paul M. Ordin, "Design and Performance of Flight-Type Liquid Hydrogen Heat Exchanger," RM E57F14 (NACA, 1957).
19. Edward Otto, Kirby W. Hiller, and Phil S. Ross, "Design and Performance of Fuel Control for Aircraft Hydrogen Fuel System," RM E57F19 (NACA, 1957).
20. Interview with William V. Gough, Jr., Alexandria, VA, 6 May 1975.
21. Glen Hennings in a group interview that included John Gibb, Harold Christenson, and David Fenn, NASA Plumbrook Station, Sandusky, OH, 28 May 1974.
22. David M. Straight and Arthur L. Smith, "Brief Studies of Turbojet Combustor and Fuel System Operation with Hydrogen Fuel at − 400°F," RM E56K27a (NACA, 1957).
23. Joseph N. Sivo and David B. Fenn, "Performance of a Short Combustor at High Altitudes Using Hydrogen Fuels," RM E56D24 (NACA, 1956).
24. W. A. Fleming, H. R. Kaufman, J. L. Harp, Jr., and L. J. Chelko, "Turbojet Performance and Operation at High Altitudes with Hydrogen and JP-4 Fuels," RM E56E14 (NACA, 1956).
25. Harold R. Kaufman, "High-Altitude Performance Investigation of J65-B-3 Turbojet Engine with Both JP-4 and Gaseous-Hydrogen Fuels," RM E57A11 (NACA, 1957); Willis M. Braithwaite, David B. Fenn, and Joseph S. Algranti, "Altitude-Chamber Evaluation of an Aircraft Liquid-Hydrogen Fuel System Used with a Turbojet Engine," RM E57F13a (NACA, 1957).
26. Gough interview, 6 May 1975; Donald R. Mulholland, Loren W. Acker, Harold H. Christenson, and William V. Gough, Jr., "Flight Investigations of a Liquid-Hydrogen Fuel System," with appendix, "Flight Instrumentation for Liquid-Hydrogen Fuel Systems," by Scott H. Simpkinson and Jacob C. Moser, RM E57F19a (NACA, 1957).

27. As deduced from their personal flight logs by Gough and Algranti, 21–22 Feb. 1978.

28. Lewis Laboratory Staff, "Hydrogen for Turbojet and Ramjet Powered Flight," Lewis Flight Propulsion Laboratory, Cleveland, 26 Apr. 1957.

29. Arnold E. Bierman and Robert C. Kohl, "Preliminary Study of a Piston Pump for Cryogenic Fluids," memo 3-6-59E (NASA, 1959). A centrifugal pump was also investigated: George W. Lewis, Edward R. Tysl, and Melvin J. Hartmann, "Design and Experimental Performance of a Small Centrifugal Pump for Liquid Hydrogen," TM X-388 (NASA, ca. 1959).

30. David B. Fenn, Loren W. Acker, and Joseph S. Algranti, "Flight Operation of a Pump-Fed Liquid Hydrogen Fuel System," TMX-242 (NASA, 1960).

31. NACA 1957 Flight Propulsion Conference, Lewis Flight Propulsion Laboratory, Cleveland, 21–22 Nov. 1957. The remainder of the chapter is based on this report.

Chapter 7:   New Initiatives in High-Altitude Aircraft

1. Col. John D. Seaberg (USAF ret.) to author, 28 June 1976.

2. U.S. Air Force, Wright Field, "Design Study Requirements," 27 Mar. 1953, declassified 6 July 1976.

3. Fairchild, Hagerstown, MD, AF 33(616)-2182, 1 July–31 Dec. 1953; Bell Aircraft, Niagara Falls, NY, AF 33(616)-2160, 1 Jul–31 Dec. 1953; G. L. Martin, Baltimore, AF 33 (600)-25825, 1 Jul.–31 Dec. 1953; information supplied by John D. Seaberg, 20 July 1976.

4. Weekly reports, New Developments Off., Bombardment Aircraft Br., Wright Air Development Center, 2, 23 Feb. 1954 (declassified by USAF 7 July 1976).

5. Weekly report, New Developments Off., 23 Mar. 1954.

6. Travel confirmation sheet, John D. Seaberg, 12 Nov. 1954; Lt. Gen. T. S. Power to Maj. John D. Seaberg, 14 June 1955, with attachments; Seaberg to author, 28 June 1976.

7. Seaberg to author, 28 June 1976.

8. Dwight D. Eisenhower, The White House Years: Waging Peace, 1956–1961 (Garden City, NY: Doubleday & Co., 1965), pp. 544–48. See also interview of Richard Bissell as quoted by George C. Wilson in the Washington Post, 6 Jan. 1975.

9. Interview with Lt. Gen. Donald L. Putt (USAF ret.), Atherton, CA, 30 Apr. 1974. Putt was deputy chief of staff for development.

10. Clarence L. Johnson to Monte D. Wright, NASA History Office, 12 May 1976.

11. Eisenhower, White House Years, 1956–1961, p. 548. Eisenhower commented that "NASA was purposely kept in the dark as to the unit's intelligence activities." NACA Press Release 7 May 1956, "NACA Announces Start of New Research Program."

12. Interviews with Randolph S. Rae, Silver Spring, MD, 15–16 Apr. 1974.

13. Interview with Homer J. Wood, Sherman Oaks, CA, 3 May 1974.

14. Lt. Col. L. F. Ayers, statement on the Rex engine project, 13 Aug. 1956, tab 27 of "Addendum to Case History of Garrett Contract (07-58669)," a file in the Archives of the Command Historian, Hq. Air Force Systems Command, Andrews AFB, MD; cited hereafter as USAF Garrett File. The author examined this classified file and excerpted a chronology which has been declassified by the USAF.

15. "REX-1, A New Aircraft System" by R. S. Rae, Summers Gyroscope Co., Santa Monica, CA, Feb. 1954. Copies are in the files of R. P. Carmichael, Dir. of Advanced Systems Design, WADC, Wright Patterson AFB, OH, and of H. J. Wood, Consultant, Sherman Oaks, CA.

16. "The REX-1 Propulsion System," Summers Gyroscope Co., Santa Monica, CA, 1 Mar. 1954; copy no. 48 is in the files of Mr. Rae.

17. To WSCP, Attn: Mr. E. B. Bell, from WCLPP, comment no. 2 by W. Worth, R. E. Roy, R. P. Carmichael on "Project REX-1 Proposal from Summers Gyroscope Co.," 21 May 1954. Draft of incomplete document in the files of R. P. Carmichael, Wright Patterson AFB, OH.

18. Interviews with Rae, 15–16 Apr. 1974.
19. J. M. Wickham, "A Hydrogen Powered Strategic Bomber," Boeing Airplane Co., Apr. 1954, as quoted by Worth, Roy, and Carmichael, 21 May 1954.
20. A. D. Baxter, "Investigation of Walter-Werke, Kiel," tech. note no. Aero 1667, Royal Aircraft Establishment, Farnborough, Aug. 1945; A. M. Nelson, "Cycle Study of Rocket Driven Turbine Compressor Engine," memo. report TSEPP-506-151, 3 Dec. 1946, in files of Dir. of Advanced Systems Design, WADC, Wright-Patterson AFB, OH. The Sept. 1953 proposal by W. C. House to the Air Force is mentioned in a WADC staff reply to a letter by the Garrett Corp. to Sec. of the Air Force Sharp on 3 July 1956 (tab 25 of USAF Garrett file). The information on House is from a letter to the author from House, 22 May 1974, and an attached, undated brochure, "Air-Turborocket ATR Supersonic Engine," Aerojet General Corp. House believes that the brochure was prepared "around 1964."
21. Tab 3 of USAF Garrett file.
22. Tab 4a, tab 25, and staff reply to Garrett's letter to Sharp, 3 July 1956, USAF Garrett file.
23. Tab 6, USAF Garrett file.
24. Statement by Lt. Col. L. F. Ayers, 13 Aug. 1956, tab 27, USAF Garrett file.
25. "Minutes of Meeting of Fuels and Propulsion Panel of the USAF Scientific Advisory Board, USAF, Pentagon"; information on hydrogen excerpted by author 17 June and declassified by USAF 8 Aug. 1974.
26. Rae interviews, 15–16 Apr. 1974.
27. Ibid.; Wood interview, 3 May 1974.
28. According to Philip J. Richie, Dir. of Procurement, Wright Air Development Center, Wright-Patterson AFB, OH, in a memo for the record 9 May 1956, Richie noted that Clifford Garrett, in discussions with Air Force officials, repeatedly stated that he did not want to buy into the Rex interest unless he could stay all the way. According to Richie, Garrett was assured "on all levels" that this would be the case: tab 21, USAF Garrett file.
29. Philip J. Richie, "Rex Program Procurement," 27 Dec. 1955, reference 13 in USAF Garrett file.
30. Ibid.
31. "Minutes of Meeting of Fuels and Propulsion Panel of the USAF Scientific Advisory Board," Mar. 1955, reproduced Apr. 1956. Excerpts related to hydrogen made by the author and declassified by the USAF 8 Aug. 1974.
32. Interview with Richard E. Horner, Washington, 13 Mar. 1974. Horner was Asst. Sec. of the Air Force for R&D.
33. Ibid.
34. Interview with Col. Norman C. Appold (USAF ret.), Marietta, GA, 4 Jan. 1974.
35. Richie, "Rex Procurement," 27 Dec. 1955.
36. Project record book of Frank Patella, Power Plant Lab., Wright Air Development Center, extracted 21 Aug. 1956, tab 26, USAF Garrett file.
37. Staff reply to Garrett letter to Sec. Sharp, 3 July 1956, tab 25, USAF Garrett file.
38. Tab 26, USAF Garrett file.
39. Ibid.
40. Tab 10, USAF Garrett file.
41. Tab 26, USAF Garrett file.
42. Lt. Col. Ayers's statement of Aug. 1956, tab 27, USAF Garrett file.
43. Tab 26, USAF Garrett file.
44. Memo by E. C. Phillips, 16 Aug. 1955, tab 12, USAF Garrett file.
45. Tabs 26, 14, 15, 27, USAF Garrett file; Richie, "Rex Procurement," 27 Dec. 1955.
46. According to Aaron Shaffer in a group interview with John L. Mason and Harvar Starck of the Garrett Corp., Los Angeles, and George E. Hlavka, Jet Propulsion Lab., Pasadena, at the Garrett Corp., 25 Apr. 1974.

47. George Hlavka, Garrett group interview, 25 Apr. 1974.

48. The Garrett problem statement of 7 Nov. 1955 is quoted in Lockheed report 11195, which is appendix F of Garrett report RD-14-R, 15 Feb. 1956.

49. Lockheed report 11195, appendix F, Garrett report RD-14-R, 15 Feb. 1956.

50. Garrett's selection of the Rex III was reported to the Air Force on 15 Feb. 1956 (report RD-14-R). However, Frank Patella, the Power Plant Laboratory manager for the Garrett contract, sensed on 21 Nov. 1955 that Garrett had chosen the Rex III. The Garrett data to Lockheed Aircraft on 7 Nov. 1955 also indicated a selection of Rex III. J. L. Bartlett, Jr., L. M. Goldsmith, and Aaron Shaffer, "Rex Engine Cycle Study and Selection," contract AF 33 (616)3143 task 30303, report RD-14-R, Rex Div., Garrett Corp., 15 Feb. 1956; Patella log, USAF Garrett file; Lockheed report 11195, included as appendix F of Garrett report RD-14-R.

51. Aaron Shaffer, Garrett group interview, 25 Apr. 1974.

52. Tab 26, USAF Garrett file.

53. Ibid.

54. Tab 20, USAF Garrett file.

55. Tab 19, USAF Garrett file.

56. Tab 21, USAF Garrett file.

57. Tabs 22 and 26, USAF Garrett file.

58. Tabs 27, 25, and 24, USAF Garrett file.

59. J. L. Bartlett, Jr., I. M. Goldsmith, and A. Shaffer, "Rex Engine Cycle Study and Selection," contract AF 33 (616)3143, task 30303, report RD-14-R, Rex Div., Garrett Corp., Los Angeles, 15 Feb. 1956.

60. G. Hlavka, "Rex Model 301-1 Engine Preliminary Design Report," contract AF 33 (616)-3143, task 30303, report RD-21-D, Rex Div., Garrett Corp., Los Angeles, 11 May 1956.

61. G. E. Hlavka and H. G. Starck, "Phase III Final Report," contract AF 33 (616)3143, report RD-97-R, Rex Div., Garrett Corp., Los Angeles, 3 Mar. 1958.

62. Tab 27, USAF Garrett file.

63. "Minutes of Meeting of Panel on Fuels and Propulsion, USAF Scientific Advisory Board," 21 Oct. 1955.

64. R. P. Carmichael, "Cycle Performance of Some Selected Engine Configurations Using Liquid Hydrogen Fuel," technical note 55-687, Power Plant Laboratory, Wright Air Development Center, Wright-Patterson AFB, 8 Nov. 1955; interview with R. P. Carmichael, Dir. of Advance Systems Design, AFSC, Wright-Patterson AFB, 6 June 1974.

Chapter 8:   Suntan

1. The author first learned about Suntan in an interview with Col. John D. Seaberg (USAF ret.) at NASA Hq. on 22 Aug. 1973. Seaberg, a principal in the project, had learned of my interest in hydrogen from Col. Norman C. Appold (USAF ret.), who had managed the project. Primary information about Suntan came from subsequent interviews with other principals, including Appold, along with some contemporary documentation. The paucity of the latter was the direct result of deliberate Air Force policy which not only employed a very high classification category, but resorted to changes in project numbers and procurements through multiple channels to disguise the activity. Although the project is now declassified, the earlier policy makes the tracing of documents difficult. Particularly elusive was a series of interviews conducted by an Air Force historian, Dr. Ernest Schweibert, in 1958, but these were located in 1976.

2. The exact date of Johnson's visit or visits and whom he saw have not been established. In an interview by Schweibert on 17 Dec. 1958, Appold placed the time of the visit between 1 Jan. 1956 and the 18th, when he was summoned to the Pentagon for a meeting on Johnson's proposal. Johnson most likely saw Richard E. Horner, who succeeded Trevor Gardner in Feb. 1956 as Sec. of the Air Force for R&D (Gardner had

left the previous November) and Lt. Gen. Donald L. Putt, Chief of Staff for Development. Both were to be closely involved in Suntan.

3. Ben R. Rich, "Lockheed CL-400 Liquid Hydrogen-Fueled Mach 2.5 Reconnaissance Vehicle," 15 May 1973, presented at a working symposium on liquid hydrogen-fueled aircraft, NASA Langley Research Center, 15–16 May 1973.

4. "A Statement by Col. Norman C. Appold of the Sun Tan Project Office—Dec. 1958," an interview by Schweibert 17 Dec. 1958, declassified 24 May 1976.

5. Ibid.

6. "Statements by Mr. Robert Miedel of Directorate of Procurement Hq ARDC, and Mr. Wm. E. Miller and Lt. Richard Doll of the Sun Tan Procurement Office—December 1958," interviews by Schweibert made available to the author 24 May 1976.

7. Interview with Col. Norman C. Appold (USAF ret.), Marietta, GA, 4 Jan. 1974; interviews with Col. John D. Seaberg (USAF ret.), 22 Aug. and 23 Nov. 1973; interview with Col. Alfred J. Gardner (USAF ret.), Washington, 19 Sept. 1973; interview with Col. Jay R. Brill, USAF-AFSC, Andrews AFB, MD, 16 Jan. 1974.

8. Interview with Col. Ralph J. Nunziato (USAF ret.), Redwood City, CA, 30 Apr. 1974.

9. Schweibert interview with Miedel, Miller, and Doll, Dec. 1958.

10. Appold interview, 4 Jan. 1974.

11. Interview with C. L. Johnson and Ben R. Rich, Lockheed-California Co., Burbank, 2 May 1974, and with C. L. Johnson, Washington, 14 Feb. 1974.

12. C. L. Johnson, Lockheed Aircraft Co., to Ernest H. Lee, Jr., Vice-Pres. for Engineering, J. H. Pomeroy Co., Los Angeles, 26 Mar. 1956, in files of Lockheed-California Co., Burbank.

13. J. H. Pomeroy Co. report on hydrogen liquefaction, prepared under Lockheed Purchase Order 1737, 1 Oct. 1956, files of Lockheed-California Co., Burbank.

14. According to Ben Rich in interview, 2 May 1974 (Scott is deceased).

15. Interview with Wesley A. Kuhrt, United Aircraft Corp., East Hartford, CT, 20 Mar. 1974. The childhood experience with hydrogen and other biographical information is also in Robert Zaiman, "The Flexible Mr. Kuhrt," *Beehive*, Jan. 1960. Kuhrt became director of research in 1964; in 1975 he was Vice-Pres. for Technology, United Technology Corp. He holds a number of patents, among them several for engines using hydrogen as a fuel.

16. Interview with C. Branson Smith, National Science Foundation, Washington, 25 Mar. 1974; interview with William Sens and John Wells, Pratt & Whitney Div., United Aircraft Corp., E. Hartford, CT, 10 Apr. 1974. Pratt & Whitney received Silverstein's NACA report on hydrogen aircraft in April and Carmichael's Wright Field analysis of hydrogen engine cycles in November 1955.

17. Perry W. Pratt to W. A. Parkins, interoffice correspondence, Pratt & Whitney Aircraft, E. Hartford, CT, 17 Feb. 1956.

18. Handwritten notes of Pratt, 20 Feb. 1956; interview with Sens and Wells, 10 Apr. 1974.

19. Interview with Sens and Wells, 10 Apr. 1974.

20. William Sens, draft memo, 24 Feb. 1974, files of Pratt & Whitney, E. Hartford, CT.

21. Patents 3 000 176, 3 241 311, 3 237 400.

22. Interview with Kuhrt, 20 Mar. 1974.

23. Interview with Richard J. Coar and Richard C. Mulready, Pratt & Whitney Aircraft Div., E. Hartford, 20 Mar. 1974. Coar graduated summa cum laude from Tufts College in 1942. Contract AF 18 (600)1616, 16 Apr. 1956.

24. Ibid. According to James Pierce, a former engineer for Herrick L. Johnston and now president of Cryovac, Columbus, OH, the liquefier sold to Pratt & Whitney in 1956 was built in 1952 or 1953 for training purposes for the Air Force–AEC hydrogen bomb program: interview, 4 June 1974.

25. Richard Mulready, "Liquid Hydrogen Engines," chap. 5 of *Technology and Uses of Liquid Hydrogen*, ed. R. B. Scott, W. H. Denton, and C. M. Nicholls (New York: Macmillan, 1965), summarized hydrogen engine developments at Pratt & Whitney from 1956 to 1963.

26. William Sens to author, 18 Apr. 1974.

27. Interview with R. J. Coar and R. C. Mulready, 20 Mar. 1974. Coar did not know whether the early liquid-hydrogen pump obtained by Pratt & Whitney was made by Aerojet.

28. A series of Pratt & Whitney charts entitled "SF-1: Model 304 Liquid Hydrogen Engine," Apr. 1958, Pratt & Whitney files.

29. Schweibert interview of Miedel, Miller, and Doll, Dec. 1958.

30. Coar and Mulready interview, 20 Mar. 1974.

31. Pages 10–12 of an unidentified Pratt & Whitney Aircraft report, shown the author by William Sens. The data identified the engine serial number, date fabrication started, date of final assembly, date of first run, date removed from the test stand, the amount of nitrogen, hydrogen, and liquid hydrogen consumed, and brief comments on the outcome of the tests.

32. Interview with Blackwell C. Dunnam, Air Force Aero Propulsion Laboratory, AFSC, Wright-Patterson AFB, OH, 6 June 1974.

33. Blackwell C. Dunnam, "Air Force Experience in the Use of Liquid Hydrogen as an Aircraft Fuel," a paper prepared for a technical meeting on hydrogen, copy sent to author 21 May 1974; interview by Schweibert of Col. N. C. Appold, Lt. Col. J. D. Seaberg, Maj. A. J. Gardner, and Capt. J. R. Brill, Feb. 1959, released with portions excised by USAF Sept. 1976.

34. Ibid.

35. Dunnam, "Air Force Experience with Hydrogen," 1974.

36. Interview by Schweibert of Appold, Seaberg, Gardner, and Brill, 9 Feb. 1959, released with portions excised by USAF, 16 Sept. 1976.

37. Ibid.

38. Ibid.; interview of Miedel, Miller, and Doll, Dec. 1958; interview with Johnson and Rich, 2 May 1974.

39. Interview with Col. A. J. Gardner (USAF ret.), 19 Sept. 1973; Schweibert interview with Appold et al., 9 Feb. 1959.

40. Schweibert interview with Appold et al., 9 Feb. 1959.

41. Ben R. Rich, "Lockheed CL-400," presented at NASA-Langley, 15–16 May 1973; interview with C. L. Johnson, Washington, 14 Feb. 1974; Johnson to Monte D. Wright, NASA History Office, 12 May 1976.

42. Johnson to Wright, 12 May 1976.

43. Rich, "Lockheed CL-400," NASA-Langley, 15–16 May 1973.

44. Schweibert interview of Appold et al., 9 Feb. 1959.

45. Johnson to Wright, 12 May 1976; Schweibert interview of Appold et al., 9 Feb. 1959; interview with Johnson and Rich, 2 May 1974.

46. Schweibert interview of Appold et al., 9 Feb. 1959.

47. Ibid.

48. Ibid.

49. Rich, "Lockheed CL-400," NASA-Langley, 15–16 May 1973; interview with C. L. Johnson, 14 Feb. 1974; Johnson to Wright, 12 May 1976.

50. Interview with Appold, 4 Jan. 1974.

51. Interview with Nunziato, 30 Apr. 1974.

52. Nunziato is emphatic that the cost was on the order of $100 million and Putt, who admits to not remembering figures very well, concurs (interviews 30 Apr. 1974). Seaberg and William Miller (procurement) also believe the $100 million figure (Seaberg to author, 1 Sept. 1976; telephone interview with Miller, 1 Sept. 1976).

53. Interview with Appold, 4 Jan. 1974; interview with Richard E. Horner, Washington, 13 Mar. 1974; Appold to Monte D. Wright, NASA History Office, 14 Apr. 1976.

54. C. L. Johnson to author, 17 Jan. 1974.

55. J. D. Seaberg to Monte D. Wright, NASA History Office, 29 Mar. 1976.

Chapter 9: The Early U.S. Space Program

1. Constance McLaughlin Green and Milton Lomask. *Vanguard, A History* (NASA SP-4202, 1970), pp. 34–36.

2. In addition to Green and Lomask, *Vanguard,* other sources for Vanguard and its precursor, Viking. are: The Martin Co., *The Vanguard Satellite Launching Vehicle: An Engineering Summary,* report 11022, Apr. 1960; Kurt K. Stehling, *Project Vanguard* (Garden City, NY: Doubleday, 1961); Milton Rosen, *The Viking Rocket Story* (New York: Harper, 1955); John P. Hagen, "The Viking and Vanguard," in Eugene M. Emme, ed., *The History of Rocket Technology* (Detroit: Wayne State Univ. Press, 1964), pp. 122–41; Alfred Rosenthal, *Venture into Space: Early Years of Goddard Space Flight Center* (NASA SP-4301, 1968), pp. 16–25.

3. Green and Lomask, *Vanguard,* pp. 165–82, 196–98, 206–09.

4. There are numerous accounts of von Braun's role in developing the A-4 (V-2) missile and the U.S. action to bring German rocket experts to the U.S. For example, see: Walter Dornberger, *V-2* (New York: Viking Press, 1958); James McGovern. *Crossbow and Overcast* (New York: William Morrow, 1964); Walter Dornberger, "The German V-2," in *History of Rocket Technology,* pp. 29–45; and Clarence Lasby, *Project Paperclip: German Scientists and the Cold War* (New York: Atheneum, 1971).

5. Wernher von Braun, "The Redstone, Jupiter, and Juno," in *History of Rocket Technology,* pp. 108–10.

6. Ibid., p. 111; John B. Medaris and Arthur Gordon, *Countdown for Decision* (New York: G. P. Putnam & Sons, 1960), p. 147.

7. Robert L. Perry, "The Atlas, Thor, Titan, and Minuteman," in *History of Rocket Technology,* pp. 142–61; interview with Karl J. Bossart, La Jolla, CA, 27 Apr. 1974.

8. Report of the Air Force Strategic Missiles Committee, Dr. John von Neumann. Chairman, 10 Feb. 1954, NASA History Office, reprinted in "Documents in the History of NASA: An Anthology" (NASA History Office, Washington, D.C., 1975), pp. 24–29; Dwight D. Eisenhower, *The White House Years: Waging Peace, 1956–1961* (Garden City, NY: Doubleday, 1965), p. 208.

9. Eisenhower, *White House Years: Waging Peace,* p. 257.

10. Eugene M. Emme, *Aeronautics and Astronautics: An American Chronology of Science and Technology in the Exploration of Space, 1915–1960* (NASA, 1961), p. 96.

11. Telephone interview with D. A. Young, Waldport, OR, 8 Aug. 1973.

12. Interview with Richard S. Cesaro. TCOM Corp., Rockville, MD, 12 Mar. 1974.

13. Memo by Herbert F. York, Chief Scientist, ARPA, 2 June 1958, copy in the retired files of A. O. Tischler, NASA, Federal Records Center, Suitland, MD, accession 20-A-3765, record group 255. The 9 panels and members were listed also in a DOD press release 1 June 1958, ARPA 69A3623(2).

14. Interview with Richard B. Canright, Pennsylvania Health Research Institute, Camp Hill, PA, 7 Mar. 1974; interview with Cesaro, 12 Mar. 1974.

15. Statement by James H. Doolittle, Chairman of the NACA, 26 Feb. 1958, as quoted by Emme in *Aeronautics and Astronautics: A Chronology,* p. 96.

16. Interview with Bruce T. Lundin, NASA-Lewis, Cleveland, 30 May 1974.

17. Interviews with Abe Silverstein, Cleveland, 29 May, and Bruce T. Lundin, 30 May 1974.

18. Bruce T. Lundin, "Some Remarks on a Future Policy and Course of Action for the NACA," 9 Dec. 1957, NASA History Office.

19. Robert Rosholt, *An Administrative History of NASA, 1958–1963* (NASA SP-4101, 1966); interview with Walter T. Olson. NASA-Lewis, 11 July 1974.

20. From retired files of A. O. Tischler, NASA, Federal Records Center, Suitland, MD, accession 70-A-3765, record group 255.

Chapter 10: Early High-Energy Upper Stages

1. Interview with C. Branson Smith, National Science Foundation, Washington, 14 Mar. 1974. Mr. Smith also loaned the author his work logs, notebooks, and other pertinent papers of the period Aug. 1953 to Aug. 1957.

2. C. Branson Smith, "Rocket Motors," TDM 1331, rev. 1, 31 July 1956, an internal document of Pratt & Whitney Aircraft Div. of United Aircraft.

3. C. B. Smith to P. W. Pratt, "Review of Our Position in the Rocket Propulsion Field," 10 July 1957, an internal Pratt & Whitney document.

4. Minutes of 14 Nov. 1956 Meeting of the Fuels and Propulsion Panel of the USAF Scientific Advisory Board; excerpts declassified for the author by USAF.

5. Air Force response to Fuels and Propulsion Panel of the USAF Scientific Advisory Board regarding recommendations made at its 14 Nov. 1956 meeting, 6 Dec. 1957.

6. Perry W. Pratt to Commander, ARDC, 4 Mar. 1958. The author has not seen this source, but it is one of five references cited in a later, undated document, "Proposal for the Development of a Hydrogen-Oxygen Rocket Engine," PWA (preliminary). The references and attachments are dated from 4 Mar. 1958 to 18 July 1958; the undated proposal was probably written close to the latter date.

7. Interviews with: Col. Norman C. Appold (USAF ret.), Lockheed Georgia Co., Marietta, GA, 4 Jan. 1974; Col. John D. Seaberg (USAF ret.), Washington, 22 Aug. and 23 Nov. 1973; Col. Alfred J. Gardner (USAF ret.), Washington, 19 Sept. 1973; and Col. Jay R. Brill, Air Research and Development Command Hq., Andrews AFB, MD, 16 Jan. 1974. These men were the Suntan management team.

8. Interviews with Seaberg, 22 Aug. and 23 Nov. 1973, and Brill, 16 Jan. 1974. Both men recalled the letter and believed it to be in USAF files. Seaberg, now a civilian working for the Air Force, and Brill, on active duty and a Brigadier General, instituted separate searches within the USAF, including the Archives of the History Office, Hq. ARDC, but the key letter was not located.

9. Interview with Seaberg, 22 Aug. 1973; interview with Richard Horner, Washington, 13 Mar. 1974 (Horner did not recall the specific meeting but conceded that his reported action was reasonable and possible); interview by Schweibert of Col. N. C. Appold, Lt. Col. J. D. Seaberg, Maj. A. J. Gardner, and Capt. J. R. Brill, 9 Feb. 1959, released with portion excised by Air Force, 16 Sept. 1976; R. B. Canright to Monte D. Wright, NASA History Office, 1 Feb. 1976.

10. "Proposals for the Development of a Hydrogen-Oxygen Rocket Engine," PWA (preliminary) with specification no. 2208 attached and dated 18 July 1958.

11. Interview with Krafft Ehricke, Rockwell International, El Segundo, CA, 26 Apr. 1974.

12. Ibid.; Wernher von Braun to Monte D. Wright, NASA History Office, 29 Dec. 1975.

13. A. G. Negro, "Small High Energy Rocket Engines," PDM 57-45, 27 May 1957, and "Small High Energy Rocket for Satellite Vehicle," PDM 57-132, 31 Oct. 1957, both internal documents of the Rocketdyne Div. of Rockwell International.

14. Statement of Krafft A. Ehricke, Advanced Studies, General Dynamics/Astronautics, on 18 May 1962 before Subcommittee on Space Sciences of the Committee on Science and Astronautics, House of Representatives, 87th Congress, 2d sess., 15–18 May 1962.

15. "Minutes of Meeting (Amended) Informal Technical Advisory Committee for Propulsion, NASA-ARPA-ARDC, 7 August 1958," by A. O. Tischler, 20 Aug. 1958, Tischler's retired files.

16. "Minutes of Second Meeting (Amended) Informal Technical Advisory Committee for Propulsion: NASA-ARDC-ARPA-AOMC. 14 August 1958," by A. O. Tischler, 4 Sept. 1958, Tischler's retired files.

17. "Minutes of Third Meeting (Amended) Informal Technical Advisory Committee for Propulsion: NASA-ARPA-USAF-AOMC, 28 Aug. 1958," by A. O. Tischler, 18 Sept. 1958, Tischler's retired files.

18. In interviews with Silverstein, Tischler, Cesaro, and Canright, the author attempted to verify whether the NASA meeting was the catalyst for action. Memories varied slightly, but there was general agreement between Silverstein, Tischler, and Canright. Cesaro neither confirmed nor denied this account, but said that the Centaur program started before the NACA-NASA got involved. While this is true, it does not rule out the competitive nature of the August 1958 meetings and actions. Interviews with Silverstein, Washington, 22 Nov. 1971, and Cleveland, 29 May 1974; Tischler, NASA, Washington, 25 June 1974; Canright, 7 Mar. 1974; and Cesaro, 12 Mar. 1974. In a letter to Monte D. Wright, NASA History Office, 1 Feb. 1976, Canright backed off somewhat, stating that ARPA orders generally took about a week to release after the technical staff defined the proper scope of work. He did not remember anything different about Order 19-59. J. D. Seaberg, in a letter to Wright, 29 Mar. 1976, pointed out the earlier negotiations between the Air Force and ARPA with Pratt & Whitney and General Dynamics–Astronautics. Seaberg felt that the NASA meeting was a coincidence. None of these later comments negate the view that the NACA-NASA sponsored meeting spurred ARPA to faster action.

19. Lt. Col. John D. Seaberg, "USAF Participation in Project CENTAUR," 1962, copy to author by Seaberg.

20. A. O. Tischler, "Minutes of Fourth Meeting (Revised) Informal Technical Advisory Committee for Propulsion: NASA-USAF-AOMC, 11 Sept. 1958," 29 Sept. 1958, Tischler's retired files.

21. Tischler, "Minutes of Fifth Meeting (Revised) Technical Advisory Committee for Propulsion: NASA-ARPA-USAF-AOMC, 25 Sept. 1958," 15 Oct. 1958, Tischler's retired files.

22. Tischler's retired files; interview with Alfred M. Nelson, Bethesda, MD, 8 Mar. 1974. Nelson prepared the document on propellant selection data, 8 Oct. 1958; it bears WADC identification 58WCLP-8970: "Some Thoughts on the Next Decade of Space Exploration," 21 Nov. 1958. Reprinted as document II-4 of "Documents in the History of NASA: An Anthology," NASA History Office, 1975. NASA Administrator Glennan sent the document to Congress and to the President "on behalf of Drs. Hugh L. Dryden, Abe Silverstein, John Hagen, Homer Newell and myself."

23. Logbook of Cell 22, Rocket Laboratory, NACA Lewis Flight Propulsion Laboratory, Cleveland (files of Frank J. Kutina, NASA–Lewis); H. W. Douglass, G. Hennings, and H. G. Price, Jr., "Experimental Performance of Liquid Hydrogen and Liquid Fluorine in Regeneratively Cooled Rocket Engines," TM X-87 (NASA, 1959); William L. Jones, Carl A. Aukerman, and John W. Gibb, "Experimental Performance of a Hydrogen-Fluorine Rocket Engine at Several Chamber Pressures and Exhaust-Nozzle Expansion Area Ratios," TM X-387 (NASA, 1960). For more on the simulated altitude testing technique, see Anthony Fortini, "Performance Investigation of a Nonpumping Rocket Ejector System for Altitude Simulation," TN D-257 (NASA, 1959); Joseph N. Sivo and Daniel J. Peters, "Comparison of Rocket Performance Using Exhaust Diffuser and Conventional Techniques for Altitude Simulation," TM X-100 (NASA, 1959).

24. Logbook of Rocket Test Facility (South 40), files of Frank J. Kutina, Lewis.

25. Edward A. Rothenberg, Franklin J. Kutina, Jr., and George R. Kinney, "Experimental Performance of Gaseous Hydrogen and Liquid Oxygen in Uncooled 20 000-Pound Thrust Rocket Engines," memo 4-8-59E (NASA, 1959).

26. William A. Tomazic, Edward R. Bartoo, and R. James Rollbuhler, "Experiments with Hydrogen and Oxygen in Regenerative Engines at Chamber Pressure from 100 to 300 Pounds per Square Inch Absolute," TM X-253 (NASA, 1960).

27. Sloop to Assoc. Dir., Lewis, "Accomplishments of Rocket Systems Branch in 1959," 8 Mar. 1960; work diary of C. Branson Smith, PWA, for a visit ca. 1957; interview with C. H. King, Jr., NASA, Washington, 18 Mar. 1974.

28. Seaberg, "USAF Participation in Project CENTAUR"; T. Keith Glennan, NASA Ad-
     ministrator, to Donald A. Quarles, Dep. Sec. of Defense, 6 May 1956, files of Abraham
     Hyatt, NASA History Office.
29. Abraham Hyatt, Asst. Dir. for Propulsion, to Glennan, "Centaur Project Manage-
     ment Arrangements," 21 May 1959; R. E. Horner to Herbert F. York, 10 June 1959;
     J. B. Macauley to R. E. Horner, 19 June 1959; Horner to York, 23 June 1959, all in
     files of Abraham Hyatt, NASA History Office.

Chapter 11:   Large Engines and Vehicles, 1958

1. Eugene M. Emme, *Aeronautics and Astronautics: An American Chronology of Science
    and Technology in the Exploration of Space 1950–1960* (NASA, 1961), p. 77.
2. Minutes of 14 Nov. 1956 Meeting of the Fuels and Propulsion Panel of the USAF
    Scientific Advisory Board and the USAF reply (excerpts declassified by USAF). A
    chronology of major ARPA events including the Air Force actions on large engines was
    part of the testimony of Dep. Sec. of Defense Donald A. Quarles, Herbert F. York,
    Dir. of Defense Research and Engineering, and Roy W. Johnson, ARPA Dir., in hearings
    before the Subcommittee on Defense Appropriations, House of Representatives, 86th
    Congress, 1st sess., 14 Apr. 1959.
3. A. O. Tischler, "Proposed Minutes of Sixth Meeting, Informal Technical Advisory
    Committee for Propulsion: NASA-USAF-ARPA-AOMC, 9 Oct. 1958," 31 Oct. 1958,
    Tischler's retired files; interviews with A. O. Tischler, NASA, Washington, 25 Jan. 1974;
    and with Abraham Hyatt, Rockwell International, El Segundo, CA, 26 Apr. 1974.
4. J. B. Medaris and Arthur Gordon, *Countdown for Decision* (New York: G. P.
    Putnam's Sons, 1960), p. 101; Thomas Watson Ray, "Apollo's Antecedents: The Con-
    ceptualization, Planning, Resource Build-Up, and Decisions That Led to the Manned
    Lunar Landing Program" (Ph.D. dissertation, Univ. of Colorado, 1974), p. 20. Ray's
    thesis contains an excellent account of early Saturn work.
5. Interview with Krafft A. Ehricke, 26 Apr. 1974.
6. Ibid.; Wernher von Braun to Monte D. Wright, NASA, 29 Dec. 1975.
7. "Proposal: A National Integrated Missile and Space Vehicle Development Program,"
    ABMA, Huntsville, AL, 10 Dec. 1957; 2d summary, 14 Mar. 1958; interview with
    Frank L. Williams, NASA, Washington, 11 Mar. 1974.
8. Wernher von Braun, ABMA, to Carl B. Palmer, NACA Hq., 10 Mar. 1958, with copy
    of teletype attached, NASA History Office.
9. Interview with Francis L. Williams, NASA, 3 Oct. 1975.
10. "A National Integrated Missile and Space Vehicle Development Program," interim report
    to the NACA Special Committee on Space Technology by the Working Group on
    Vehicular Program, 1 Apr. 1958, NASA History Office.
11. The NASA Hq. copy, with red tag and comment still attached, is in the NASA History
    Office.
12. Dryden to Commander ABMA, Attn: Mr. F. L. Williams, 15 July 1958, files of F. L.
    Williams, NASA Hq.
13. "Report to NACA Special Committee on Space Technology by Working Group on
    Vehicular Program," 18 July 1958, NASA History Office.
14. "Recommendations Regarding a National Civil Space Program," by the NASA Special
    Committee on Space Technology, NASA, 28 Oct. 1958, NASA History Office.
15. "Report to NACA Special Committee on Space Technology by Working Group on
    Vehicular Program," 18 July 1958, NASA History Office.
16. ARPA to C. G., U.S. AOMC Scientific Satellite Program, 17 Apr. 1958, cited in R. C.
    Callaway, D. M. Hammock, H. H. Koelle, and F. L. Williams, "Preliminary Information
    on an Advanced Orbital Carrier and Space Vehicle (Juno IV)," report DSP-TM-4-58,
    Future Projects Design Br., Structures & Mechanics Lab., Development Operations

Div., ABMA, Redstone Arsenal. AL. 16 June 1958. This report, in the files of F. L. Williams of NASA, contains a brief history of Juno IV. For another account of Juno IV, see: Wernher von Braun, "Redstone. Jupiter and Juno," in Eugene M. Emme, ed.. *The History of Rocket Technology* (Detroit: Wayne State Univ. Press, 1964), pp. 107-21.

17. Sources on ARPA for 1958–1959 include: ARPA chronology in H. R. hearings (note 2); "Saturn Chronology," in ARPA files, accession 68-A-6057, box 1. folder titled "Order No. 14-59, Large Space Vehicle Booster Approx. 1.5 mil lb Thrust." Federal Records Center, Suitland. MD. apparently written in 1959 by an ARPA staff member familiar with the events; "Background of Saturn Development and Management." an attachment to "Report to Administrator, NASA, on Saturn Vehicle Development Plan by Saturn Vehicle Team," 15 Dec. 1959, NASA History Office; David S. Akens. *Saturn Illustrated Chronology*, 5th ed. (NASA–Marshall Space Flight Center, 1971); interview with Richard Canright, Penn. Health Research Inst., Camp Hill, 7 Mar. 1974; and interview with Richard S. Cesaro, TCOM Corp., Rockville. MD. 12 Mar. 1974.

18. Interview with Canright, 7 Mar. 1974.

19. Ibid.

20. Interview with Cesaro, 12 Mar. 1974.

21. ARPA, "Saturn Chronology." The remainder of this chapter, except the last two sentences, is based on this source.

22. "Agreement on the AOMC between NASA and the Department of the Army," 3 Dec. 1958, as quoted by Thomas Ray, "Apollo's Antecedents." See also Robert L. Rosholt. *Administrative History of NASA*, p. 47.

Chapter 12: Saturn, 1959

1. "Report of the Joint ARPA-NASA Committee on Large Clustered Booster Capabilities." 8 Jan. 1959, with endorsement by Hyatt, files of Abraham Hyatt. NASA History Office.

2. "A National Space Vehicle Program," report to the President presented by the National Aeronautics and Space Administration. 27 Jan. 1959, NASA History Office.

3. "Saturn Chronology," ARPA, ca. 1959, accession 68-A-6057, box 1, Federal Records Center, Suitland, MD.

4. Herbert F. York, San Diego, CA. to Eugene M. Emme, NASA, regarding "Saturn Chronology," 10 June 1974.

5. Ibid.

6. "Saturn Chronology," ARPA. ca. 1959.

7. Telephone interview with James Powell, formerly of Glenn L. Martin Co. and responsible for the Titan C concept, now at the Energy Research and Development Administration. Washington, 17 Sept. 1975.

8. York to Emme, 10 June 1974.

9. "Saturn Chronology," ARPA. ca. 1959.

10. Ibid.

11. Interview of York by Emme, 12 June 1973.

12. "Saturn Chronology," ARPA. ca. 1959.

13. Ibid.; York to Emme, 10 June 1974.

14. Interview with Richard E. Horner. Washington. 13 Mar. 1974.

15. Wernher von Braun to Monte D. Wright, 29 Dec. 1975.

16. Interviews, Francis L. Williams. NASA. Washington, 11, 26 Mar. 1974.

17. "Project Horizon Report: A U.S. Army Study for the Establishment of a Lunar Outpost." vol. 2, "Technical Considerations and Plans." 9 June 1959.

18. E. W. Hall to Silverstein, "Proposed Saturn Development Schedule," 23 Oct. 1959. E. W. Hall file, NASA History Office.

19. E. W. Hall, "Report on Technical Survey of ABMA Activities," 2 Nov. 1959; also

incorporated in "Technical Survey of ABMA Activities: Propulsion Development & Saturn Vehicles." A. Hyatt, A. O. Tischler, and E. W. Hall, 10 Nov. 1959. The 30 Oct. meeting at the Pentagon is on Hall's appointment calendar. Hall and Hyatt files, NASA History Office.

20. E. W. Hall, notes titled "Saturn Upper Stage Selection," 3 Nov. 1959, files of Hall, NASA History Office.

21. Homer J. Stewart to Administrator, NASA, "Considerations of Vega, Agena B, and Centaur," 2 Oct. 1959, document IV-1 in "Documents in the History of NASA, An Anthology." HHR-43, NASA History Office.

22. Richard Horner to Herbert York, 9 Nov. 1959, files of E. W. Hall, NASA History Office; Richard Horner, Assoc. Admin., to Abe Silverstein, Dir. of Space Flight Dev., "Establishment of a Saturn Vehicle Team," 7 Nov. 1959, files of E. W. Hall, NASA History Office; "Report to the Administrator, NASA, on NASA Development Plan," by Saturn Vehicle Team, 15 Dec. 1959, NASA History Office. The report, without attachments, is in "Documents in the History of NASA."

23. Comments of Col. N. C. Appold in "Minutes of the Meetings of Saturn Vehicle Team," NASA Hq., Washington, 28 Nov. 1959, files of Eldon W. Hall, NASA History Office.

24. Telephone interview with T. C. Muse, Arlington, VA, 22 Sept. 1975.

25. Wernher von Braun to Monte D. Wright, 29 Dec. 1975.

26. Von Braun to the author, 8 Nov. 1973; von Braun to Wright, 29 Dec. 1975; interview with Krafft Ehricke, 26 Apr. 1974; interview with Richard Canright, 7 Mar. 1974.

27. Interview with Eldon Hall, 28 Feb. 1974.

28. Ibid.; interview with Hyatt, 26 Apr. 1974.

29. Minutes of the meetings of 27, 28 Nov. and 3, 4, 14, and 15 Dec. are in the files of Eldon W. Hall, NASA History Office. Reference is made to meetings of 10, 11 Dec., but no record of minutes has been located. A working copy of the report, dated 10 Dec., is also in Hall's file. The final report, "Report to the Administrator, NASA, on Saturn Development Plan by Saturn Vehicle Team," 15 Dec. 1959, is reprinted without appendices as document IV-7 of "Documents in the History of NASA."

30. Interviews of Francis L. Williams, 7, 11 Mar. 1974.

31. Eldon W. Hall to the author, 14 Dec. 1971; interviews with Hall 28 Feb. 1974, with Abraham Hyatt 26 Apr. 1974, with Alfred Nelson 23 Sept. 1975.

32. "Report to Administrator, NASA, on Saturn Development Plans," 15 Dec. 1959.

33. Interview with Francis C. Schwenk, NASA, Washington, 8 Mar. 1974.

34. Horner to Ostrander and Silverstein, 29 Dec. 1959, E. W. Hall files, NASA History Office.

Appendix A-1

1. The written history of sulfuric acid dates from the sixteenth century; its preparation and reaction with iron to form hydrogen before that time is speculation but plausible.

2. J. R. Partington, *A History of Chemistry*, 4 vols. (London: Macmillan, 1961), 2:526; *Philosophical Transactions* (1766): 141; Cavendish's papers are in *The Scientific Papers of the Honourable Henry Cavendish, F.R.S.*, Sir Edward Thorpe, ed., 2 vols. (Cambridge: University Press, 1921). Volume II, *Chemical and Dynamical*, was the source used. See also Partington, *History of Chemistry*, 3:302–62.

3. Tiberius Cavallo, *The History and Practice of Aerostation* (London, 1785), pp. 51–61, 97–107; *Encyclopaedia Brittanica*, 11th ed., s.v. "Aeronautics"; E. Charles Vivian, *History of Aeronautics* (New York: Harcourt, Brace & Co., 1921), pp. 319–25; L. T. C. Rolt, *The Aeronauts: A History of Ballooning, 1783–1903* (New York: Walker & Co., 1966), pp. 31–35, 49–54; Robert Jackson, *Airships: A Popular History of Dirigibles, Zeppelins, Blimps and Other Lighter-Than-Air Craft* (Garden City: Doubleday & Co., 1973), pp. 7–10, 17–18.

4. *Enc. Brit.*, 11th ed., s.v. "Aeronautics"; Rolt, *Aeronauts*, pp. 90–93.

5. Partington, *History of Chemistry*, 3:445–48; Rolt, *Aeronauts*, p. 246; *Enc. Brit.*, 11th ed., s.v. "Hydrogen."

6. Some early measurements of hydrogen's heat of combustion, in 10^9 J/kg, are: Crawford (1788), 174; Dalton (1808), 106; Rumford (1812–1814), 79; Despretz (1828), 97; Dulong (1838), 145; and Berthelot (ca. 1869), 122; Partington, *History of Chemistry*, 3:426–34, 370–71, 156–57. For the modern value: Sanford Gordon, NASA-Lewis, to author, 9 Dec. 1974.

7. *Enc. Brit.*, 11th ed., s.v. "Gas Engine."

8. Partington, *History of Chemistry*, 4:289–91, 630–31.

9. Francis Bacon, "Sylva Sylvarum or a Naturall History. In Ten Centuries," 1627, as quoted by James Dewar in his "History of Cold and the Absolute Zero," Presidental Address before the British Association for the Advancement of Science, Belfast, 1902; rev. for the *Annual Report of the Smithsonian Institution for 1902* (Washington: 1903), pp. 207–08.

10. Georges Claude, *Liquid Air, Oxygen, Nitrogen* (Philadelphia: P. Blakiston's Son, 1913), pp. 75–79, fn. p. 80.

11. Ibid., fn. p. 88.

12. James Dewar, *Collected Papers of Sir James Dewar*, ed. Lady Dewar (Cambridge: University Press, 1927), pp. 678–91.

13. *Scientific American Supplement*, 29 Jan. 1910, p. 69.

Appendix A-2

1. The material on Tsiolkovskiy's life and works is drawn from the following sources: K. E. Tsiolokovskiy, *Works on Rocket Technology*, ed. M. K. Tikhonravov, NASA Technical Translation F-243 (Washington, 1965); K. E. Tsiolkovskiy, *Collected Works of K. E. Tsiolkovskiy*, ed. A. A. Blagonravov, vol. 2: *Reactive Flying Machines*, NASA Technical Translation F-237 (Washington, 1965); N. A. Rynin, *Interplanetary Flight and Communication*, vol. 3, no. 7, *K. E. Tsiolkovskii: Life, Writings, and Rockets*, translated by Israel Programs for Scientific Translations (Jerusalem: IPST, 1971); Prof. Kosmodemyansky, *Konstantin Tsiolkovsky: His Life and Work* (Moscow: Foreign Languages Publishing House, 1956); K. E. Tsiolkovskii, "An Autobiography," *Astronautics* 4 (May 1958): 48, which is also in Rynin, *Tsiolkovskii: Life, Writings*, pp. 2–8; and A. A. Blagonravov, ed., *Transactions for the First Lectures Dedicated to the Development of the Scientific Heritage of K. E. Tsiolkovskiy*, NASA Technical Translation F-544 (Washington, 1970).

2. Rynin, *Tsiolkovskii: Life, Writings*, p. 4.

3. Tsiolkovskiy, *Reactive Flying Machines*, pp. 82–84.

4. The first two series of numbers can be found in Tsiolkovskiy, *Reactive Flying Machines*, pp. 9, 170.

5. Kosmodemyansky, *Tsiolkovsky*, p. 13.

6. Tsiolkovskiy, *Reactive Flying Machines*, pp. 72–117; Tsiolkovskiy, *Works on Rocket Technology*, pp. 24–59.

7. Tsiolkovskiy, *Reactive Flying Machines*, pp. 78–79.

8. Ibid., p. 119; Rynin, *Tsiolkovskii: Life, Writings*, p. 32.

9. Tsiolkovskiy, *Work on Rocket Technology*, p. 35, which appears in shorter form in Tsiolkovskiy, *Reactive Flying Machines*, p. 86.

10. Ibid. He seems to have used oil and petroleum as synonyms.

11. Tsiolkovskiy, *Reactive Flying Machines*, p. 122.

12. Ibid., p. 330.

13. Ibid., pp. 452, 459.

14. Ibid., p. 516.

15. Ibid., pp. 503–15.

16. Ibid., p. 513.

17. Milton Lehman, *This High Man* (New York: Farrar. Straus, 1963); Robert H. Goddard, *The Papers of Robert H. Goddard*, 3 vols., ed. Esther Goddard and G. Edward Pendray (New York: McGraw-Hill, 1970); Eugene M. Emme, "Yesterday's Dream, Today's Reality," *The Airpower Historian*, Oct. 1960, pp. 216–22.

18. Goddard noted the anniversary of his cherry tree dream for four decades. Like Tsiolkovskiy, Goddard's first space ships involved centrifugal force. Milton Lehman, *This High Man*, pp. 28, 263.

19. Goddard's undergraduate thesis: "On Some Peculiarities of Electrical Conductivity Exhibited by Powders and a Few Solid Substances" (1908); his Master's thesis: "Theory of Diffraction" (1910); his doctor's thesis: "On the Conductivity at Contacts of Dissimilar Solids" (1911); his post-doctoral thesis: "The Positive Result of Force on a Material Dielectric Carrying a Displacement Current" (1913). Lehman, *This High Man*, pp. 50, 56–59.

20. Goddard, *Papers*, pp. 14, 22, 34.

21. Ibid., p. 121.

22. Ibid., pp. 162–74; Robert H. Goddard, "A Method for Reaching Extreme Altitudes," Smithsonian Miscellaneous Collections for 1919. Hydrogen-oxygen was not mentioned in the 1916 report but was included in supplementary notes when published in 1919. Goddard, *Papers*, pp. 22–23.

23. Goddard, *Papers*, p. 416.

24. Lehman, *This High Man*, p. 125; Goddard, *Papers*, pp. 499–506; Emme, "Yesterday's Dream," p. 219.

25. In a conversation with the author on 6 Mar. 1974, Mrs. Goddard stated that Goddard definitely did not work with liquid hydrogen; he had a hard enough time getting liquid oxygen for his initial experiments with it. He often told her he would like to have some liquid hydrogen. His first liquid oxygen came from a professor at Harvard or MIT who was conducting experiments involving liquefying air and throwing away the liquid oxygen. Goddard went to get the excess liquid oxygen and hurried back before it evaporated. Sometimes he sent a mechanic for the oxygen. Later, Linde opened a plant in Massachusetts and Goddard began obtaining his liquid oxygen from that source.

26. For Oberth's life and work: Hermann Oberth, "From My Life," *Astronautics* 4 (June 1959): 38–39; Hermann Oberth, *Rockets in Planetary Space*, NASA trans. TTF-9227, 1965; Hermann Oberth, *Ways to Spaceflight*, NASA trans. TTF-622, 1972; Willy Ley, *Rockets, Missiles, and Space Travel* (New York: Viking, 1951, 1961).

27. M. Subotowicz, "The Rocket Conceptions of K. Siemienowicz, 1650," *Jour. British Interplanetary Soc.* 14 (Sept.–Oct. 1955): 245–47.

28. Oberth, *Ways to Spaceflight*, p. 42.

29. Ibid., pp. 38–41.

30. Ibid., pp. 94–95, 343.

31. Ibid., pp. 319, 332–36.

32. Ibid., pp. 50, 369, 22.

33. Ibid., pp. 411–12.

34. Lehman, *This High Man*, p. 132.

35. Goddard, *Papers*, pp. 497–517.

Appendix A-3

1. John Toland, *The Great Dirigibles: Their Triumphs and Disasters* (New York: Dover, 1972); Douglas H. Robinson, *Giants in the Sky* (Seattle: Univ. of Washington Press, 1973).

2. R. W. Knight, "The Hindenburg Accident," report 11, Safety & Planning Div., Bureau of Air Commerce, Dept. of Commerce, Aug. 1938. The report included a translation of the German Commission's report as Sec. II, which came to the same conclusion, pp. 63–64.

3. "Hydrogen Safety Manual," NASA TMX-52454, 1968, pp. 3, 30.

4. U.S. Patent 130 915, 27 Aug. 1872; Paul Haenlein, *Zeitschrift für Luftschriffahrt* 1, no. 8 (1888): 84.

5. "Aernonautical Engineering Supplement," *The Aeroplane* 18 (18 Feb. 1920): 362.

6. Richard K. Smith, *First Across: The U.S. Navy's Transatlantic Flight of 1919* (Annapolis: Naval Institute Press, 1973), pp. 87–94; Robinson, *Giants in the Sky*, p. 300; Kurt H. Weil, "The Hydrogen I.C. Engine—Its Origins and Future in the Emerging Energy-Transportation-Environment System," *Proceedings of the Intersociety Energy Conversion Conference*, San Diego, Sept. 1972, pp. 1355–63; G. F. Mucklow, "The Effect of Reduced Intake-Air Pressure and of Hydrogen on the Performance of a Solid Injection Oil Engine," *Journal of the Royal Aeronautical Society* 31 (1927): 17–51; *Enc. Brit.*, 1961, s.v. "Lighthouses"; Chief of Naval Airship Training and Experimentation to Chief, Bureau of Aernonautics, signed by B. May, 19 Mar. 1953; Harold C. Gerrish and Hampton H. Foster, "Hydrogen as an Auxiliary Fuel in Compression-Ignition Systems," report 535 in *Twenty-First Annual Report of the National Advisory Committee for Aeronautics, 1935* (Washington, 1936), pp. 495–510.

7. Georges Claude, *Liquid Air, Oxygen, Nitrogen* (Philadelphia: P. Blakiston's Son, 1913), p. 92.

8. Rexmond C. Cochrane, *Measures for Progress: A History of the National Bureau of Standards* (Washington, 1966), pp. 83–84.

9. Manne Siegbahn, "The Physics Prize," in *Nobel: The Man and His Prizes*, by the Nobel Foundation (New York: Elsevier Publishing Co., 1962), pp. 492–97.

10. Adalbert Farkas, *Orthohydrogen, Parahydrogen and Heavy Hydrogen* (Cambridge: University Press, 1935), pp. 1–2; *Technology and Uses of Liquid Hydrogen*, ed. R. B. Scott, W. H. Denton, and C. M. Nicholls (New York: Macmillan Co., 1964), pp. 2–3.

11. The heat of normal-to-para conversion at 20.4 K and heat of vaporization were taken from table I, p. 4, of Guenther von Elbe and Howard T. Scott, Jr., "Hazards of Liquid Hydrogen in Research and Development Facilities," ASD-TDR-62-1027, Dir. of Aero-Mechanics, AFSC, Wright-Patterson AFB, OH, Dec. 1962; *The Kirk-Othmer Encyclopedia of Chemical Technology*, 2d ed. (New York: John Wiley & Sons, 1966), 2:338–39, gives 168 cal/gm as the heat of normal-to-para hydrogen conversion and 21.4 cal/gm as the heat of vaporization of normal hydrogen at 20.4 K. The 1 percent loss per hour is from Scott, Denton, and Nicholls, *Technology of Hydrogen*, pp. 2–3.

12. Farkas, *Hydrogen*, p. 115; Arne Westgren, "The Chemistry Prize," in *Nobel: The Man and His Prizes*, pp. 379, 391; Eduard Farber, *Nobel Prize Winners in Chemistry, 1901–1961*, rev. ed. (New York: Abelard-Schuman, 1963), pp. 207–12.

13. Westgren, "The Chemistry Prize," pp. 377–79.

14. Farkas, *Hydrogen*, p. 116; Farber, *Nobel Prize Winners*, p. 138.

15. Westgren, "Chemistry Prize," pp. 379–80; Farber, *Nobel Prize Winners*, pp. 137–41.

16. All three methods of separating deuterium from hydrogen are described by Farkas, *Hydrogen*.

17. *Enc. Brit.*, 1961, s.v. "Hydrogen."

18. Walter Dornberger, *V-2* (New York: Viking Press, 1958).

19. Wernher von Braun to author, 8 Nov. 1973. All information about Thiel is based on this source.

20. Ibid.

Bibliography

This brief bibliography is intended as a general guide for those who might wish to dig deeper into one of the many interesting topics the author has barely touched and for those interested in contemporary space history in general.

Bibliographies and Chronologies

Among the benefits of the space age are computers with enormous memory capacities. Some have stored a vast amount of references pertaining to aeronautics and space, including fuels such as hydrogen. The considerable interest in hydrogen today, from studies of energy alternatives, means that one must be quite restrictive in a bibliographical search to avoid being buried in computer read-out paper. Much of the computer-stored information on hydrogen is of fairly recent origin, however, and may not be applicable to an earlier period of interest. The author made use of selective bibliographical searches by NASA's RECON system, the Department of Defense's Documentation Center (Cameron Station, Alexandria, VA) and the Chemical Propulsion Information Agency (Applied Physics Laboratory, Johns Hopkins Univ., Silver Spring, MD). These are invaluable for locating specific reports on government-sponsored research not generally in the published literature.

The chronology of greatest general use to the author was Eugene M. Emme, *Aeronautics and Astronautics: An American Chronology of Science and Technology in the Exploration of Space, 1915-1960* (NASA, 1961). This chronology provided many clues for digging deeper and also helped to relate other events with those pertaining to the subject. Several unpublished chronologies were found during the course of the research and are given in the notes; their scope was more restricted but equally valuable.

U.S. Propulsion Developments, 1945-1959

Histories. R. Cargill Hall, "Early U.S. Satellite Proposals," *Technology and Culture,* 4 (Fall 1963):410-34, and unpublished material on the same subject by Hall, in the NASA History Office, give an excellent account of satellites and rockets in 1945-1950. One of the contractors involved was the Aerojet Corporation; that company's rocket engine work is described by George H. Osborn, Robert Gordon, and Herman L. Coplen, "Liquid Hydrogen Rocket Engine Development, 1944-1950," presented at the 21st International Astronautical Congress, West Germany, 9 Oct. 1970. The "1944" in the title appears to be incorrect; actual research on hydrogen at Aerojet appears to have started late in 1945.

Cargill Hall's paper and others on the early development of rockets, plus papers on missile developments in the 1950s are in *The History of Rocket Technology,* Eugene M. Emme, ed. (Detroit: Wayne State University Press, 1964). Willy Ley's *Rockets, Missiles and Space Travel* (New York: Viking Press, 1961) is another source for rocket developments through the 1950s. John D. Clark's *Ignition! An Informal History of Liquid Rocket Propellants* (New Brunswick, NJ: Rutgers University Press, 1972) is a refreshing and candid view on propellants and research on them by a participant. Although he considers all propellants, most of Clark's attention was on storable or non-cryogenic propellants.

311

Other highly useful histories are Robert L. Rosbolt, *An Administrative History of NASA, 1958-1963* (NASA SP-4101, 1966); Alfred Rosenthal, *Venture into Space: Early Years of Goddard Space Flight Center* (NASA SP-4301, 1968); Constance McLaughlin Green and Milton Lomask, *Vanguard, A History* (NASA SP-4202, 1970); and Thomas A. Sturm, *The USAF Scientific Advisory Board: Its First Twenty Years, 1944-1964* (Washington, 1967).

Three PhD dissertations are also very useful: Arthur Levine, "United States Aeronautical Research Policy, 1915-1958; A Study of the Major Policy Decisions of the National Advisory Committee for Aeronautics," Columbia Univ., 1963; James Arthur Dewar, "Project Rover: A Study of the Nuclear Rocket Development Program, 1953-1963," Kansas State Univ., 1974; and Thomas Wilson Ray, "Apollo's Antecedents: The Conceptualization, Planning, Resource Build-Up, and Decisions That Led to the Manned Lunar Landing Program," Univ. of Colorado, 1974.

Books by Participants. Accounts by key participants in aeronautical and missile developments are invaluable for obtaining a good understanding of the subject. Among those found most helpful were Dwight D. Eisenhower, *The White House Years: Waging Peace, 1956-1961* (Garden City, NY: Doubleday & Co., 1965); H. H. Arnold, *Global Mission* (New York: Harper & Bros., 1949); Herbert York, *Race to Oblivion: A Participant's View of the Arms Race* (New York: Simon & Schuster, 1970); J. B. Medaris with Arthur Gordon, *Countdown for Decision* (New York: G. P. Putnam's Sons, 1960); and Theodore von Kármán with Lee Edson, *The Wind and Beyond* (Boston: Little Brown & Co., 1967).

Technical Books. Technology and Uses of Liquid Hydrogen, ed. R. B. Scott, W. H. Denton, and C. M. Nicholls (New York: Macmillan & Co., 1964), goes right to the subject of this history of hydrogen and is one of the most useful technical sources. Within it is a chapter by Richard Mulready on Pratt & Whitney Aircraft's development of liquid hydrogen engines. Other very useful sources on liquid hydrogen during the 1950s are the first six volumes of *Advances in Cryogenic Engineering*, ed. K. D. Timmerhaus (New York: Plenum Press, 1960-1961); these contain papers presented at six conferences held during the 1950s.

The following text on propulsion systems was useful: George P. Sutton, *Rocket Propulsion Elements* (New York: John Wiley & Sons, 1949 and 1956), the classic for the period of interest. There are numerous texts on thermodynamics, turbomachinery, and aircraft engines. Those found helpful were: John F. Sandfort, *Heat Engines* (Garden City, NY: Doubleday, 1962), available in paperback; P. J. McMahon, *Aircraft Propulsion* (n.p.: Harper & Row, 1971); Mark W. Zemansky, *Heat and Thermodynamics*, 5th ed. (New York: McGraw-Hill, 1968); George F. Babits, *Applied Thermodynamics* (Boston: Allyn & Bacon, 1968); and D. G. Shepherd, *Principles of Turbomachinery* (New York: Macmillan & Co., 1956). A good reference on nuclear rockets is R. W. Bussard and R. D. DeLauer, *Nuclear Rocket Propulsion* (New York: McGraw-Hill, 1958).

Research and development reports. The reports of government laboratories and contractors were major sources for this history. These are found in a variety of ways: by prior knowledge of the activity, a bibliography of the organization doing the work or sponsoring it, references in other reports, literature searches, suggestions from colleagues or participants, and sometimes sheer luck. If there is an easy or systematic way, the author has yet to discover it. Sometimes an investigation may have been highly classified and no reports can be found. In other cases periodic progress reports were required regardless of results obtained, and there is a plethora of reports with a modicum of useful information in each. In investigating the hydrogen work at Ohio State University between 1945 and 1951, 91 references were found, ranging from bimonthly progress reports to one that summarized all the rocket work for the entire period. Seventy-one reports were identified for the hydrogen work of Aerojet Corporation between 1945 and 1949; these were briefly summarized by the Osborn, Gordon, and Coplen paper mentioned above. Clearly the best way to locate research and development reports sponsored by the government is with the help of government libraries. The author found the library services of the NASA indispensable. The task is sometimes more difficult than necessary, for many of the old reports still bear their original classification markings and the services of a security official are necessary. Fortunately, the NASA officials I dealt with were knowledgeable and cooperative.

Other documents. Official correspondence, minutes of meetings, internal memoranda, patents, and other documents of government agencies and aerospace corporations are invaluable sources. Unfortunately, the pressure to clear files often results in the destruction of such material, but much is placed in retired files. The government maintains a *retired records facility* at Suitland, MD; it and the retired files of the Garrett Corporation proved very useful. Other helpful sources are those participants who keep their own records, especially the few rare ones with work diaries, but locating these depends on persistent inquiry and some luck.

Interviews. Interviews of participants in this history were major sources of information; the importance of oral history for recent events cannot be overemphasized. This was particularly true in the case of the Air Force's supersecret Suntan program, where only a small amount of documentation has been located. In other cases, interviews furnished clues to other sources, helped to make sense of disconnected fragments, and gave color and background to otherwise dull events. Sometimes information from interviewees conflicts, but cross-checking and locating additional information or other clues help to clarify many of these conflicts. For those discrepancies that remain, perhaps a reader will come forward with better documentation or information. No one expects a participant's memory to be faultless or his view completely unbiased, but the author was impressed by the overall quality of the information from interviews, as well as the cooperativeness and objectiveness of those interviewed.

Pre-1945 Interest in Hydrogen

The best single source on hydrogen to the twentieth century is J. R. Partington, *History of Chemistry*, 4 vols. (London: Macmillan Co., 1961–). Partington was a master at digging out details. The 2d, 3d, and 4th volumes are rich with references for those who wish more details. (Vol. 1 had not appeared in 1975.)

The works of several key contributors to the science and technology of hydrogen are especially helpful. One is the *Scientific Papers of the Honourable Henry Cavendish*, ed. T. Edward Thorpe (Cambridge: University Press, 1921). Another is the *Collected Papers of Sir James Dewar*, ed. Lady Dewar (Cambridge: University Press, 1927). Dewar's "History of Cold and the Absolute Zero," a paper he gave before the British Association for the Advancement of Science in 1902 and published in the *Annual Report* of the Smithsonian Institution for the same year, was very helpful. Two excellent accounts of cryogenics are: K. Mendelssohn, *The Quest for Absolute Zero* (New York: McGraw-Hill, 1966), and Michael McClintock, *Cryogenics* (New York: Reinhold, 1964).

Balloons and dirigibles, the first users of hydrogen in flight, have many excellent histories extending from Tiberius Cavallo, *The History and Practice of Aerostation* (London, 1785), which entranced the author, to Douglas H. Robinson, *Giants in the Sky* (Seattle: Univ. of Washington Press, 1973), an excellent treatise. The use of hydrogen in balloons and dirigibles is a subject in itself with much source material. The library of the National Air and Space Museum of the Smithsonian is a good place to start.

Rocket Pioneers

The most convenient English sources on both Tsiolkovskiy and Oberth are the translations of their works by the NASA (TT F-237, TT F-243, and TT F-622). NASA has also translated the multiple-volume works of N. A. Rynin; vol. 3, no. 7, is on Tsiolkovskiy.

The best single source on Goddard is the *Papers of Robert H. Goddard*, 3 vols., ed. Esther Goddard and G. Edward Pendray (New York: McGraw-Hill, 1970). An interesting and readable biography is Milton Lehman, *This High Man* (New York: Farrar, Strauss, 1963). To go deeper, consult the National Air and Space Museum and the Goddard Memorial Library, the latter at Worcester, Mass.

Aeronautics, Engines, and Rockets through World War II

The literature on early aeronautics is only of general interest for the subject of hydrogen. The *Annual Reports* of the National Advisory Committee for Aeronautics and other NACA publications are a rich source of aeronautical technical information: for a general picture of early aeronautics. E. Charles Vivian and W. Lockwood Marsh. *History of Aeronautics* (New York: Harcourt, Brace & Co., 1921) is useful.

The best sources on aircraft engines—piston and jet—and aviation fuels through World War II are Robert Schlaifer. "Development of Aircraft Engines." and S. D. Heron. "Development of Aviation Fuels." published by Harvard Univ. in one volume in 1950 and reprinted by the Maxwell Reprint Co. in 1970.

Two excellent sources on the contributions of the Office of Scientific Research and Development during World War II are James Phinney Baxter III. *Scientists against Time* (Cambridge: MIT Press. 1946). and the Office of Scientific Research and Development series. *Science in World War II* (Boston: Little Brown & Co.. 1948). in which the volume *Rockets, Guns, and Targets*. ed. John E. Buchard. is of special interest.

German rocket developments before and during World War II have been well documented. The author used W. G. A. Perring. "A Critical Review of German Long-Range Rocket Development." *Journal of the Royal Aeronautical Society* 50 (July 1946): 483–525. for early postwar analysis of the V-2; Walter Dornberger. *V-2* (New York: Viking Press. 1958). for the inside story of the German rocket build-up by a major participant: and James McGovern, *Crossbow and Overcast* (New York: William Morrow & Co., 1964). for an absorbing account of how the Allies obtained the services of German rocket experts.

Index

322

The Author

John L. Sloop retired from government service in 1972 after 31 years of aeronautical and space research and its management. He joined the National Advisory Committee for Aeronautics at its Langley laboratory in 1941, was transferred to its new engine laboratory in Cleveland in 1942, and headed a group working on aircraft engine ignition problems during World War II. After the war, he was placed in charge of cooling research in a newly formed section on rockets; he concentrated on internal film cooling using porous walls and other techniques. In 1949, he was made head of the laboratory's rocket research; during the 1950s his group made many contributions in theoretical and experimental research on high-energy propellants, ignition, combustion, and cooling. Over 150 technical reports were published by the rocket group by 1960.

Abe Silverstein, who initially headed all of the National Aeronautics and Space Administration's spaceflight programs, brought Sloop to Washington in 1960 as one of his technical assistants. Sloop served on a number of internal management committees on launch vehicles and spacecraft and participated in the planning that led to the Saturn vehicle and Apollo missions. A year later, he became deputy director of the group that managed NASA's small and medium launch vehicles (Scout, Delta, Atlas-Agena, and Atlas-Centaur). In 1962, he was named director of propulsion and power generation in NASA's office of advanced research and technology, where his responsibilities included solid- and liquid-propellant rockets and on-board power using chemical and solar energy. In 1964, Sloop became assistant associate administrator for advanced research and propulsion, of the office that managed research in NASA laboratories in the fields of aeronautics, space vehicles, propulsion, electronics, human factors, and basic research.

Sloop is the author of 45 publications and over a hundred unpublished papers and talks. He is a Fellow of the American Institute of Aeronautics and Astronautics and in 1974 shared its Goddard award with two others "for significant contributions to the development of practical lox-hydrogen rocket engines which have played an essential role in the Nation's space program and in the advancement of space technology." He is also a member of the Society for the History of Technology. He has held various offices in the American Rocket Society and the AIAA.

Sloop was born in Charlotte, N.C., in 1916, earned a B.S. in electrical engineering from the University of Michigan in 1939, and is a registered engineer in Ohio.

Mr. and Mrs. Sloop (the former Atlasse Yeargin) live in Bethesda, Maryland. They have four children: Linda Carr (b. 1942), Lt. (jg) William Locke (1944–1969), Judith Farrell (b. 1946), and John Robert (b. 1948).

NASA HISTORICAL PUBLICATIONS

HISTORIES

Frank W. Anderson, Jr., *Orders of Magnitude: A History of NACA and NASA, 1915–1976*, NASA SP-4403, 1976, GPO.*

William R. Corliss, *NASA Sounding Rockets, 1958–1968: A Historical Summary*, NASA SP-4401, 1971, NTIS.†

Edward C. and Linda N. Ezell, *The Partnership: A History of the Apollo-Soyuz Test Project*, NASA SP-4209, 1978, GPO.

Constance McL. Green and Milton Lomask, *Vanguard: A History*, NASA SP-4202, 1970, GPO; also Washington: Smithsonian Institution Press, 1971.

Barton C. Hacker and James M. Grimwood, *On the Shoulders of Titans: A History of Project Gemini*, NASA SP-4203, 1977, GPO.

R. Cargill Hall, *Lunar Impact: The History of Project Ranger*, NASA SP-4210, 1977, GPO.

Edwin P. Hartman, *Adventures in Research: A History of the Ames Research Center, 1940–1965*, NASA SP-4302, 1970, NTIS.

Mae Mills Link, *Space Medicine in Project Mercury*, NASA SP-4003, 1965, NTIS.

Alfred Rosenthal, *Venture into Space: Early Years of Goddard Space Flight Center*, NASA SP-4301, 1968, NTIS.

Robert L. Rosholt, *An Administrative History of NASA, 1958–1963*, NASA SP-4101, 1966, NTIS.

Loyd S. Swenson, James M. Grimwood, and Charles C. Alexander, *This New Ocean: A History of Project Mercury*, NASA SP-4201, 1966, GPO.

REFERENCE WORKS

The Apollo Spacecraft: A Chronology, NASA SP-4009: vol. 1, 1969, NTIS; vol. 2, 1973, GPO; vol. 3, 1976, GPO; vol. 4, in press.

Astronautics and Aeronautics: A Chronology of Science, Technology, and Policy, annual volumes 1961–1974, with an earlier summary volume, *Aeronautics and Astronautics, 1915–1960*. Early volumes available from NTIS, recent volumes from GPO.

Documents in the History of NASA: An Anthology, looseleaf, NASA HHR-43, Aug. 1975, NASA History Office.

Skylab: A Chronology, NASA SP-4011, 1977, GPO.

Jane Van Nimmen and Leonard C. Bruno, *NASA Historical Data Book, 1958-1968*, vol. 1, *NASA Resources*, NASA SP-4012, 1976, NTIS.

Helen T. Wells, Susan H. Whiteley, and Carrie E. Karegeannes, *Origins of NASA Names*, NASA, SP-4402, 1976, GPO.

---

* GPO: Order from Superintendent of Documents, Government Printing Office, Washington, DC 20402.
† NTIS: Order from National Technical Information Service, Springfield, VA 22161.

☆ U. S. GOVERNMENT PRINTING OFFICE : 1978 O - 251-594

www.ingramcontent.com/pod-product-compliance
Lightning Source LLC
Chambersburg PA
CBHW080234180526

45167CB00006B/2274